AF894630

SOTHIS PRESS

Sothis Press aims to completely reimagine the public domain book with an emphasis on universal access to captivatingly designed print editions. We believe that technology will enable every single book ever published to be perpetually accessible in beautiful print form. Sothis is beauty in ubiquity.

www.sothis.press
1 Waterhouse Square, London, ECIN 2ST

Standard Building Code, Also Known as Southern Standard Building Code

1973 Edition

Southern Building Code Congress Intl

SOTHIS PRESS

ISBN 978-1-0365-2317-6

Standard Building Code, Also Known as Southern Standard Building Code is in the public domain.

This edition is published by Sothis Press © 2023. The company is registered in England and Wales. The Company Number is 14900389.

SOUTHERN STANDARD BUILDING CODE

SOUTHERN BUILDING CODE CONGRESS

1973 Edition

STANDARD BUILDING CODE

ALSO KNOWN AS SOUTHERN STANDARD BUILDING CODE

1973 EDITION

THIRD PRINTING

AUTHORIZED EDITION

SOUTHERN BUILDING CODE CONGRESS INTERNATIONAL, INC.

3617 EIGHTH AVENUE, SOUTH

BIRMINGHAM, ALABAMA 35222

PRINTED IN THE U.S.A.

Adopted November 16, 1945, at the Annual Research Conference, Birmingham, Alabama the Standard Building Code has been kept current thereafter through revisions and changes officially approved at Annual Conferences held at

ORLANDO, FLA., November 10th-14th, 1947
GALVESTON, TEXAS, November 12th-16th, 1948
NASHVILLE, TENN., November 13th-16th, 1949
ST. PETERSBURG, FLA., November 12th-15th, 1950
MOBILE, ALA., November 4th-7th, 1951
GREENVILLE, S. C., November 15th-19th, 1952
SHREVEPORT, LA., November 1st-4th, 1953
DAYTONA BEACH, FLA., November 14th-18th, 1954
BIRMINGHAM, ALA., November 6th-9th, 1955
AUGUSTA, GA., November 11th-15th, 1956
ST. PETERSBURG, FLA., November 10th-13th, 1957
BILOXI, MISS., November 9th-12th, 1958
CHATTANOOGA, TENN., November 1st-4th, 1959
BIRMINGHAM, ALA., October 30th-November 2nd, 1960
SHREVEPORT, LA., November 5th-9th, 1961
CLEARWATER, FLA., November 11th-15th, 1962
DALLAS, TEXAS, November 10th-14th, 1963
MOBILE, ALA., October 25th-29th, 1964
MIAMI BEACH, FLA., November 7-11, 1965
MEMPHIS, TENN., November 13-17, 1966
NORFOLK, VA., November 12-16, 1967
NEW ORLEANS, LA., October 27-31, 1968
JACKSONVILLE, FLA., November 2-6, 1969
ATLANTA, GA., November 1-5, 1970
NEW ORLEANS, LA., October 31-November 4, 1971
HOUSTON, TEXAS, October 28-November 3, 1972

PREFACE

The purpose of this Code is to provide minimum requirements to safeguard life, health and public welfare and the protection of property as it relates to these safeguards by regulating and controlling the design, construction, alteration, repair, equipment, use and occupancy, location, maintenance, removal and demolition of all buildings or structures and appurtenances thereto.

The Standard Building Code is dedicated to the development of better building construction and greater safety to the public and uniformity in building laws; to the granting of full justice to all building materials on a fair basis of the true merits of each material; and to development on a sound economic basis for the future growth of our Nation through unbiased and equitable dealing with building construction.

This 1973 Edition represents the many revisions and changes officially approved at Annual Conferences for the years 1945-1972, and is a comprehensive document providing for the use of all safe materials or methods of construction.

RECOMMENDED PROCEDURES TO FOLLOW IN THE USE OF THE STANDARD BUILDING CODE

1. Determine Occupancy Classification of the structure
   (a) Select classification which most accurately fits the use of the structure. (Chapters IV - V)

2. Fire District provisions, if any.
   (a) Chapter III

3. Determine minimum Type of Construction necessary to accommodate proposed structure. (Chapter VI)
   (a) Determine maximum allowable heights and floor areas for Types of Construction and Occupancy. (Tables—Chapter IV)
   (b) Check allowable area increases permitted. (Chapter IV)

4. Check detailed Occupancy requirements. (Note that each code section dealing with Occupancy Classification—Chapter IV—provides a complete check list that may be utilized for the particular occupancy.)

5. Check detailed Construction requirements.
   (a) Fire Protection of Structural Members (Chapter VI)
   (b) Fire Protection Requirements (Chapter VII)
   (c) Means of Egress Requirements (Chapter XI)

6. Review design as related to standards. (Chapters XII - XX)

7. Check other requirements as necessary.
   (a) Use of public property—Chapter XXII
   (b) Elevators—Chapter XXIV
   (c) Sprinklers and Standpipes—Chapter IX
   (d) Use of Combustible Materials—Interior—512.3, 704
   (e) Roof coverings—301, 706
   (f) Light, ventilation and sanitation—2001, 2002
   (g) Other

These steps are naturally varied in sequence by individual preferences, however the first three are Basic Steps which should be determined in proper order to assist in design of buildings.

SOUTHERN BUILDING CODE CONGRESS INTERNATIONAL, INC.
STANDARD BUILDING CODE

Ordinance No.............................

An ordinance regulating the erection, construction, enlargement, alteration, repair, moving, removal, conversion, demolition, occupancy, equipment, use, height, area, and maintenance of buildings or structures in the City of...; providing for the issuance of permits and collection of fees therefor; declaring and establishing a Fire District; providing penalties for the violation thereof, and repealing all ordinances and parts of ordinances in conflict therewith.

Be it ordained by the..

of the City of..
as follows:

TABLE OF CONTENTS

CHAPTER I—ADMINISTRATION

Page

SECTION 108 — INSPECTIONS

SECTION 109 — CERTIFICATE OF OCCUPANCY

SECTION 110 — POSTING FLOOR LOADS

SECTION 111 — BOARD OF ADJUSTMENTS AND APPEALS

SECTION 112 — APPEALS

SECTION 113 — DECISIONS OF THE BOARD OF ADJUSTMENTS AND APPEALS

SECTION 114 — VIOLATIONS AND PENALTIES

SECTION 115 — VALIDITY

CHAPTER II—DEFINITIONS

SECTION 201 — DEFINITIONS

CHAPTER III—FIRE DISTRICT

SECTION 301 — GENERAL BUILDING RESTRICTIONS — WITHIN THE FIRE DISTRICT

SECTION 302 — CHANGES TO BUILDINGS

Page

Page

SECTION 406 – GROUP "C" – SCHOOLS

SECTION 407 – GROUP "D" – INSTITUTIONAL

SECTION 408 – GROUP "E" – ASSEMBLY

SECTION 409 – GROUP "F" – STORAGE

SECTION 410 – GROUP "G" – INDUSTRIAL

SECTION 411 – GROUP "H" – SPECIAL HAZARDOUS

SECTION 412 – MIXED OCCUPANCY AND SEPARATION REQUIREMENTS

Page

CHAPTER V—SPECIAL OCCUPANCY REQUIREMENTS

Page

SECTION 604 — TYPE III

SECTION 605 — TYPE IV

SECTION 606 — TYPE V

SECTION 607 — TYPE VI

SECTION 608 — EXCEPTIONS TO FIRE PROTECTION

SECTION 609 — MIXED TYPES OF CONSTRUCTION

SECTION 610 — BUILDINGS LOCATED ON THE SAME LOT

Page

CHAPTER VII—FIRE PROTECTION REQUIREMENTS

SECTION 701 – PROTECTION OF VERTICAL OPENINGS, STAIRS AND ELEVATORS

SECTION 702 – PARTITIONS

SECTION 703 – PROTECTION OF WALL OPENINGS

SECTION 704 – RESTRICTIONS ON INTERIOR USE OF COMBUSTIBLE MATERIALS

SECTION 705 – FIRESTOPPING

SECTION 706 – ROOF COVERINGS

SECTION 707 – SKYLIGHTS

SECTION 708 – MANSARD OR SLANTING ROOFS

SECTION 709 – DORMER WINDOWS

Page

SECTION 806 — CHIMNEY CONNECTORS AND VENT CONNECTORS

SECTION 807 — DRAFT REGULATORS

SECTION 808 — FIREPLACES

SECTION 809 — INCINERATORS

SECTION 810 — BARBECUES

SECTION 811 — VENTING OF GAS APPLIANCES

SECTION 812 — FURNACE AND BOILER ROOM

SECTION 813 — STEAM AND HOT WATER

SECTION 814 — FIRE PROTECTION OF MISCELLANEOUS HEATING DEVICES

Page

SECTION 815 — AUXILIARY EQUIPMENT

SECTION 816 — COMBUSTION EQUIPMENT — GENERAL

SECTION 817 — WATER HEATERS AND HOT WATER STORAGE TANKS — BURNING SOLID OR LIQUID FUEL

SECTION 818 — ROOM OR SPACE HEATERS SOLID OR LIQUID FUEL

SECTION 819 — CENTRAL HEATING BOILERS AND FURNACES SOLID OR LIQUID FUEL

SECTION 820 — FLOOR FURNACES (SOLID OR LIQUID FUEL)

Page

CHAPTER X—FIRE RESISTANCE STANDARDS FOR MATERIALS AND CONSTRUCTION

CHAPTER XI—MEANS OF EGRESS REQUIREMENTS (EXITS AND EXIT ACCESS)

Page

SECTION 1302 – FOOTINGS AND FOUNDATIONS

SECTION 1303 – PILES

SECTION 1304 – CAISSONS

CHAPTER XIV—MASONRY CONSTRUCTION

SECTION 1401 – GENERAL

SECTION 1402 – MATERIALS

SECTION 1403 – WORKING STRESSES

SECTION 1404 – WALL THICKNESS

Page

SECTION 1405 — LATERAL SUPPORT

SECTION 1406 — PARAPET WALLS

SECTION 1407 — BONDING

SECTION 1408 — ANCHORAGE

SECTION 1409 — MISCELLANEOUS DETAILS

SECTION 1410 — PLAIN GROUTED MASONRY

SECTION 1411 — REINFORCED MASONRY

SECTION 1412 — REINFORCED GYPSUM CONCRETE

Page

SECTION 1602 — CONCRETE QUALITY

SECTION 1603 — MINIMUM SLAB THICKNESS

SECTION 1604 — VERMICULITE CONCRETE

CHAPTER XVII—WOOD CONSTRUCTION

SECTION 1700 — GENERAL

SECTION 1701 — CONSTRUCTION PRACTICES

SECTION 1702 — PROTECTION AGAINST DECAY AND TERMITES

SECTION 1703 — FIRE PROTECTION

SECTION 1704 — FASTENINGS

Page

SECTION 1705 — FLOOR FRAMING

SECTION 1706 — VERTICAL FRAMING

SECTION 1707 — ROOF AND CEILING FRAMING

SECTION 1708 — FRAMING AT CHIMNEYS AND FIREPLACES

Page

CHAPTER XX—LIGHT, VENTILATION AND SANITATION

SECTION 2001 – LIGHT AND VENTILATION

SECTION 2002 – SANITATION

CHAPTER XXI—SAFEGUARDS DURING CONSTRUCTION

SECTION 2101 – GENERAL

CHAPTER XXII—USE OF PUBLIC PROPERTY

SECTION 2201 – GENERAL

SECTION 2202 – OTHER PROJECTIONS

SECTION 2203 – SPACE UNDER PUBLIC PROPERTY

Page

SECTION 2204 – MOVING OF BUILDINGS

CHAPTER XXIII—SIGNS AND OUTDOOR DISPLAYS

SECTION 2301 – GENERAL

SECTION 2302 – STRUCTURAL REQUIREMENTS

SECTION 2303 – CONSTRUCTION

SECTION 2304 – USE OF PLASTIC MATERIALS

CHAPTER XXIV—ELEVATORS AND ESCALATORS

Page

CHAPTER XXV—PREFABRICATED CONSTRUCTION AND MOBILE HOMES OVER EIGHT (8) FEET WIDE

CHAPTER XXVI—PLASTICS

SECTION 2601 — GENERAL

SECTION 2602 — WALL PANELS

SECTION 2603 — GLAZING OF UNPROTECTED OPENINGS

SECTION 2604 — ROOF PANELS

SECTION 2605 — SKYLIGHTS

SECTION 2606 — LIGHT-TRANSMITTING PANELS IN MONITORS AND SAWTOOTH ROOFS

Reference Index

The purpose of this Reference Index is to catalog by name and address those agencies, associations, institutes and others who are referred to in this code by name, initials or symbols. Further there are those listed, for convenience, whose technical and other services are made available to the Building Official.

AA	Aluminum Association 750 Third Avenue, New York, New York 10017.
ACI	American Concrete Institute P. O. Box 4754, Redford Sta., Detroit, Michigan 48219.
AHA	American Hardboard Association 20 North Wacker Drive, Chicago, Illinois 60606.
AHA	American Hospital Association 840 N. Lakeshore Drive, Chicago, Illinois 60611.
AIA	American Insurance Association 85 John Street, New York, New York 10038
AIMA	Acoustical and Insulating Materials Association 205 West Touhy Avenue, Park Ridge, Illinois 60068.
AISC	American Institute of Steel Construction, Inc. 101 Park Avenue, New York, New York 10017.
AISI	American Iron and Steel Institute 150 East 42nd Street, New York, New York 10017.
AITC	American Institute of Timber Construction. 333 West Hampden Avenue, Englewood, Colorado 80110.
ANSI	American National Standards Institute 1430 Broadway, New York, New York 10018
APA	American Plywood Association 1119 A Street, Tacoma, Washington 98401.
API	American Petroleum Institute 1801 K Street, N.W., Washington, D.C. 20006.
ARMA	Asphalt Roofing Manufacturers Association 2990 Richmond Avenue, Houston, Texas 77006.
ASHRAE	American Society of Heating, Refrigeration and Air Conditioning Engineers United Engineering Center 345 East 47th Street, New York, New York 10017.
ASTM	American Society for Testing and Materials 1916 Race Street, Philadelphia, Pennsylvania 19103.

AWPA	American Wood Preservers Association 1012 - 14th Street, N.W., Washington, D.C. 20005.
AWPI	American Wood Preservers Institute 1651 Old Meadow Road, McLean, Virginia 22101.
AWPB	American Wood Preservers Bureau P. O. Box 5824, Spartanburg, South Carolina 29301
AWS	American Welding Society, Inc. 345 East 47th Street, New York, New York 10017.
BIA	Brick Institute of America 1750 Old Meadow Road, McLean, Virginia 22101.
CPA	Clay Products Association 111 West Washington Street, Chicago, Illinois 60602.
CRSI	Concrete Reinforcing Steel Institute 228 North LaSalle Street, Chicago, Illinois 60601.
CS	Commercial Standards (U. S. Dept. of Commerce) Superintendent of Documents, Government Printing Office, Washington, D. C. 20402.
FPL	Forest Products Laboratory 1619 Massachusetts Ave., N.W., Washington, D.C. 20036.
FS	Federal Specifications Supt. of Documents, Government Printing Office Washington, D. C. 20234.
GA	Gypsum Association 201 North Wells St., Room 2510, Chicago, Illinois 60606.
GOAA	General Outdoor Advertising Association 22 Grand Street, New Rochelle, New York 10801.
HPMA	Hardwood Plywood Manufacturer's Association P. O. Box 6246, Arlington, Virginia 22206.
IBRM	Institute of Boiler and Radiator Manufacturers 393 Seventh Avenue, New York, New York 10001.
IMIAWC	International Masonry Industry All-Weather Council 208 South LaSalle Street Chicago, Illinois 60604
MBMA	Metal Building Manufacturers Association 2130 Keith Building, Cleveland, Ohio 44115.
MCA	Manufacturing Chemists' Association, Inc. 1825 Connecticut Ave., N.W., Washington, D.C. 20009.
MHMA	Mobile Homes Manufacturers Association 1800 North Kent Street, Arlington, Virginia 22209.

MLA	Metal Lath Association 221 N. LaSalle St., Suite 1507, Chicago, Illinois 60601.
NASFCA	National Automatic Sprinkler and Fire Control Association, Inc. 2 Holland Avenue. White Plains. New York 10603.
NBS	National Bureau of Standards (U. S. Dept. of Commerce) Supt. of Documents, Government Printing Office, Washington, D. C. 20402.
NCMA	National Concrete Masonry Association P. O. Box 9185, Rosslyn Sta., Arlington, Virginia 22209.
NEMI	National Elevator Manufacturing Industry, Inc. 101 Park Avenue, New York, New York 10017.
NFPA	National Fire Protection Association 60 Batterymarch Street, Boston, Massachusetts 02110.
NForPA	National Forest Products Association 1619 Massachusetts Avenue, N.W., Washington, D. C. 20036.
NLA	National Lime Association 4000 Brandywine Street, N.W., Washington, D. C. 20016.
NPA	National Particleboard Association 711 14th St., N.W., Room 720, Washington, D.C. 20005.
NPCA	National Pest Control Association 250 West Jersey Street, Elizabeth, New Jersey 07202.
NWAH-ACA	National Warm Air Heating & Air Conditioning Association 640 Engineers Building, Cleveland, Ohio 44114.
OSU	Ohio State University, Engineering Experiment Station 2070 Neil Avenue, Columbus, Ohio 43210.
PCA	Portland Cement Association 5420 Old Orchard Road, Skokie, Illinois 60076.
PCI	Prestressed Concrete Institute 205 West Wacker Drive, Chicago, Illinois 60606.
PHMI	Prefabricated Home Manufacturers Institute Shoreham Building, Washington, D. C. 20005.
PI	Perlite Institute 45 West 45th Street, New York, New York 10036.
PTL	Pittsburgh Testing Laboratory 850 Poplar Street, Pittsburgh, Pennsylvania 15220.

RCSHSB	Red Cedar Shingle and Handsplit Shake Bureau 5510 White Building, Seattle, Washington 98101.
SGA	Southern Gas Association 924 Life Building, Dallas, Texas 75202
SJI	Steel Joist Institute 2001 Jefferson Davis Highway, Suite 707, Arlington, Virginia 22202.
SMACCNA	Sheet Metal & Air Conditioning Contractor's Nat. Assn. 1611 No. Kent Street, Arlington, Virginia 22209.
SFPA	Southern Forest Products Association P. O. Box 52468 New Orleans, La. 70152
SPI	Society of Plastics Industry, Inc. 250 Park Avenue, New York, New York 10017.
SPIB	Southern Pine Inspection Bureau P. O. Box 846, Pensacola, Florida 32502.
SWI	Steel Window Institute 2130 Keith Building, Cleveland, Ohio 44115.
SWRI	Southwest Research Institute 8500 Culebra Road, San Antonio, Texas 78228.
TCA	Tile Council of America 360 Lexington Avenue, New York, New York 10017.
TPI	Truss Plate Institute 919 18th Street, N.W., Washington, D.C. 20006.
ULI	Underwriters' Laboratories, Inc. 207 East Ohio Street, Chicago, Illinois 60611.
VI	Vermiculite Institute 141 West Jackson Boulevard, Chicago, Illinois 60604.
WWPA	Western Wood Products Association 700 Yeon Building, Portland, Oregon 97204.

SOUTHERN BUILDING CODE CONGRESS
SOUTHERN STANDARD BUILDING CODE

Ordinance No.........................

An ordinance regulating the erection, construction, enlargement, alteration, repair, moving, removal, conversion, demolition, occupancy, equipment, use, height, area, and maintenance of buildings or structures in the City of...; providing for the issuance of permits and collection of fees therefor; declaring and establishing a Fire District; providing penalties for the violation thereof, and repealing all ordinances and parts of ordinances in conflict therewith.

Be it ordained by the...

of the City of..
as follows:

CHAPTER I—ADMINISTRATION

SECTION 101 — TITLE AND SCOPE

101.1 — TITLE

The provisions embraced within the following chapters and sections shall constitute and be known and may be cited as "The Building Code" hereinafter referred to as "this code."

101.2 — CODE REMEDIAL

This code is hereby declared to be remedial, and shall be construed to secure the beneficial interests and purposes thereof—which are public safety, health, and general welfare—through structural strength, stability, sanitation, adequate light and ventilation, and safety to life and property from fire and other hazards incident to the construction, alteration, repair, removal, demolition, use and occupancy of buildings, structures, or premises.

101.3 — SCOPE

(a) The provisions of this code shall apply to the construction, alteration, repair, equipment, use and occupancy, location, maintenance, removal and demolition, of every building or structure or any appurtenances connected or attached to such buildings or structures.

(b) No provision of this code shall be held to deprive any federal or state agency, or any applicable governing body having jurisdiction, of any power or authority which it had on the effective date of this act or of any remedy then existing for the enforcement of its orders, nor shall it deprive any individual or corporation of its legal rights as provided by law.

101.4 — EXISTING BUILDINGS

(a) If, within any twelve-month period, alterations or repairs costing in excess of fifty percent of the then physical value of the building are made to an existing building, such building shall be made to conform to the requirements of this code for new buildings, also that for buildings located in fire districts the provisions of Sections 302.1 and 302.2 shall apply.

(b) If an existing building is damaged by fire or otherwise in excess of fifty percent of its then physical value before such damage is repaired, it shall be made to conform to the requirements of this code for new buildings.

(c) If the cost of such alterations or repairs within any twelve-month period or the amount of such damage as referred to in paragraph (b) is more than twenty-five but not more than fifty percent of the then physical value of the building, the portions to be altered or repaired shall be made to conform to the requirements of this code for new buildings to such extent as the Building Official may determine.

(d) For the purpose of this section physical value of the building shall be determined by the Building Official.

(e) If the occupancy of an existing building is entirely changed the building shall be made to conform to the requirements of this code for the new occupancy. If the occupancy of only a portion of an existing building is changed and that portion is separated from the remainder as stipulated in Section 412, then only such portion need be made to conform.

(f) Repairs and alterations, not covered by the preceding paragraphs of this section, restoring a building to its condition previous to damage or deterioration, or altering it in conformity with the provisions of this code or in such manner as will not extend or increase an existing non-conformity or hazard, may be made with the same kind of materials as those of which the building is constructed; but not more than twenty-five percent of the roof covering of a building shall be replaced in any period of twelve months unless the entire roof covering is made to conform with the requirements of this code for new buildings.

101.5 — MAINTENANCE

All buildings or structures, both existing and new, and all parts thereof, shall be maintained in a safe and sanitary condition. All devices or safeguards which are required by this code in a building when erected, altered, or repaired, shall be maintained in good working order. The owner, or his designated agent, shall be responsible for the maintenance of buildings and structures.

SECTION 102 — ORGANIZATION

102.1 — BUILDING OFFICIAL

(a) There is hereby established a department to be called the "Building Department," which shall be in charge of a Building Official.

(b) The Building Official shall have had at least ten years' experience or equivalent, as an architect, engineer, building inspector, building contractor, or superintendent of building construction, or any combination of these for five years of which he shall have been in responsible charge of work. He shall be appointed by the Chief Appointing Authority of the applicable governing body. His appointment shall continue during good behavior and satisfactory service. He shall not be removed from office except for cause after full opportunity has been given him to be heard on specific charges before such Chief Appointing Authority.

102.2 — INSPECTORS

The Building Official, with the approval of the Chief Appointing Authority, may appoint such number of officers, inspectors, assistants, and other employees as shall be authorized from time to time. No person shall be appointed as inspector of construction who has not had at least five years' experience as a building inspector, builder, engineer, architect, or as a superintendent, foreman, or competent mechanic in charge of construction.

102.3 — DEPUTY

The Building Official may designate as his deputy an employee in the department who shall, during the absence or disability of the Building Official exercise all the powers of the Building Official.

102.4 — RESTRICTIONS ON EMPLOYEES

No officer or employee connected with the department, except one whose only connection is as a member of the board, established by this act, shall be financially interested in the furnishing of labor, material, or appliances for the construction, alteration, or maintenance of a building, or in the making of plans or of specifications therefor, unless he is the owner of such building. No such officer or employee shall engage in any work which is inconsistent with his duties or with the interests of the department.

102.5 — RECORDS

The Building Official shall keep, or cause to be kept, a record of the business of the department. The records of the department shall be open to public inspection.

SECTION 103 — POWERS AND DUTIES OF BUILDING OFFICIAL

103.1 —RIGHT OF ENTRY

The Building Official shall enforce the provisions of this code, and he, or his duly authorized representative, may enter any building, structure, or premises to perform any duty imposed upon him by this code.

103.2 — STOP WORK ORDERS

Upon notice from the Building Official, work on any building or structure that is being done contrary to the provisions of this code or in a dangerous or unsafe manner, shall be immediately stopped. Such notice shall be in writing and shall be given to the owner of the property, or to his agent, or to the person doing the work, and shall state the conditions under which work may be resumed. Where an emergency exists, no written notice shall be required to be given by the Building Official.

103.3 — REVOCATION OF PERMITS

The Building Official may revoke a permit or approval, issued under the provisions of this act, in case there has been any false statement or misrepresentation as to a material or fact in the application or plans on which the permit or approval was based.

103.4 — UNSAFE BUILDINGS

(a) All buildings or structures which are unsafe, unsanitary, or not provided with adequate egress, or which constitute a fire hazard, or are otherwise dangerous to human life, or which in relation to existing use constitute a hazard to safety or health by reason of inadequate maintenance, dilapidation, obsolescene, or abandonment, are severally in contemplation of this section, unsafe buildings. All such unsafe buildings are hereby declared illegal and shall be abated by repair and rehabilitation or by demolition in accordance with the following procedure:

(1) Whenever the Building Official shall find any building or structure or portion thereof to be unsafe, as defined in this section, he shall, in accordance with established procedure for legal notices, give the owner, agent, or person in control of such building or structure written notice stating the defects thereof. This notice shall require the owner within a stated time either to complete specified repairs or improvements, or to demolish and remove the building or structure or portion thereof.

(2) If necessary, such notice shall also require the building, structure or portion thereof to be vacated forthwith and not reoccupied until the specified repairs and improvements are completed, inspected and approved by the Building Official. The Building Official shall cause to be posted at each entrance to such building a notice stating: THIS BUILDING IS UNSAFE AND ITS USE OR OCCUPANCY HAS BEEN PROHIBITED BY THE BUILDING OFFICIAL." Such notice shall remain posted until the required repairs are made or demolition is completed. It shall be unlawful for any person, firm or corporation or their agents, or other servants, to remove such notice without written permission of the Building Official, or for any person to enter the building except for the purpose of making the required repairs or of demolishing same.

(3) The owner, agent or person in control shall have the right, except in cases of emergency, to appeal from the decision of the Building Official, as provided hereinafter, and to appear before the Board of Adjustments and Appeals at a specified time and place to show cause why he should not comply with said notice.

(4) In case the owner, agent, or person in control cannot be found within the stated time limit, or, if such owner, agent, or person in control shall fail, neglect, or refuse to comply with notice to repair, rehabilitate, or to demolish and remove said building or structure or portion thereof, the Building Official, after having ascertained the cost, shall cause such building or structure or portion thereof, to be demolished, secured, or required to remain vacant.

(5) The decision of the Building Official shall be final in cases of emergency which, in his opinion, involve imminent danger to human life or health. He shall promptly cause such building, structure, or portion thereof to be made safe or cause its removal. For this purpose he may at once enter such structure or land on which it stands, or abutting land or structures, with such assistance and at such cost as he may deem necessary. He may order the vacation of adjacent structures and may require the protection of the public by

appropriate fence or such other means as may be necessary, and for this purpose may close a public or private way.

(6) Costs incurred under paragraphs 103.4(a)(4) and 103.4(a)(5) shall be charged to the owner of the premises involved and shall be collected in the manner provided by law.

103.5 — REQUIREMENTS NOT COVERED BY CODE

Any requirement necessary for the strength or stability of an existing or proposed building or structure, or for the safety or health of the occupants thereof, not specifically covered by this code, shall be determined by the Building Official subject to appeal to the Board of Adjustments and Appeals.

103.6 — ALTERNATE MATERIALS AND ALTERNATE METHODS OF CONSTRUCTION

The provisions of this code are not intended to prevent the use of any material, or method of construction not specifically prescribed by this code, provided any such alternate has been approved and its use authorized by the Building Official. The Building Official shall approve any such alternate, provided he finds that the proposed design is satisfactory and complies with the provisions of Chapter XII, and that the material, method, or work offered is, for the purpose intended, at least the equivalent of that prescribed in the code in quality, strength, effectiveness, fire-resistance, durability, and safety. The Building Official shall require that sufficient evidence or proof be submitted to substantiate any claim that may be made regarding its use. If, in the opinion of the Building Official, the evidence and proof are not sufficient to justify approval, the applicant may refer the entire matter to the Board of Adjustments and Appeals as stipulated in Section 111.

103.7 — LIABILITY

Any officer or employee, or member of the Board of Adjustments and Appeals, charged with the enforcement of this code, acting for the applicable governing body in the discharge of his duties, shall not thereby render himself liable personally, and he is hereby relieved from all personal liability for any damage that may accrue to persons or property as a result of any act required or permitted in the discharge of his duties. Any suit brought against any officer or employee because of such act performed by him in the enforcement of any provision of this code shall be defended by the Department of Law until the final termination of the proceedings.

103.8 — REPORTS

The Building Official shall annually submit a report to the Chief Administrator covering the work of the department during the preceding year. He shall incorporate in said report a summary of the decisions of the Board of Adjustments and Appeals during said year.

SECTION 104 — TESTS

The Building Official may require tests or test reports as proof of compliance. Tests, if required, are to be made at the expense of the owner, or his agent, by an approved testing laboratory or other approved agency. Copies of such test reports or the results of all such tests shall be kept on file in the office of the Building Official.

SECTION 105 — APPLICATION FOR PERMIT

105.1 — WHEN REQUIRED

(a) Any owner, authorized agent, or contractor who desires to construct, enlarge, alter, repair, move, demolish, or change the occupancy of a building or structure, or to erect, or construct a sign of any description, or to install or alter fire-extinguishing apparatus, elevators, engines, or to install a steam boiler, furnace, heater, incinerator, or other heat producing apparatus, or other appurtenances, the installation of which is regulated by this code, or to cause any such work to be done, shall first make application to the Building Official and obtain the required permit therefor.

(b) A general permit shall carry with it the right to install in any building or structure, or part thereof, heating apparatus, elevators, sidewalk elevators, vaults, chutes, coal holes, lifts, cranes, derricks, steam power boilers, steam, oil, gas or vapor engines, provided the same are shown on the drawings and set forth in the specifications filed with the application for the permit; but where these are not shown on the drawings and covered by the specifications submitted with said application, special permits shall be required.

(c) Ordinary minor repairs may be made with the approval of the Building Official without a permit; provided that such repairs shall not violate any of the provisions of this code.

(d) See Section 504 for Temporary Structures.

105.2 — FORM

(a) Each application for a permit with the required fee, shall be filed with the Building Official, on a form furnished for that purpose, and shall contain a general description of the proposed work and its location. The application shall be signed by the owner, or his authorized agent.

(b) Each application for a permit shall indicate the proposed occupancy of all parts of the building and of that portion of the site or lot, if any, not covered by the building or structure, and shall contain such other information as may be required by the Building Official.

105.3 — DRAWINGS AND SPECIFICATIONS

(a) When required by the Building Official, two or more copies of specifications, and of drawings drawn to scale with sufficient clarity and detail to indicate the nature and character of the work, shall accompany every application. Such drawings and specifications shall contain information, in the form of notes or otherwise, as to the quality of materials, where quality is essential to conformity with this code. Such information shall be specific, and this code shall not be cited as a whole or in part, nor shall the term "legal" or its equivalent be used, as a substitute for specific information.

(b) The Building Official may require details, computations, stress diagrams, and other data necessary to describe the construction and basis of calculations and they shall bear the signature of the person responsible for the design.

(c) All drawings, specifications, and accompanying data shall bear the name and address of the designer. In the case of buildings or structures of Group C, D and E Occupancy, and all buildings or structures exceeding two stories in height or 5000 square feet in area, except one and two family dwellings, such designer shall be an architect or engineer legally registered under the laws of this state regulating the practice of architecture or engineering and shall affix his official seal to said drawings, specifications and accompanying data.

105.4 — PLOT DIAGRAM

The Building Official shall require drawings showing the location of the proposed building or structure and of every existing building or structure on the site or lot. He may also require a boundary line survey, if necessary, prepared by a qualified surveyor.

105.5 — LIMITATION

An application for a permit for any proposed work shall be deemed to have been abandoned six months after the date of filing, unless before then a permit shall have been issued; provided that, for cause, one or more extensions of time for periods of not exceeding ninety days each may be allowed by the Building Official.

105.6 — EXAMINATION OF DRAWINGS

(a) The Building Official shall examine or cause to be examined each application for permit and the drawings and computations filed therewith and shall ascertain by such examinations whether the construction indicated and described is in accordance with the requirements of this code and all other pertinent laws or ordinances.

(b) **Permits issued upon Architects or Engineers affidavits.** The Building Official may accept a sworn affidavit from a registered architect or engineer stating that the plans submitted conform to the laws as to egress, type of construction and general arrangement and if accompanied by drawing showing the structural design, and by a statement that the plans and design conform to the requirements of this code as to strength, stresses, strains, loads and stability, he may without any examination or inspection accept such affidavit, provided the Architect or Engineer who made such affidavit agrees to submit to the Building Official, on the completion of the structure a certification that the structure has been erected in accordance with the requirements of this code. Where the Building Official relies upon such affidavit, the architect or engineer shall assume full responsibility for the compliance with all provisions of this code and/or other pertinent laws or ordinances.

105.7 — STREET LINES

No permit shall be given by the Building Official for the construction of any building, or for the alteration of any building where said building is to be changed and such change will affect the exterior walls, bays, balconies, or other appendages or projections fronting on any street, alley or public lane, or for the placing on any lot or premises of any building or structure removed from another lot or premises, unless the applicant has made application at the office of the Director of Public Works for the lines of the public street on which he proposes to build, erect or locate said building; and it shall be the duty of the Building Official to see that the street lines are not encroached upon in any manner whatsoever, except as provided for in Chapter XXII.

SECTION 106 — PERMITS

106.1 — ACTION ON APPLICATION

(a) No person, firm or corporation shall erect, construct, enlarge, alter, repair, move, improve, remove, convert or demolish any building or structure in the applicable jurisdiction, or cause the same to be done, without first obtaining a separate building permit for such buildings or structures from the Building Official.

(b) If the Building Official is satisfied that the work described in an application for permit and the drawings filed therewith conform to the requirements of this code and other pertinent laws and ordinances, he shall issue a permit therefor to the applicant.

(c) If the application for a permit and the drawings filed therewith describe work which does not conform to the requirements of this code or other pertinent laws or ordinances, the Building Official shall not issue a permit, but shall return the drawings to the applicant with his refusal to issue such permit. Such refusal shall, when requested, be in writing and shall contain the reasons therefor.

106.2 — CONTRACTORS LICENSE AND BOND REQUIRED

It shall be the duty of every contractor or builder, who shall make contracts for the erection or construction or repair of buildings for which a permit is required, and every contractor or builder making such contracts and subletting the same, or any part thereof, to pay a license tax as provided in the general license ordinance, and to register his name in a book provided for that purpose, with the Building Official, giving full name, residence and place of business, and, in case of removal from one place to another to have made corresponding change in said register accordingly; and it shall be the further duty of every such person to give good and sufficient bond in a sum prescribed and as required by the administrative authority and approved by the Department of Law, conditioned to conform to the building regulations, the regulations of this section, and other ordinances or laws of the applicable governing body in reference to buildings.

106.3 — CONDITIONS OF THE PERMIT

The Building Official shall act upon an application for a permit with plans as filed, or as amended, without unreasonable or unnecessary delay. A permit issued shall be construed to be a license to proceed with the work and shall not be construed as authority to violate, cancel, alter, or set aside any of the provisions of this code, nor shall such issuance of a permit prevent the Building Official from thereafter requiring a correction of errors in plans or in construction, or of violations of this code. Every permit issued shall become invalid if the work authorized by such permit is not commenced within six months after its issuance, or if the work authorized by such permit is suspended or abandoned for a period of one year after the time the work is commenced; provided, that, for cause, one or more extensions of time, for periods not exceeding ninety days each, may be allowed in writing by the Building Official.

106.4 — DRAWINGS TO BE KEPT AT SITE

When the Building Official issues a permit, he shall endorse, in writing, or stamp, both sets of plans "Approved". One set of drawings so approved shall be retained by the Building Official and the other set shall be returned to the applicant. The approved drawings shall be kept at the site of work and shall be open to inspection by the Building Official or his authorized representative.

106.5 — PERMITS ISSUED UPON AFFIDAVITS

Whenever a permit is to be issued in reliance upon an affidavit as provided in Section 105.6 or whenever the work to be covered by a permit involves construction under conditions which, in the opinion of the Building Official, are hazardous or complex, the Building Official shall require that the architect or engineer who signed the affidavit or made the drawings or computations shall supervise such work, be responsible for its conformity with the approved drawings, and forthwith upon its completion make and file with the Building Official written affidavit that the work has been done in conformity with the approved plans and with the structural provisions of the code. In the event such architect or engineer is not available, the owner shall employ in his stead a competent person or agency whose qualifications are approved by the Building Official.

106.6 — FOUNDATION PERMITS

When application for permit to erect or enlarge a building has been filed and pending issuance of such permit, the Building Official may, at his discretion, issue a special permit for the foundations of such building. The holder of such a special permit shall proceed at his own risk and without assurance that a permit for the superstructure will be granted.

SECTION 107 — FEES

107.1 — GENERAL

No permit shall be issued until the fees prescribed in this section shall have been paid. Nor shall an amendment to a permit be approved until the additional fee, if any, due to an increase in the estimated cost of the building or structure, shall have been paid.

107.2 — FAILURE TO OBTAIN A PERMIT

(a) If any person commences any work on a building or structure before obtaining the necessary permit from the applicable governing body, he shall be subject to the penalty prescribed herein.

(b) Where construction is commenced before a permit is obtained, the permit fees shall be doubled.

107.3 — ACCURATE RECORDS

The Building Official shall keep a permanent and accurate accounting of all permit fees and other monies collected, the names of all persons upon whose account the same was paid, the date and amount thereof.

107.4 — SCHEDULE OF PERMIT FEES

On all buildings, structures or alterations requiring a building permit, as set forth in Section 105, fee shall be paid as required at the time of filing application, in accordance with the following schedule:

(a) Permit Fees

1. Where the valuation does not exceed $100.00, no fee shall be required, unless an inspection is necessary, in which case there shall be a $1.50 fee.

2. For a valuation over $100.00 up to and including $15,000.00 the fee shall be $3.00 per thousand or fraction thereof.

3. For a valuation over $15,000.00 up to and including $100,000.00, the fee shall be $45.00 for the first fifteen thousand plus $2.00 for each additional thousand or fraction thereof.

4. For a valuation over $100,000.00 up to and including $500,000.00, the fee shall be $215.00 for the first one hundred thousand plus $1.00 for each additional thousand or fraction thereof.

5. For a valuation over $500,000.00 up to and including $1,000,000.00, the fee shall be $615.00 for the first five hundred thousand plus 40¢ for each additional thousand or fraction thereof.

6. For a valuation over $1,000,000.00, the fee shall be $815.00 for the first million plus 15¢ for each additional thousand or fraction thereof.

(b) Moving of Building or Structures

For the moving of any building or structure, the fee shall be $10.00.

(c) Demolition of Building or Structures

For the demolition of any building or structure, the fee shall be $4.00.

107.5 — BUILDING PERMIT VALUATIONS

If, in the opinion of the Building Official, the valuation of building, alteration, or structure appears to be underestimated on the application, permit shall be denied, unless the applicant can show detailed estimated cost to meet the approval of the Building Official. Permit valuations shall include total cost, such as plumbing, electrical, mechanical equipment and other systems.

SECTION 108 — INSPECTIONS

108.1 — INSPECTIONS — GENERAL

(a) Before issuing a permit the Building Official may examine or cause to be examined any building for which an application has been received for permit to enlarge, alter, repair, move, demolish, or change the occupancy thereof. He shall inspect all buildings and structures, from time to time, during and upon completion of the work for which a permit was issued. He shall make a record of every such examination and inspection and of all violations of this code.

(b) When deemed necessary by the Building Official, he shall make an inspection of materials or assemblies at the point of manufacture or fabrication. He shall make a record of every such examination and inspection and of all violations of this code.

(c) The Building Official may make, or cause to be made, the inspections required by this Section. He may accept reports of inspectors of recognized inspection services provided that after investigation he is satisfied as to their qualifications and reliability. No certificate called for by any provision of these requirements shall be based on such reports unless the same are in writing and certified by a responsible officer of such service.

108.2 — INSPECTIONS REQUIRED

(a) The Building Official shall inspect or cause to be inspected at various intervals all construction or work for which a permit is required, and a final inspection shall be made of every building or structure upon completion, prior to the issuance of the Certificate of Occupancy, as required in Section 109.

(b) Work requiring a building permit shall not be commenced until the permit holder or his agent shall have posted the building permit card in a conspicuous place on the front of the premises. The permit shall be protected from the weather and in such position as to permit the Building Official to conveniently make the required entries thereon. This permit card shall be maintained in such position by the permit holder until the Certificate of Occupancy has been issued by the Building Official.

(c) The Building Official upon notification from the permit holder or his agent shall make the following inspections of buildings and such other inspections as may be necessary, and shall either approve that portion of the construction as completed or shall notify the permit holder or his agent wherein the same fails to comply with the law:

Foundation Inspection: To be made after trenches are excavated and forms erected.

Frame Inspection: To be made after the roof, all framing, fire-blocking and bracing is in place and all pipes, chimneys, and vents are complete.

Final Inspection: To be made after the building is completed and ready for occupancy.

(d) No work shall be done on any part of a building or structure beyond the point indicated in each successive inspection without first obtaining the written approval of the Building Official. Such written approval shall be given only after an inspection shall have been made of each successive step in the construction as indicated by each of the foregoing three inspections.

(e) No reinforcing steel or structural frame work of any part of any building or structure shall be covered or concealed in any manner whatsoever without first obtaining the approval of the Building Official, the designing architect or engineer.

(f) In all buildings where plaster is used for fire protection purposes, the permit holder or his agent shall notify the Building Official after all lathing and backing is in place. No plaster shall be applied until the approval of the Building Official has been received (See Chapter X).

SECTION 109 — CERTIFICATE OF OCCUPANCY

109.1 — WHEN REQUIRED

No new building shall be occupied and no change in occupancy of a building or part of a building shall be made until after the Building Official shall have issued a certificate of occupancy therefor.

109.2 — CONTENTS OF CERTIFICATE

Upon completion of a building hereafter erected in accordance with approved plans, and after the final inspection herein referred to, and upon application therefor, the Building Official shall issue a certificate of occupancy stating the nature of the occupancy permitted, the number of persons for each floor when limited by law, the allowable load per square foot for each floor in accordance with the provisions of this code.

109.3 — TEMPORARY OCCUPANCY

A temporary certificate of occupancy may be issued for a portion or portions of a building which may safely be occupied prior to final completion of the building.

109.4 — EXISTING BUILDINGS

A certificate of occupancy for any existing building may be obtained by applying to the Building Official and supplying the information and data necessary to determine compliance with this code for the occupancy intended. Where necessary, in the opinion of the Building Official, two sets of detailed drawings, or a general inspection, or both, may be required. When, upon examination and inspection, it is found that the building conforms to the provisions of this code for such occupancy, a certificate of occupancy shall be issued.

SECTION 110 — POSTING FLOOR LOADS

110.1 — FLOOR LOADS

(a) No existing or new building shall be occupied for any purpose which will cause the floors thereof to be loaded beyond their safe capacity. The Building Official may permit occupancy of a building for mercantile, commercial, or industrial purposes, by a specific business, when he is satisfied that such capacity will not thereby be exceeded.

(b) It shall be the responsibility of the owner, agent, proprietor or occupant of Group F and G Occupancies, or any occupancy where excessive floor loading is likely to occur, to employ a competent architect or engineer in computing the safe load capacity. All such computations shall be accompanied by an affidavit from the architect or engineer stating the safe allowable floor load on each floor in pounds per square foot uniformly distributed; it shall thereupon be filed as a permanent record of the department of building.

110.2 — SIGNS REQUIRED

In every building or part of a building used for business storage, industrial or hazardous purposes, the safe floor loads, as approved by the Building Official, shall be marked on plates of approved design which shall be supplied and securely affixed by the owner of the building in a conspicuous place in each story to which they relate. Such plates shall not be removed or defaced, and if lost, removed or defaced, shall be replaced by the owner of the building.

110.3 — LOADS IN EXCESS OF POSTED CAPACITY

No such owner shall place, or permit to be placed, on any floor of a building a greater load than the safe load so determined and posted.

SECTION 111 — BOARD OF ADJUSTMENTS AND APPEALS

111.1 — APPOINTMENT

There is hereby established a board to be called the Board of Adjustments and Appeals, which shall consist of five (5) members. Such Board shall be composed of one Architect, one General Contractor or Engineer and three Members at large from the building industry. The said Board shall be appointed by the Chief Appointing Authority.

111.2 — TERM OF OFFICE

Of the members first appointed two shall be appointed for a term of one year, two for a term of two years, one for a term of three years, and thereafter they shall be appointed for terms of four years. Vacancies shall be filled for an unexpired term in the manner in which original appointments are required to be made. Continued absence of any member from regular meetings of the Board shall, at the discretion of the Chief Appointing Authority, render any such member liable to immediate removal from office.

111.3 — QUORUM

Three members of the board shall constitute a quorum. In varying the application of any provisions of this code or in modifying an order of the Building Official, affirmative votes of the majority present, but not less than three affirmative votes shall be required. No board member shall act in a case in which he has a personal interest.

111.4 — RECORDS

The Building Official shall act as Secretary of the Board of Adjustments and Appeals and shall make a detailed record of all its proceedings, which shall set forth the reasons for its decisions, the vote of each member participating therein, the absence of a member, and any failure of a member to vote.

111.5 — PROCEDURE

The board shall establish rules and regulations for its own procedure not inconsistent with the provisions of this code. The board shall meet at regular intervals, to be determined by the Chairman, or in any event, the board shall meet within ten days after notice of appeal has been received.

SECTION 112 — APPEALS

112.1 — TIME LIMIT

(a) Whenever the Building Official shall reject or refuse to approve the mode or manner of construction proposed to be followed or materials to be used in the erection or alteration of a building or structure, or when it is claimed that the provisions of this code do not apply, or that an equally good or more desirable form of construction can be employed in any specific case, or when it is claimed that the true intent and meaning of this code or any of the regulations thereunder have been misconstrued or wrongly interpreted, the owner of such building or structure, or his duly authorized agent, may appeal from the decision of the Building Official to the Board of Adjustments and Appeals. Notice of appeal shall be in writing and filed within 90 days after the decision is rendered by the Building Official. A fee of $10.00 shall accompany such notice of appeal.

(b) In case of a building or structure which, in the opinion of the Building Official, is unsafe or dangerous, the Building Official may, in his order, limit the time for such appeal to a shorter period. Appeals hereunder shall be on forms provided by the Building Official.

SECTION 113 — DECISIONS OF THE BOARD OF ADJUSTMENTS AND APPEALS

113.1 — VARIATIONS AND MODIFICATIONS

(a) The Board of Adjustments and Appeals, when so appealed to and after a hearing, may vary the application of any provision of this code to any particular case when, in its opinion, the enforcement thereof would do manifest injustice, and would be contrary to the spirit and purpose of this code or public interest, or when, in its opinion the interpretation of the Building Official should be modified or reversed.

(b) A decision of the Board of Adjustments and Appeals to vary the application of any provision of this code or to modify an order of the Building Official shall specify in what manner such variation or modification is made, the conditions upon which it is made and the reasons therefor.

113.2 — DECISIONS

(a) Every decision of the Board of Adjustments and Appeals shall be final, subject, however, to such remedy as any aggrieved party might have at law or in equity. It shall be in writing and shall indicate the vote upon the decision. Every decision shall be promptly filed in the office of the Building Official, and shall be open to public inspection; a certified copy shall be sent by mail or otherwise to the appellant and a copy shall be kept publicly posted in the office of the Building Official for two weeks after filing.

(b) The Board of Adjustments and Appeals shall, in every case, reach a decision without unreasonable or unnecessary delay.

(c) If a decision of the Board of Adjustments and Appeals reverses or modifies a refusal, order, or disallowance of the Building Official, or varies the application of any provision of this code, the

Building Official shall immediately take action in accordance with such decision.

SECTION 114 — VIOLATIONS AND PENALTIES

Any person, firm, corporation or agent who shall violate a provision of this code, or fail to comply therewith, or with any of the requirements thereof, or who shall erect, construct, alter, demolish or move any structure, or has erected, constructed, altered, repaired, moved or demolished a building or structure in violation of a detailed statement or drawing submitted and approved thereunder, shall be guilty of a misdemeanor. Each such person shall be deemed guilty of a separate offense for each and every day or portion thereof during which any violation of any of the provisions of this code is committed, or continued and upon conviction of any such violation such person shall be punished within the limits and as provided by State Laws.

SECTION 115 — VALIDITY

If any section, sub-section, sentence, clause or phrase of this Ordinance is for any reason held to be unconstitutional, such decision shall not affect the validity of the remaining portions of this Code.

CHAPTER II—DEFINITIONS

SECTION 201 — DEFINITIONS

201.1

For the purpose of this Code, certain abbreviations, terms, phrases, words, and their derivatives, shall be construed as set forth in this Section.

201.2

Words used in the present tense include the future. Words in the masculine gender include the feminine and neuter. Words in the feminine and neuter gender include the masculine. The singular number includes the plural and the plural number includes the singular.

ADDITION is an extension or increase in floor area or height of a building or structure.

ALLEY—means any public space or thoroughfare twenty (20) feet or less in width which has been dedicated or deeded for public use.

ALTER OR ALTERATION—means any change or modification in construction or occupancy.

AMUSEMENT DEVICE—means a mechanically operated device which is used to convey persons in any direction as a form of amusement.

APARTMENT shall mean a dwelling unit as defined in this code.

APARTMENT HOUSE—means any building, or portion thereof, which is designed, built, rented, leased, let or hired out to be occupied, or which is occupied as the home or residence of more than two (2) families living independently of each other and doing their own cooking in the said building, and shall include flats and apartments.

APPLICABLE GOVERNING BODY—a city, county, state, state agency or other political government subdivision or entity authorized to administer and enforce the provisions of this code, as adopted or amended.

APPROVED means approved by the building official or other authority having jurisdiction.

ARCHITECT—within the meaning of this Code, shall be deemed to be a duly registered and licensed architect.

AREA (building) is the maximum horizontally projected area of the building at or above grade, exclusive of court and vent shafts.

AREA (floor) is the usable area of each story of a building, or portion thereof, within surrounding exterior walls.

A. S. T. M.—means American Society for Testing and Materials.

ASSEMBLY OCCUPANCY—(Defined in Section 408.1).

ATTIC STORY—means any story situated wholly or partly in the roof, so designated, arranged or built as to be used for business, storage or habitation.

AUTOMATIC as applied to fire protection devices, is a device or system providing an emergency function without the necessity for human intervention and activated as a result of a predetermined temperature rise, rate of rise of temperature, or combustion products, such as incorporated in an automatic sprinkler system, automatic fire door, automatic fire shutter, or automatic fire vent.

AUTOMOTIVE SERVICE STATION—(Defined in Section 505).

BALCONY—means that portion of the seating space of an assembly room, the lowest part of which is raised four (4) feet or more above level of the main floor.

BASEMENT—means that portion of a building between floor and ceiling, which is partly below and partly above grade (as defined in this Section), but so located that the vertical distance from grade to the floor below is less than the vertical distance from grade to ceiling, provided, however, that the distance from grade to ceiling shall be at least four (4) feet six (6) inches. (See STORY).

BEAM—a primary structural member supporting secondary structural members, floor, roof, joists, and the like.

BOILER is a heating appliance intended to supply hot water or steam.

BUILDING—means any structure built for the support, shelter or enclosure of persons, animals, chattels, or property of any kind which has enclosing walls for 50% of its perimeter. The term "building" shall be construed as if followed by the words "or part thereof". (For the purpose of this Code each portion of a building separated from other portions by a fire wall shall be considered as a separate building.) For the purpose of area and height limitations this definition shall be applicable to sheds and open sheds.

SHED—means any structure built for the support, shelter or enclosure of persons, animals, chattels, or property of any kind which has enclosing walls for less than 50% of its perimeter.

OPEN SHED—means any structure that has no enclosing walls.

BUILDING (existing) is any structure erected prior to the adoption of this Code, or one for which a legal building permit has been issued.

BUILDING OFFICIAL is the officer or other designated authority charged with the administration and enforcement of this code, or his duly authorized representative.

BUILDING LINE—means the line, established by law, beyond which a building shall not extend, except as specifically provided by law.

BUSINESS OCCUPANCY—(Defined in Section 405.1).

CAST STONE—is a building stone manufactured from cement concrete precast and used as a trim, veneer or facing on or in buildings or structures.

CELLAR—means that portion of a building, the ceiling of which is entirely below grade or less than four (4) feet six (6) inches above grade. (See STORY.)

CHIMNEY CONNECTOR is the pipe which connects a fuel burning appliance to a chimney.

CITY—(See definition APPLICABLE GOVERNING BODY.)

COMBUSTIBLE MATERIAL—A material which cannot be classified as non-combustible in accordance with that definition.

COMMON-PROPERTY LINE—means a line dividing one lot from another when said lots are not of one ownership.

COMPLIANCE APPROVAL—means approval as being in compliance with the provisions of the Southern Standard Building Code as analyzed by the Committee on Compliance of the Southern Building Code Congress.

CONCRETE—(See Chapter XVI.)

CONSTRUCTION TYPES

Type I — See Section 602
Type II — See Section 603
Type III — See Section 604
Type IV — See Section 605
Type V — See Section 606
Type VI — See Section 607

CURB LEVEL—referring to a building, means the elevation at that point of the street grade that is opposite the center of the wall nearest to and facing the street line.

DEAD LOAD—(See Section 1202.)

DISPLAY SIGN—means a structure that is arranged, intended, designed or used as an advertisement announcement or direction, and includes a sign, sign screen, billboard and advertising devices of every kind.

DORMITORY is a space in a unit where group sleeping accommodations are provided with or without meals for persons not members of the same family group, in one room, or in a series of closely associated rooms under joint occupancy and single management, as in college dormitories, fraternity houses, military barracks, and ski lodges.

DWELLING—when used in this Code without other qualifications, means a structure occupied exclusively for residential purpose by not more than two families.

DWELLING UNIT is a single unit providing complete, independent living facilities for one or more persons including permanent provisions for living, sleeping, eating, cooking and sanitation.

ENGINEER—within the meaning of this Code, shall be deemed to be a duly registered and licensed engineer.

EXISTING BUILDING—(See BUILDING—EXISTING).

EGRESS, MEANS OF—See Definition, Section 1102.

EXIT—See Definition, Section 1102.

EXIT ACCESS—See Definition, Section 1102.

FAMILY—means one or more persons living together, whether related to each other by birth or not, and having common housekeeping facilities.

FIRE DISTRICT—(See Section 301.)

FIRE DOOR—means a door and its assembly, so constructed and assembled in place as to give the specified protection against the passage of fire.

FIRE-RESISTANCE RATING—means the time in hours that the material or construction will withstand the standard fire exposure as determined by a fire test made in conformity with the "Methods of Fire Tests of Building Construction and Materials, ASTM E119-71."

FIRE RETARDANT (Pressure treated) WOOD—means wood chemically impregnated in accordance with A.W.P.A. Specification C1-61. Which must have a flame spread rating not exceeding 25, with no evidence of significant progressive combustion when tested for 30 minutes in accordance with "Methods of Test for Surface Burning Characteristics of Building Materials, ASTM E 84-68." All material shall bear identification showing the fire performance rating thereof issued by an approved agency having a re-examination service and when exposed to the weather or sustained high humidity shall be identified as "Exterior". Exterior grade shall comply with the requirements of the Method of Test for Durability of Fire Retardant Treatment of Wood ASTM D 2898-70T.

FIRE WALL is a fire resistive wall, having protected openings, which restricts the spread of fire and extends continuously from the foundation to or through the roof.

FLAME SPREAD is the propagation of flame over a surface.

FLAME SPREAD RATING—Is that numerical value assigned to a material tested in accordance with "Methods of Test for Surface Burning Characteristics of Building Materials, ASTM E84-68."

FLOOR AREA—See Area (floor).

FRONT OF LOT—means the front boundary line of a lot bordering on the street, and in the case of a corner lot, may be either frontage.

GALLERY—means that portion of the seating space of an assembly room having a seating capacity of more than ten (10) located above a balcony.

GARAGE—PRIVATE GARAGE—(Defined in Section 506).

GARAGE—PUBLIC GARAGE—means any garage other than a private garage.

GRADE is a reference plane representing the average of finished ground level adjoining the building at all exterior walls.

GRADE—with reference to lumber, means the division of sawn lumber into quality classes with respect to its physical and mechanical properties as defined in published lumber manufacturers' standard grading rules.

HABITABLE SPACE is space in a structure for living, sleeping, eating or cooking. Bathrooms, toilet compartments, closets, halls, storage or utility space, and similar areas are not considered habitable space.

HEATING—(All definitions in Chapter VIII and Mechanical Code.)

HEIGHT—as applied to a building, means the vertical distance from grade to the highest finished roof surface in the case of flat roofs or to a point at the average height of roofs having a pitch of more than one (1) foot in four and one-half (4½) feet; HEIGHT of a building in stories does not include basements and cellars, except as specifically provided otherwise.

HEIGHT—as applied to a story, means the vertical distance from top to top of two successive finished floor surfaces.

HEIGHT—as applied to a wall, means the vertical distance to the top measured from the foundation wall, or from a girder or other intermediate support of such wall.

HOTEL is any building containing six or more guest rooms intended or designed to be used, or which are used, rented or hired out to be occupied or which are occupied for sleeping purposes by guests.

INCOMBUSTIBLE MATERIAL—is synonymous with NON-COMBUSTIBLE MATERIAL.

INDUSTRIAL OCCUPANCY—(Defined in Section 410.1).

INNER COURT—an open unoccupied space bounded by the walls of the building, but located within the exterior walls of the building.

INSTITUTIONAL OCCUPANCY—(Defined in Section 407.1).

INTERIOR LOT LINE—is synonymous with COMMON-PROPERTY LINE.

LINTEL—means the beam or girder placed over an opening in a wall which supports the wall construction above.

LIVE LOAD—(See Section 1203).

LOAD DURATION is the period of continuous application of a given load, or the aggregate of periods of intermittent applications of the same load.

LODGING HOUSE is any building or portion thereof containing not more than five guest rooms which are used by not more than five guests where rent is paid in money, goods, labor or otherwise. A lodging house shall comply with all the requirements for dwellings.

LOT—a parcel of land considered as a unit.

MASONRY—means that form of construction, composed of stone, brick, concrete, gypsum, hollow clay tile, concrete block or tile, or other similar building units or materials or a combination of these materials laid up unit by unit and set in mortar. For the purpose of this Code, plain monolithic concrete shall be considered as masonry. (See Section 1402.6).

SOLID MASONRY—means masonry built without hollow spaces.

MEZZANINE is an intermediate level between the floor and ceiling of any story, and covering less than 33⅓ percent of the floor area immediately beneath.

MIXED TYPES OF CONSTRUCTION—has the meaning as set forth in Section 609 of this Code.

MOTEL shall mean hotel as defined in this Code.

MULTIPLE DWELLING—has the same meaning as APARTMENT HOUSE.

NON-COMBUSTIBLE MATERIAL—A material which, in the form in which it is used, meets any of the following requirements:

(1) Materials that are judged to be non-combustible shall have been successfully tested in accordance with the "Method of Test for Determining Non-Combustiblity of Elementary Materials, ASTM E136-65."

(2) Materials having a structural base of non-combustible material as defined above in paragraph 1, with a surfacing not more than one-eighth (⅛) inch thick which in addition has a flame spread rating not greater than fifty (50) when tested in accordance with the "Method of Test for Surface Burning Characteristics of Building Materials, ASTM E84-68."

(3) Materials, other than described in 1. or 2., having a surface flame-spread rating no greater than twenty-five (25) without evidence of continued progressive combustion on any exposed surface that may be exposed by cutting through the material in any way.

The term non-combustible does not apply to the flamespread characteristics of interior finish or trim materials. No material shall be classed as a non-combustible building material which is subject to increase in combustibility or flamespread beyond the limits herein established through the effects of age, moisture or other atmospheric conditions.

OCCUPANCY is the purpose for which a building, or part thereof, is used or intended to be used.

MIXED OCCUPANCY—means mixed occupancy as set forth in Section 412 of this Code.

SPECIAL OCCUPANCY—means Group H Occupancy, as set forth in Section 411 of this Code.

OWNER is any person, agent, firm or corporation having a legal or equitable interest in the property.

PARTITION—means an interior wall, other than folding or portable, that subdivides spaces within any story, attic or basement of a building.

PENTHOUSE is an enclosed structure above the roof of a building, other than a roof structure or bulkhead, occupying not more than one-third of the roof area.

PERMIT is an official document or certificate issued by the authority having jurisdiction authorizing performance of a specified activity.

PERSON—means a natural person, his heirs, executors, administrators, or assigns, and also includes a firm, partnership, or corporation, its or their successors or assigns, or the agent of any of the aforesaid.

PLENUM is an air compartment or chamber to which one or more ducts are connected and which forms part of an air distribution system.

PUBLIC PARKING DECKS—means a special structure limited in use only to the temporary parking of motor vehicles.

PUBLIC SPACE is a legal open space on the premises, accessible to a public way or street, such as yards, courts or open spaces permanently devoted to public use which abuts the premises, and that is permanently maintained accessible to the Fire Department and free of all incumbrances that might interfere with its use by the Fire Department.

REPAIR is the reconstruction or renewal of any part of an existing building for the purpose of its maintenance.

REQUIRED—means required by some provision of this Code.

RESIDENTIAL OCCUPANCY—(Defined in Section 404.1).

ROOF STRUCTURE—means a structure above a roof or any part of a building enclosing a stairway, tank, elevator machinery or ventilating apparatus, or such part of a shaft as extends above the roof.

ROOM CAPACITY—(See Chapter XI).

SCHOOL OCCUPANCY—(Defined in Section 406.1).

SEATING CAPACITY—(See Chapter XI).

SELF-CLOSING as applied to a fire door or other opening, means normally closed and equipped with an approved device which will insure closing after having been opened for use.

SERVICE STATION—(Defined in Section 505).

SHAFT—means a vertical opening extending through one or more stories of a building, for elevators, dumbwaiter, light, ventilation, or similar purpose.

SHALL—as used in this Code, is mandatory.

SIGNS—(See Chapter XXIII).

SPECIAL OCCUPANCY—(Defined in Section 411.1).

SPRINKLERED—means equipped with an approved automatic sprinkler system properly maintained. See Chapter IX.

STAGE—GENERAL—a stage is a partially enclosed portion of an Assembly Building, cut off from the audience section by a proscenium wall, which is designed or used for the presentation of plays, demonstrations, or other entertainment. "Stages" shall be classified as "working stage" and "non-working stage".

STAGE—WORKING—(Also Theatrical Stage — See Section 512.11) — a working stage is a partially enclosed portion of an Assembly Building, cut off from the audience section by a proscenium wall of masonry of not less than 4 hour fire-resistance construction, and which is equipped with scenery loft, gridiron, fly-gallery, and lighting equipment, and the proscenium opening shall be equipped with a fire-proof and smoke-proof curtain, and the depth from the proscenium curtain to the back wall shall be not less than fifteen (15) feet.

STAGE—NON-WORKING—a non-working stage is a partially enclosed portion of an Assembly Building, cut off from the audience section by a proscenium wall of not less than one-hour fire-resistive construction, without the equipment common to the Working Stage (such as fly gallery and gridiron) and of such dimensions that such equipment cannot be installed (but flat scenery may be used on such stage).

A fireproof curtain is not required for a non-working stage, but if there is a fabric or other curtain it shall be of incombustible materials or treated with an approved fire retardant. The depth of the stage may be more or less than fifteen (15) feet.

STAGE, PLATFORM—a platform is a raised section of floor within the assembly hall or auditorium area, and setting on the floor thereof, not enclosed above the platform floor level, and usually or relatively small area as compared to the auditorium seating area. A platform may be of permanent, temporary, or portable construction; it may have "flat" movable scenery and draw curtains.

STAGE, ROSTRUM—(See "platform"). Usually used for single or small group of persons such as lecturers, no scenery or curtains. May be permanent, temporary, or portable.

STAGE, DAIS—a small and low "platform" or "rostrum" may be placed on a "stage". May be permanent, temporary, or portable.

STAGE, PODIUM—a small "dais" of size sufficient to accommodate one or two persons, such as a band or orchestra conductor or a soloist. A "podium" may be located on a Stage, Platform, Rostrum, or Dais, or the floor of the audience section of a place of assembly. A podium is almost always a portable construction.

STAIRWAY—means one or more flights of stairs and the necessary landings and platforms connecting them, to form a continuous and uninterrupted passage from one story to another in a building or structure.

STANDARD FIRE TEST—means the fire test formulated under the procedure of the United States of America Standards Institute as the "Methods of Fire Tests of Building Construction and Materials, ASTM E119-71," ANSI A2.1-1972.

STORY is that portion of a building included between the upper surface of a floor and upper surface of the floor or roof next above.

STREET—means any public thoroughfare (street, avenue, boulevard, park) or space more than twenty (20) feet in width which has been dedicated or deeded to the public for public use.

STREET LINE—means a lot line dividing a lot from a street.

STRUCTURE—means that which is built or constructed, an edifice or building of any kind, or any piece of work artificially built up or composed of parts joined together in some definite manner. The term "structure" shall be construed as if followed by the words "or part thereof."

SURVEYOR—within the meaning of this Code, shall be deemed to be duly registered and licensed surveyor or Civil Engineer.

THEATER—means a building, or part thereof, which contains an assembly hall with or without stage which may be equipped with curtains and permanent stage scenery or mechanical equipment adaptable to the showing of plays, operas, motion pictures, performances, spectacles and similar forms of entertainment. (See Section 408.)

VALUATION OR VALUE—as applied to a building, means the estimated cost to replace the building in kind.

VENEER—means a facing attached to a wall for the purpose of providing ornamentation, protection, or insulation, but not counted as adding strength to the wall.

VERTICAL OPENING is an opening through a floor or roof.

WALL (bearing) is a wall supporting any vertical load in addition to its own weight.

WALL (nonbearing) is a wall which supports no vertical load other than its own weight.

WALL, CAVITY—means a wall built of masonry units or of plain concrete, or a combination of these materials, so arranged as to provide an air space within the wall, and in which the inner and outer parts of the wall are tied together with metal ties.

WALL, CURTAIN—means a non-bearing wall between columns or piers and which is not supported by girders or beams, but is supported on the ground.

WALL, FACED—means a wall in which the masonry facing and backing are so bonded as to exert common action under load.

WALL, EXTERIOR—means a wall, bearing or non-bearing, which is used as an enclosing wall for a building, but which is not necessarily suitable for use as a Party Wall or Fire Wall.

WALL, FOUNDATION—means a wall below the first floor extending below the adjacent ground level and serving as support for a wall, pier, column or other structural part of a building.

WALL OF MASONRY, HOLLOW—means a wall built of masonry units so arranged as to provide an air space within the wall, and in which the inner and outer parts of the wall are bonded together with masonry units or steel.

WALL, PANEL—means a non-bearing wall in skeleton or framed construction, built between columns or piers and wholly supported at each story.

WALL, PARAPET—means that part of any wall entirely above the roof line.

WALL, PARTY—A wall on an interior lot line, used or adapted for joint service between two (2) buildings.

WALL (retaining) is a wall designed to prevent the lateral displacement of soil or other material.

WRITING—includes printing and typewriting.

WRITTEN NOTICE is a notification in writing delivered in person to the individual or parties intended, or delivered at, or sent by certified or registered mail to the last residential or business address of legal record.

CHAPTER III

FIRE DISTRICT

SECTION 301 — GENERAL BUILDING RESTRICTIONS— WITHIN THE FIRE DISTRICT

301.1 — GENERAL

For the purpose of this Code there shall be established a Fire District and such Fire District shall be in accordance with the provisions of Section 301.4—SCOPE.

301.2 — TYPES OF CONSTRUCTION PERMITTED

Within the Fire District every building hereafter erected shall be one of the following Types (except as permitted in Section 304).

Type I

Type II

Type III

Type IV

Type V

301.3 — OTHER SPECIFIC REQUIREMENTS

(a) Exterior Walls

Exterior walls of buildings located in the Fire District shall comply with the requirements specified in Chapter VI—except as set forth in Section 608.3.

(b) Group "H" Special Hazardous Not Permitted

Every Group "H" Occupancy shall be prohibited from location within the Fire District.

(c) Fire Protection

Every building shall be fire protected throughout as specified for the various Types of Construction in Chapter VI.

(d) Roof Coverings

Roof covering in the Fire District shall conform to the requirements of Type A or B—Roof Coverings as defined in Section 706.

(e) Interior Fire Protection Within Fire District

In buildings over one (1) story in height (unless of Type III Construction; Sprinklered throughout; a Public Parking Deck; or surrounded on all sides by a permanently open space of not less than thirty (30) feet) all walls, floors, roofs and their supporting structural members shall provide not less than one (1) hour fire resistance. (Temporary partitions are set forth in Section 702.)

301.4 — SCOPE

(a) The Fire District shall include such territory or portion as outlined in an Ordinance or law entitled "An Ordinance (Resolution) Creating and Establishing A Fire District." Whenever, in such ordinance creating and establishing a Fire District, reference is made to the Fire District, it shall be construed to mean the Fire District designated and referred to in this Chapter. (See Appendix "G" for recommended method of establishing said Fire District.)

(b) The Fire District complying with the provisions of this Section shall be shown on a map, which shall be available to the public.

SECTION 302 — CHANGES TO BUILDINGS

302.1 — EXISTING BUILDINGS WITHIN THE FIRE DISTRICT

No existing building shall be hereafter increased in height unless it is of a type of construction permitted for new buildings within the Fire District or is altered to comply with the requirements for such type of construction. Nor, shall any existing building be hereafter extended on any side unless such extensions are of a type of construction permitted for new buildings within the Fire District.

302.2 — ALTERATIONS — GENERAL

Nothing in this Section, however, shall prohibit other alterations within the Fire District provided there is no change of Occupancy that is otherwise prohibited and provided the fire hazard is not increased by such alteration.

302.3 — MOVING BUILDINGS

No buildings shall hereafter be moved into the Fire District or to another lot in the Fire District unless it is of a type of construction permitted in the Fire Disrict. (See Section 2204 — Regulations for Moving Buildings.)

SECTION 303 — BUILDINGS LOCATED IN AND OUT OF THE FIRE DISTRICT

Any building located partially in the Fire District shall be of a type of construction required for the Fire District, unless the major portion of such building lies outside of the Fire District and no part is more than ten (10) feet inside the boundaries of the Fire District.

SECTION 304 — EXCEPTIONS TO RESTRICTIONS IN FIRE DISTRICT

The preceding provisions of this Chapter shall not apply in the following instances:

(a) Temporary Buildings used in connection with duly authorized construction.

(b) A private garage used exclusively as such, not more than one (1) story in height, nor more than six hundred and fifty (650) square feet in area, located on the same lot with a dwelling.

(c) Fences not over eight (8) feet in height.

(d) Coal Tipples, Material Bins, Trestles conforming to Section 503.

(e) Water Tanks and Cooling Towers conforming to Section 713 and Section 714.

(f) Greenhouses less than fifteen (15) feet high.

(g) Porches on dwellings not over one (1) story in height and not over ten (10) feet wide from the face of the building, provided such porch does not come within five (5) feet of any property line.

(h) Display signs as provided in Chapter XXIII.

(i) Sheds open on a long side not over fifteen (15) feet high five hundred (500) sq. ft. in area.

(j) One and two family dwellings when of a type of construction not permitted in the Fire District may be extended 25% of the floor area existing at the time of inclusion in the Fire District by any type of construction permitted by this Code.

SECTION 305 — OUTSIDE THE FIRE DISTRICT

Outside the Fire District, all types of construction are permitted provided they comply with the provisions prescribed elsewhere in this Code that apply regardless of location. Roof Coverings shall conform to the requirements as defined in Section 706.

CHAPTER IV

CLASSIFICATION OF BUILDING BY OCCUPANCY

SECTION 401 – CLASSIFICATION BY OCCUPANCY OR USE

401.1 — GENERAL

Every new and existing building, structure or part thereof shall, for the purpose of this Code, be classified according to its use, or occupancy as a building or structure of one of the following OCCUPANCY GROUPS:

GROUP A—RESIDENTIAL
GROUP B—BUSINESS
GROUP C—SCHOOL
GROUP D—INSTITUTIONAL
GROUP E—ASSEMBLY
GROUP F—STORAGE
GROUP G—INDUSTRIAL
GROUP H—HAZARDOUS

Each occupancy group is intended to embrace buildings as hereinafter defined and those of similar character or use. Wherever there is any uncertainty as to the classification of a building, the Building Official shall fix the classification within which it falls, according to the relative fire hazard involved.

SECTION 402 — HEIGHT AND AREA RESTRICTIONS

402.1 — APPLICATION

In the protection of each occupancy, the maximum height and area for buildings or structures of the different types of construction, shall be governed by the intended use of the buildings or structures, or occupancy, as provided for in this Chapter.

402.2 — DEFINITIONS

For the purpose of this Code, "height" and "area," as applied to a building, has the meaning designated in Chapter II, Definitions.

402.3 — EXCEPTIONS — HEIGHT

(a) Church spires, chimneys, tanks and supports, aerial supports, parapet walls not over four (4) feet high, bulkheads and penthouses used solely to enclose stairways, tanks, elevator machinery or shafts, or ventilation or air-conditioning apparatus, need not be considered in determining the highest point of the building; provided that the highest point shall be taken to be the highest point of the roof of the highest penthouse when the aggregate area of all penthouses and other roof structures exceeds one-third (⅓) of the area of the roof upon which they stand. (See requirements of Section 712 and 713).

(b) Where a 1-story automobile parking area (enclosed or open) of Type I or Type II Construction, with grade entrance is provided under a building of Group A Occupancy, the number of stories to be used in determining the minimum type of construction may be measured from the roof slab of such parking area.

402.4 — MEZZANINES

Mezzanine floors or galleries shall not be regarded as a story unless they exceed thirty-three and one-third (33⅓) percent of the aggregate ground floor area.

402.5 — LIMITATIONS, HEIGHT

The basement or cellar of a building shall not count as a story if the first floor above such basement or cellar is less than seven (7) feet above grade.

402.6 — HEIGHT INCREASE FOR SPRINKLERS.

The maximum allowable number of stories may be increased by one story if the building is provided with automatic sprinklers throughout in accordance with Section 901 of Chapter IX, except such height increase shall not be permitted in buildings where the installation of automatic sprinkler equipment is a mandatory requirement of this code, or when the provisions of Section 403.6 are used.

402.7 — EXISTING BUILDINGS

No existing building may be increased in height unless the entire building is altered as needed to meet the requirements of this Code for a new building of such increased height and number of stories.

SECTION 403 — EXCEPTIONS TO AREA RESTRICTIONS

403.1 — GENERAL

The exceptions and requirements of this Section shall modify the area limits of this Chapter, as herein provided.

403.2 — AREA INCREASES NOT PERMITTED (WITH EXCEPTION)

The increase of floor areas permitted by this Section may be additive when applicable, except that in buildings where, because of occupancy, type construction or fire district, one-hour fire-resistance construction is a requirement, or automatic sprinkler equipment is required, as specified in Section 901, no increase of area shall be permitted because of such construction or equipment. Except as provided in Section 403.7, the total permissive area shall not exceed 450% of the basic area.

403.3 — EXCEPTIONS — AREA — FIRE DIVISION WALLS

(a) General

For the purpose of this Code, each part of a building included within fire walls shall be deemed to be a separate building.

(b) New Buildings

No building shall be limited in area when divided into sections by fire walls having not less than four hour fire-resistance ratings, as specified in Chapter VI, provided no section exceeds the maximum allowable floor area in this Chapter.

(c) Existing Buildings

No building hereafter erected shall be extended to exceed the maximum floor area set forth in this chapter, governed by the occupancy and type of construction. However, a building heretofore lawfully erected, which already exceeds such maximum area, may be extended, provided such extension does not exceed the maximum area prescribed and provided such extension is separated from the existing building by a fire wall having a fire-resistance rating of not less than four hours.

403.4 — AREA INCREASE FOR ONE-HOUR FIRE-RESISTANCE

Except in buildings where the occupancy or conditions are such that one-hour fire-resistance is a requirement of this Code, buildings of Type IV, Type V and Type VI Constructions, interiors of which are provided throughout with not less than one-hour fire-resistive construction may have the maximum allowable areas of this Chapter increased fifty (50) percent.

403.5 — AREA INCREASE FOR SEPARATION ON TWO OR MORE SIDES OF A BUILDING

Where streets or public spaces of minimum width not less than 20 feet extend along two or more sides of a building of any occupancy classification, except Group "H" Hazardous, the maximum areas specified in this Chapter and modified as provided in this Section for such buildings may be increased by the percentage specified in Table 403.5 for each foot by which the minimum width of such streets or public spaces exceeds 20 feet, but such increase shall not exceed the maximum percentage shown in Table 403.5.

TABLE 403.5 — RATE OF AREA INCREASE FOR SEPARATION

	Rate for Areas Increase for Separation Over 20' Wide	Max. Area Increase
Separation along 2 sides but along not less than 50% of perimeter of building	1% per Ft. over 20 Ft.	50%
Separation along 3 sides but along not less than 75% of perimeter of building	2% per Ft. over 20 Ft.	100%
Separation along all sides or along 100% of perimeter of building	3% per Ft. over 20 Ft.	100%

403.6 — AREA INCREASE FOR SPRINKLERS

The maximum allowable floor and attic area may be increased by 200% for one story buildings, and by 100% for buildings over one story in height if the building is provided with automatic sprinklers throughout in accordance with Section 901 of Chapter IX, except such area increase shall not be permitted in buildings where the installation of automatic sprinkler equipment is a mandatory requirement of this code, or when the provisions of Section 402.6 are used.

403.7 — UNLIMITED AREAS

The area of a one-story building of Group B Business, Group F Storage, or Group G Industrial occupancy, shall not be limited provided the building is equipped with an approved automatic sprinkler system throughout, in accordance with Section 901 or other automatic extinguishing systems as approved by the Building Official, and is surrounded on all sides by a permanent open space of not less than sixty (60) feet.

Where water may cause or increase a fire, other fire extinguishing systems shall be required in rooms or buildings used for the manufacture or storage of hazardous materials including but not limited to, aluminum powder, calcium carbide, calcium phosphate, metallic sodium and potassium, quicklime, magnesium powder and sodium peroxide.

403.8 — ASSEMBLY OCCUPANCY AREA INCREASE

Refer to Sections 512.2 and 514.6.

403.9 — SCHOOL OCCUPANCY AREA INCREASE

Refer to Section 406.5.

SECTION 404 — GROUP "A" — RESIDENTIAL

404.1 — SCOPE

Buildings in which families or households live or in which sleeping accommodations are provided, and all dormitories, shall be classified as Group A—Residential Occupancy. Group A—Residential Occupancy—shall include, among others, the following:

- Dwellings
- Multiple Dwellings (more than two families)
- Hotels and Motels
- Dormitories
- Lodging Houses
- Convents
- Monasteries

404.2 — PROTECTIVE REQUIREMENTS — GROUP "A" OCCUPANCY

	SECTION
1. Allowable Height and Area	404.4
2. Types of Construction	Chapter VI
3. Exit Requirements	Chapter XI
4. Protection of Vertical Openings	701
5. Protection of Wall Openings	703
6. Sprinklers and Standpipes	901 to 902, inclusive
7. Mixed Occupancy and Separation Requirements	412
8. Light, Ventilation and Sanitation	2001 to 2002, inclusive
9. Heating Requirements	Chapter VIII

404.3 — SPECIAL REQUIREMENTS, GROUP "A" OCCUPANCY

1. Separation of Furnace and Boiler Room .... Section 812
2. Storage and handling of flammable liquids shall be prohibited in every Group "A" Occupancy. Not more than one (1) gallon of flammable liquid, used for cleaning purposes only, may be kept in a residence, provided such flammable liquid is kept in an approved container, used especially for that purpose.

404.4 — GROUP "A" RESIDENTIAL OCCUPANCY—HEIGHT AND AREA RESTRICTIONS

ALLOWABLE HEIGHTS Type Construction Used		ALLOWABLE AREAS*** Square Feet per Floor		
	Story Height	First Floor	Second Floor	Third Floor and Above
Type I	No Limit	No Limit	No Limit	No Limit
Type II	80 ft.	No Limit	No Limit	No Limit
Type III	Three	18,000	15,000	9,750
Type IV	Five*	12,000	10,000	6,500
Type V	Five*	12,000	10,000	6,400
Type VI	Three	7,000	7,000	6,400**

*When five stories or more in height, two-hour fire-resistive floors shall be required over basement or cellar.

**One-hour fire-resistive construction shall be required throughout.

***See Section 403 for Allowable Area Increases.

SECTION 405 — GROUP "B" — BUSINESS

405.1 — SCOPE

Buildings which are occupied for business or rendering of professional services shall be classified in Group B-1; buildings which are occupied for the sale or display of merchandise, or the supplying of food or drink, shall be classified in Group B-2.

Group B—Business Occupancy includes, among others, the occupancies listed below, but does not include buildings used for any purpose involving highly combustible, inflammable or explosive materials.

Group B-1—Office buildings, greenhouses, service stations, banks, undertaking parlors, temporary structures, libraries (other than school).

Group B-2—Stores, shops, markets, restaurants, bowling alleys (See exception in Section 405.2).

405.2 — EXCEPTION

Restaurants or places supplying food or drink that accommodate 75 or more people, or that have a stage, or that provide dancing or entertainment features, shall be classified in Group E—Assembly and not in Group B — Business Occupancy (See Section 408).

405.3 — PROTECTIVE REQUIREMENTS—GROUP "B" OCCUPANCY

	SECTION
1. Allowable Height and Area	405.5
Heights and Areas are based upon type of construction used.	
2. Types of Construction	Chapter VI
3. Exit Requirements	Chapter XI
4. Protection of Vertical Openings	701
5. Protection of Wall Openings	703
6. Sprinklers and Standpipes	901 to 902, inclusive
7. Mixed Occupancy and Separation Requirements	412
8. Light, Ventilation and Sanitation	2001 to 2002, inclusive
9. Heating Requirements	Chapter VIII

405.4 — SPECIAL REQUIREMENTS—GROUP "B" OCCUPANCY

	SECTION
1. Separation of Boiler and Furnace Room	812
2. Special Exit Doorway Requirements	Chapter XI
3. Temporary Structures	504
4. Service Stations	505
5. Greenhouses	509
6. Rat-proof Construction	Chapter XIX

7. Storage and handling of flammable liquids shall be prohibited; except in Group B-2 (retail stores) that are not in excess of Section 411.2 (b); except service stations which shall conform to Section 501.1 (f).

403.5 — GROUP "B" BUSINESS OCCUPANCY — HEIGHT AND AREA RESTRICTIONS

ALLOWABLE HEIGHTS		ALLOWABLE AREAS**		
Type Construction		Square Feet per Floor		
	Story Height	First Floor	Second Floor	Third Floor and Above
Type I	No Limit	No Limit	No Limit	No Limit
Type II	80 ft. (a)	No Limit	No Limit	No Limit
Type III				
B-1—Offices, etc.	Five	25,500	18,000	13,500
B-2—Mercantile	Five	13,500	12,000	8,000
Type IV*				
B-1—Offices, etc.	Five	17,000	12,000	6,000
B-2—Mercantile	Five	9,000	8,000	4,000
Type V*				
B-1—Offices, etc.	Five	14,000	11,000	6,000
B-2—Mercantile	Five	9,000	8,000	4,000
Type VI				
B-1—Offices, etc.	Two	9,000	7,000	Not Permitted
B-2—Mercantile	Two	6,000	4,000	Not Permitted

*When five or more stories in height, two-hour fire-resistive floors shall be constructed over basement or cellar.

**For allowable increase in areas for additional protection, see Section 403.

(a) The height of Type II Construction for buildings of Group "B" Business Occupancies shall not be limited provided the fire-resistance of all columns shall be not less than 3 hours, and of the other structural members, including floors, shall be not less than that shown in Table 603.5, but in no case less than 2 hours, except that roofs shall be of not less than 1½-hour fire resistance construction.

(b) In all "B-1" occupancies—Office Buildings—of Type I Construction, the partitions, columns, trusses, girders, beams and floors may be reduced by one (1) hour if the building is equipped with an approved automatic sprinkler system throughout, but no component or assembly shall be less than one (1) hour fire resistance.

SECTION 406 — GROUP "C" — SCHOOLS

406.1 — SCOPE

Buildings in which people come together for education or instructional purposes shall be classified in Group "C" — School Occupancy.

Group C — School Occupancy shall include, among others, the following:

Schools	Universities
Colleges	Academies

406.2 — EXCEPTION

(a) Parts of buildings used for the congregating or gathering of 75 or more persons in one room shall be classified as in Group E — Assembly Occupancy—(See Section 408), regardless of whether such gathering is of an educational or instructional nature or not.

(b) Schools for business or vocational training shall be classified in the same occupancies and conform to the same requirements as the trade, vocation or business taught.

406.3 — PROTECTIVE REQUIREMENTS, GROUP "C" OCCUPANCY

	SECTION
1. Allowable Height and Area	406.5
Heights and Areas are based upon type of construction used.	
2. Types of Construction	Chapter VI
3. Exit Requirements	Chapter XI
4. Protection of Vertical Openings	701
5. Protection of Wall Openings	703
6. Sprinklers and Standpipes	901 to 902, inclusive
7. Mixed Occupancy and Separation Requirements	412
8. Light, Ventilation and Sanitation	2001 to 2002, inclusive
9. Heating Requirements	Chapter VIII

406.4 — SPECIAL REQUIREMENTS, GROUP "C" OCCUPANCY

	SECTION
1. Separation of Boiler and Furnace Room	812
2. Non-combustible Stairways Required	Chapter XI
3. Corridors	Chapter XI
4. Unilateral Light	2001.4

5. No classroom shall occupy a basement area which is fifty (50) percent or more below ground level.
6. Every heating appliance which produces an unprotected open flame shall be prohibited.
7. Storage and handling of flammable liquids shall be prohibited.
8. Where permanent motion picture projectors are used, booths shall be provided, as set forth in Section 512.16.
9. Rooms used for day care nurseries, kindergarten or first grade pupils shall not be located above or below the floor of exit discharge. Rooms used for second grade pupils shall not be located more than one story above the floor of exit discharge.

406.5 — GROUP "C" SCHOOL OCCUPANCY — HEIGHT AND AREA RESTRICTIONS

ALLOWABLE HEIGHTS Type Construction Used**	Story Height	ALLOWABLE AREAS Square Feet per Floor First Floor***	Second Floor	Third Floor and Above
Type I****	No Limit	No Limit	No Limit	No Limit
Type II	80 Ft.	No Limit	No Limit	No Limit
Type III*	Two	18,000	12,000	Not Permitted
Type IV*	Two	12,000	12,000	Not Permitted
Type V*	Two	12,000	12,000	Not Permitted
Type VI*	Two	8,000	8,000	Not Permitted

*Floors located immediately above usable space in basements or cellars shall have a fire-resistance rating of not less than one-hour except where an approved automatic sprinkler system is provided; provided, however, that where basements or cellars are used as classrooms or assembly rooms they shall be counted as a story.

**At least one-hour interior fire-resistive construction shall be used throughout in all Group "C" (schools), two or more stories in height.

***The Area of a one-story Type III, IV, or V building may be increased one hundred percent (100%) if the building is surrounded on all sides by a permanent open space of not less than sixty (60) feet, and there are not less than two exits provided from each classroom, one of which opens directly to the exterior of the building. For other allowable area increases, see Section 403.

****In all Group "C" occupancies of Type I Construction, the partitions, columns, trusses, girders, beams and floors may be reduced by one hour if the building is equipped with an approved automatic sprinkler system throughout but no component or assembly may be less than one hour fire-resistance.

SECTION 407 — GROUP "D" — INSTITUTIONAL

407.1 — SCOPE

Buildings in which more than six people are detained for penal or correctional purposes, or in which the liberty of the inmates is restricted, or places of involuntary detention, shall be classified in Group D-1.

Buildings in which more than ten people are harbored for medical, charitable or other care or treatment shall be classified in Group D-2.

Group D-1 — Institutional Occupancy — shall include, among others, the following:

- Insane Asylums
- Reformatories
- Jails
- Prisons

Group D-2 — Institutional Occupancy — shall include, among others, the following:

- Hospitals
- Sanitariums
- Orphanages
- Homes for the Aged
- Nursing Homes

407.2 — EXCEPTION

Dormitories for doctors, nurses, and able-bodied employees (not for patients or inmates) of Institutional buildings shall be classified as Group A — Residential Occupancy.

407.3 — PROTECTIVE REQUIREMENTS — GROUP "D" OCCUPANCY

		SECTION
1.	Allowable Height and Area Heights and Areas are based upon type of construction used.	407.5
2.	Types of Construction	Chapter VI
3.	Exit Requirements	Chapter XI
4.	Protection of Vertical Openings	701
5.	Protection of Wall Openings	703
6.	Sprinklers and Standpipes	901 to 902, inclusive
7.	Mixed Occupancy and Separation Requirements	412
8.	Light, Ventilation and Sanitation	2001 to 2002, inclusive
9.	Heating Requirements	Chapter VIII

407.4 — SPECIAL REQUIREMENTS — GROUP "D" OCCUPANCY

SECTION

1. Separation of Boiler and Furnace Room 812
2. Special Exit Doorway Requirements ....Chapter XI
3. Special Exit Requirements for Sanitoriums ....Chapter XI
4. Non-Combustible Stairways Required ....Chapter XI
5. Handling and Storage of Combustible Film 501.3
6. Storage and Handling of Flammable Liquids shall be prohibited.

407.5 — GROUP "D" INSTITUTIONAL OCCUPANCY — HEIGHT AND AREA RESTRICTIONS

	ALLOWABLE HEIGHTS Type Construction Used*	ALLOWABLE AREAS** Square Feet per Floor		
	Story Height	First Floor	Second Floor	Third Floor and Above
Type I	No Limit	No Limit	No Limit	No Limit
Type II	80 Ft.	No Limit	No Limit	No Limit
Type III				
Group D-1	Not Permitted			
Group D-2	Two	12,000	8,000	Not Permitted
Type IV				
Group D-1	Not Permitted			
Group D-2	Two	7,000	4,000	Not Permitted
Type V				
Group D-1	Not Permitted			
Group D-2	Two	7,000	4,000	Not Permitted
Type VI				
Group D-1	Not Permitted			
Group D-2	One	5,000	Not Permitted	Not Permitted

*At least one-hour fire-resistive construction shall be provided throughout all buildings.

**See Section 403 for Allowable Area Increases.

SECTION 408 – GROUP "E" – ASSEMBLY

408.1 — SCOPE

Buildings in which provision is made for the congregation or gathering of seventy-five (75) or more persons in one room or space shall be classified in Group E — Assembly Occupancy. Such room or space shall include any occupied connecting room or space in the same story, or in a story or stories above or below, where entrance is common to the rooms or spaces. Group E occupancy includes buildings having an auditorium and a stage provided for the use of movable scenery, or having an auditorium for viewing motion pictures or for theatrical purposes.

Group E — Assembly Occupancy shall include, among others, the following:

Amusement Park Buildings	Passenger Depots
Auditoriums	Public Assembly Halls
Churches	Recreation Halls
Dance Halls	Restaurants (large)
Gymnasiums	Stadiums and Grandstands
Motion Picture Houses	Tents (Assembly)
Museums	Theaters

408.2 — SUB-CLASSIFICATIONS

Group E shall be divided into two sub-classifications as set forth in this Section, both of which shall comply with the requirements for Group E Occupancy unless otherwise specified:

(a) Group E-1 — Large Assembly

Group E-1 shall include theaters and places of public assembly having a working stage (see definition) and having a capacity of seven hundred (700) or more persons; also, Group E-1 shall include theaters or places of assembly having a non-working stage but having a capacity of 1,000 or more persons.

(b) Group E-2 — Small Assembly

Group E-2 shall include theaters and places of assembly having a capacity of 75 or more persons but having a capacity less than designated for Group E-1.

408.3 — METHOD OF DETERMINING CAPACITY

The capacity of occupant content of places of assembly shall be as set forth in Chapter XI.

408.4 — PROTECTIVE REQUIREMENTS — GROUP "E" OCCUPANCY

	SECTION
1. Allowable Height and Area Heights and Areas are based upon type of construction used.	408.6
2. Types of Construction	Chapter VI
3. Exit Requirements	Chapter XI
4. Protection of Vertical Openings	701
5. Protection of Wall Openings	703
6. Sprinklers and Standpipes	901 to 902, inclusive
7. Mixed Occupancy and Separation Requirements	412
8. Light, Ventilation and Sanitation	2001 to 2002, inclusive
9. Heating Requirements	Chapter VIII

408.5 — SPECIAL REQUIREMENTS — GROUP "E" OCCUPANCY

	SECTION
1. Separation of Boiler and Furnace Room	812
2. Special Exit Doorway Requirements	Chapter XI
3. Non-Combustible Stairway Requirements	Chapter XI

4. Special requirements governing the necessary features for total protection of Group "E" Assembly Occupancies, shall be in accordance with Section 512.

5. Tents	504.2
6. Stadiums and Grandstands	510
7. Amusement Park Buildings	511
8. Restaurants, Ratproof Construction	Chapter XIX

9. Storage and handling of flammable liquids shall be prohibited.

408.6 — GROUP "E" ASSEMBLY OCCUPANCY — HEIGHT AND AREA RESTRICTIONS

ALLOWABLE HEIGHTS		ALLOWABLE FLOOR AREA†		
Type Construction Used		Square Feet per Floor		
	Story Height	First Floor	Second Floor	Third Floor and Above
Type I*****	No Limit	No Limit	No Limit	No Limit
Type II	80 Ft.	No Limit	No Limit	No Limit
Type III*				
Group E 1 Working Stage	Not Permitted			
Group E 1 Non-Working Stage*	One	12,000	Not Permitted	
Group E 2 Working Stage	One	10,000****	Not Permitted	
Group E 2 Non-Working Stage*	Two	12,000****	12,000****	
Type IV**				
Group E 1 Working Stage	Not Permitted			
Group E 1 Non-Working Stage	One	8,000	Not Permitted	
Group E 2 Working Stage	One	6,000****	Not Permitted	
Group E 2 Non-Working Stage	Two	8,000****	8,000****	
Type V**				
Group E 1 Working Stage	Not Permitted			
Group E 1 Non-Working Stage	One	8,000	Not Permitted	
Group E 2 Working Stage	One	6,000****	Not Permitted	
Group E 2 Non-Working Stage	Two	8,000****	8,000****	
Type VI				
Group E 1	Not Permitted			
Group E 2 Working Stage	One	3,000	Not Permitted	
Group E 2 Non-Working Stage	One	5,000***	Not Permitted	

*Buildings which are surrounded on all sides by a permanent open space of not less than sixty (60) feet and provided with an approved automatic sprinkler system shall not be limited in area.

**One-hour fire-resistive floors shall be required. ***May be increased 33⅓ percent for places of worship.

****See Section 512.2 and Section 514.6 for exception to area limitation.

†See Section 403 for Allowable Area Increases.

*****In all Group "E" occupancies of Type I Construction, the columns, trusses, girders, beams and floors may be reduced by one (1) hour if the building is equipped with an approved automatic sprinkler system throughout, but no component or assembly shall be less than one (1) hour fire resistance.

SECTION 409 — GROUP "F" — STORAGE

409.1 — SCOPE

Buildings which are used for the storage of goods, wares or merchandise, excepting limited storage incidental to the display, sale or manufacture of such goods, wares or merchandise, shall be classified in Group F — Storage Occupancy.

Group "F" — Storage Occupancy — shall include, among others, the occupancies listed in this Section, but does not include buildings used to store highly combustible, inflammable or explosive products or materials (See Section 411):

- Airplane Hangars
- Garages
- Warehouses
- Storage Buildings
- Freight Depots

409.2 — PROTECTIVE REQUIREMENTS—GROUP "F"—STORAGE

	SECTION
1. Allowable Height and Area	409.4
Heights and Area are based upon type of construction used.	
2. Types of Construction	Chapter VI
3. Exit Requirements	Chapter XI
4. Protection of Vertical Openings	701
5. Protection of Wall Openings	703
6. Sprinklers and Standpipes	901 to 902, inclusive
7. Mixed Occupancy and Separation Requirements	412
8. Light, Ventilation and Sanitation	2001 to 2002, inclusive
9. Heating Requirements	Chapter VIII

409.3 — SPECIAL REQUIREMENTS — GROUP "F" — STORAGE

	SECTION
1. Separation of Boiler and Furnace Room	812
2. Non-combustible Stairway Requirements	Chapter XI
3. Occupancy Permit for Changed Floor Loads	1207
4. Posting of Floor Loads	110
5. Airplane Hangars	502
6. Coal Pockets	503
7. Garages	
Private	506
Public	508
8. Parking Lots and Public Parking Decks	507
9. Storage and handling of flammable liquids	501.1 (f)
10. Ratproof Construction	Chapter XIX

409.4 — GROUP "F" — STORAGE OCCUPANCY —HEIGHT AND AREA RESTRICTIONS

ALLOWABLE HEIGHT Type Construction Used		ALLOWABLE FLOOR AREA** Square Feet Per Floor		
	Story Height	First Floor	Second Floor	Third Floor and Above
Type I	No Limit	No Limit	No Limit	No Limit
Type II	Six	30,000	30,000	20,000
Type III*	Six	24,000	24,000	15,000
Type IV*	Four	16,000	16,000	10,000
Type V*	Four	16,000	16,000	10,000
Type VI	One	6,000	Not Permitted	

*When three stories or more in height, an approved automatic sprinkler system shall be installed throughout the building. (Does not apply to Public Parking Decks.)

**May be increased beyond the areas set forth in the above table, see Section 403.

SECTION 410 — GROUP "G" — INDUSTRIAL

410.1 — SCOPE

Buildings in which work or labor is performed in connection with the fabrication, assembly, processing, etc., of products or materials shall be classified in Group G — Industrial Occupancy. Group G — Industrial Occupancy — shall include, among others, the occupancies listed in this Section, but does not include buildings used for any purpose involving highly combustible, inflammable, or explosive products or materials (See Section 411):

Manufacturing Plant

Factory

Assembly Plant

Processing Plant

Mill

410.2 — PROTECTIVE REQUIREMENTS — GROUP "G" — INDUSTRIAL

	SECTION
1. Allowable Height and Area	410.4
(a) Exception to Area Restrictions	403
(b) Unlimited Areas	403.7

Heights and Areas are based upon type of construction used.

2. Types of Construction	Chapter VI
3. Exit Requirements	Chapter XI
4. Protection of Vertical Openings	701
5. Protection of Wall Openings	703
6. Sprinklers and Standpipes	901 to 902, inclusive
7. Mixed Occupancy and Separation Requirements	412
8. Light, Ventilation and Sanitation	2001 to 2002, inclusive
9. Heating Requirements	Chapter VIII

410.3 — SPECIAL REQUIREMENTS—GROUP "G"—INDUSTRIAL

	SECTION
1. Occupancy Permit Required for Changed Floor Loads	1207
2. Posting of Floor Loads Required	110
3. Storage and Handling of Flammable Liquids	501.1 (f)
4. Ratproof Construction	Chapter XIX

410.4 — GROUP "G" — INDUSTRIAL OCCUPANCY — HEIGHT AND AREA RESTRICTIONS

Type Construction	ALLOWABLE HEIGHT	ALLOWABLE FLOOR AREA* (Square Feet Per Floor)		
	Story Height	First Floor	Second Floor	Third Floor and Above
Type I****	No Limit	No Limit	No Limit	No Limit
Type II	80 Ft.	No Limit	30,000	20,000
Type III**	Six	31,500	22,500	15,000
Type IV***	Four	21,000	15,000	10,000
Type V***	Four	15,000	12,000	9,000
Type VI	One	10,000	Not Permitted	

*For Unlimited areas see Section 403.6. Other allowable area increases see Section 403.

**In Type III buildings over 3 stories, sprinklers shall be required.

***In buildings three stories or more in height, an approved automatic sprinkler system shall be installed.

****In all Group "G" occupancies of Type I Construction, the partitions, columns, trusses, girders, beams and floors may be reduced by one hour if the building is equipped with an approved automatic fire sprinkler system throughout but no component or assembly may be less than one hour fire resistance.

SECTION 411 — GROUP "H" — SPECIAL HAZARDOUS

411.1 — SCOPE

Buildings or structures used for purposes that involve highly combustible, inflammable or explosive products or materials or that constitute exceptional fire hazards, because of the form, character or volume stored, processed or manufactured, shall be classified in Group H — Special Hazardous Occupancy.

GROUP H — Special Hazardous Occupancy — shall include among others, the following:

Dry Cleaning Establishments
Grain Elevators
Storage or use of Highly Combustible Materials
Storage of Combustible Film

411.2 — GROUPING OF SPECIAL HAZARDOUS MATERIALS

The processing, manufacturing or storing of the following materials, among others, shall be classified as a special fire hazard, (Group "H" — Special Hazardous Occupancy), because of the highly combustible and explosive quality of the materials involved. They shall be classified as set forth below:

(a) The following materials or products stored shall be classified as Group "H" — Special Hazardous Occupancy, regardless of the volume or quantity stored or handled:

Acids: Sulphuric, Nitric or Hydroflouric — Calcium Carbide — Asphaltum — Celluloid. Chemicals: Poisonous, combustible, explosive —Cereal, flour or feed mills — Cork — Excelsior — Petroleum Products (Not retail gasoline service stations) — Pyroxylin and Pyroxylin Plastic Products — Shoddy Mills — Loose Spices and Vegetable Stocks (not retail stores) — Explosives — Feather Renovating — Films (see Section 501.3) — Fireworks — Gas: Poisonous irritating and flammable — Jute — Kapok — Munitions — Naval Stores — Nitrate of Soda — Paper; Baled Waste — Oakum or Hemp Processing — Paints: Manufacturing — Starch Mills — Waste Paper —Rubber Manufacturing Plants.

(b) Materials or products stored or handled in quantities in excess of the cubic foot areas set forth below shall be classified as Group "H"—Special Hazardous Occupancy:

	In Excess of
Acetylene gas in pressure containers	2000 cu. ft.
Artificial flowers	1200 cu. ft.
Artificial leather	600 cu. ft.
Bags—Burlap, paper or cotton	1000 cu. ft.
Barrels—second hand	1800 cu. ft.
Brooms and Broom corn	1200 cu. ft.
Cotton (Loose) Wading or Waste	400 cu. ft.
Drugs	6500 cu. ft.
Enameling	2400 cu. ft.
Fertilizer (Bags only)	2700 cu. ft.

Lacquers (Separate containers only)	3600 cu. ft.
Matches	300 cu. ft.
Mattresses	4800 cu. ft.
Paints (Separate metal containers only)	6000 cu. ft.
Spray painting shops	3200 cu. ft.
Tires (Recapping)	See Section 517
Tires—storage	12,000 cu. ft.
Varnish—Turpentine (Separate metal containers only)	2400 cu. ft.

411.3 — PROTECTIVE REQUIREMENTS — GROUP "H" — SPECIAL HAZARDOUS

	SECTION
1. Prohibited in Fire District	301
2. Allowable Height and Area Heights and Areas are based upon types of construction used.	411.5
3. Types of Construction	Chapter VI
4. Exit Requirements	Chapter XI
5. Protection of Vertical Openings	701
6. Protection of Wall Openings	703
7. Sprinklers and Standpipes	501.2 (d), and 901 to 902, inclusive
8. Light, Ventilation and Sanitation	2001 to 2002, inclusive
9. Heating Requirements	Chapter VIII
10. Separation of Boiler and Furnace Room	812
11. Mixed Occupancy and Separation Requirements	412

411.4 — SPECIAL REQUIREMENTS — GROUP "H" — SPECIAL HAZARDOUS

The hazards involved under Group "H"—Special Hazardous—require provisions that give adequate protection for these special occupancies. The occupancies in this group shall conform to the special requirements as set forth in Section 501.

411.5 — GROUP "H" — SPECIAL HAZARDOUS OCCUPANCIES, HEIGHT AND AREA RESTRICTIONS

ALLOWABLE HEIGHT Type Construction Used		ALLOWABLE FLOOR AREAS* Square Feet Per Floor		
	Story Height	First Floor	Second Floor	Third Floor and Above
Type I	Four	11,500	7,500	5,000
Type II	Three	8,300	6,000	4,000
Type III	Two	7,500	6,000	Not Permitted
Type IV	One	5,000	Not Permitted	
Type V	One	5,000	Not Permitted	
Type VI	Not Permitted			

*No modification in areas shall be permitted in Group "H" Special Hazardous Occupancies.

Note:—See Chapter V for special and detail requirements of Group "H" Special Hazardous Occupancies.

SECTION 412 — MIXED OCCUPANCY AND SEPARATION REQUIREMENTS

412.1 — DEFINITION

When a building is used for two or more occupancies, classified within different occupancy groups, it shall be considered a mixed occupancy.

412.2 — LIMITATIONS

A mixed Occupancy building shall be governed by the Height and Area limitations applying to the principal use therein. Accessory occupancies shall not exceed the area limitations or be located at a story height greater than that permitted for such accessory occupancy and type of construction being used.

412.3 — MINIMUM SEPARATION

Unless otherwise specifically prescribed in this Section, the separation of mixed occupancies shall provide not less than one (1) hour fire protection, except that portions of buildings used as accessory offices or for customary non-hazardous uses necessary for transacting the principal business of Group F Storage and Group G Industrial occupancies may be separated by partitions of non-combustible construction without fire-protection or by partitions constructed of materials as permitted in the type of construction used.

412.4 — FIRE-RESISTANCE RATING OF SEPARATION

(a) The minimum fire-resistance of constructions separating any two occupancies in a building of mixed occupancy, shall be the higher rating required for the occupancies being separated, as specified below:

MINIMUM REQUIREMENTS*

Group A—Residential	1 Hr. fire-resistance separation
Group B—Business	1 Hr. fire-resistance separation
Group C—School	2 Hr. fire-resistance separation
Group D—Institutional	2 Hr. fire-resistance separation
Group E-1—Large Assembly	4 Hr. fire-resistance separation
Group E-2—Small Assembly	2 Hr. fire-resistance separation
Group F—Storage	4 Hr. fire-resistance separation
Group G—Industrial	2 Hr. fire-resistance separation
Group H—Special Hazardous	4 Hr. fire-resistance separation

(b) A separation between a private garage and any occupancy shall be the minimum fire-resistance specified above for the occupancy except in the case of a one or two-family dwelling, no fire-resistance separation shall be required. No separation shall be required between a Sunday School room, or rooms, and a Church Auditorium of Group E2 Small Assembly.

(c) A separation between a public garage used exclusively for the storage of passenger vehicles that will accommodate not more than nine (9) passengers and any other occupancy other than Group E-1, Group F and Group H shall be two (2) hours.

*For materials and assemblies to provide the required fire-resistance, see Chapter X and Appendix B.

412.5 — PROTECTION OF HORIZONTAL OPENINGS

For requirements governing the protection of door openings in walls and partitions separating mixed occupancies, see Section 703.4.

412.6 — PROTECTION OF VERTICAL OPENINGS

For requirements governing the protection of vertical openings, see Section 701.

412.7 — TENANT SEPARATION

In a building, or portion of a building of a single occupancy classification, when enclosed spaces are provided for separate tenants, such spaces shall be separated by not less than one-hour fire resistance.

CHAPTER V

SPECIAL OCCUPANCY REQUIREMENTS

SECTION 501 – GENERAL

501.1 — SCOPE AND REFERENCES

(a) Buildings of occupancies in Group H not specifically provided for in this Code, which involve the storage, manufacture, or use of highly combustible or flammable materials shall be constructed to provide a degree of fire protection adequate for the hazard involved. Such protection may exceed the fire-resistive requirements prescribed for Type I Construction, if deemed necessary by the Building Official, but in all cases the construction shall meet the minimum requirements specified for Group H occupancies.

(b) In all buildings of Group H occupancy, approved automatic sprinklers shall be installed throughout, except that where the nature of the fire hazard is such that application of water is not effective as a means of protection, other approved means of protection shall be provided.

(c) Buildings of Group H occupancies shall not be located in the Fire District.

(d) Warehouses used to store combustible fibres such as cotton, sisal, jute, hemp, kapok, excelsior and similar materials having a flash fire hazard, shall be limited to story heights of not over 12 feet, floor to ceiling, and no single storage compartment shall exceed 5,000 square feet in floor area or 36,000 cubic feet in capacity.

(e) Stables, for storing hay, which do not exceed 1 story and storage loft, or a maximum of 20 feet in height, and do not exceed 3,000 square feet in floor area, may be of Type VI Construction if located 30 feet or more from adjoining property lines and other structures.

(f) Buildings or structures of occupancies involving the use of highly combustible material or processes, and their equipment, shall be erected, altered, and installed in accordance with safe practice. Except as otherwise provided in this Code, the provisions of the various regulations or standards of the National Fire Protection Association governing the particular occupancy shall be considered as constituting safe practice. Those standards include, among others, the following:

N.F.P.A. Standards for the storage, handling and use of flammable liquids, "Flammable & Combustible Liquids Code, NFPA No. 30-1972."

N.F.P.A. Standards for spray finishing using flammable materials, "Spray Finishing Using Flammable Materials, NFPA No. 33-1969."

N.F.P.A. Standards for dip tanks containing flammable or combustible liquids, "Dip Tanks, NFPA No. 34-1971."

N.F.P.A. Standards for the installation and operation of oxygen-fuel gas systems for welding and cutting, "Oxygen Fuel Gas Systems for Welding and Cutting, NFPA No. 51-1969."

N.F.P.A. Standards for the storage and handling of liquefied petroleum gases, "Storage and Handling of Liquified Petroleum Gases, NFPA No. 58-1972."

N.F.P.A. Recommended good practice requirements for the installation and use of combustion engines and gas turbines, "Combustion Engines and Gas Turbines, NFPA No. 37-1970."

N.F.P.A. Standards for the storage and handling of pyroxylin plastic in warehouses and wholesale jobbing and retail stores, "Pyroxlin Plastic in Warehouses and Wholesale, Jobbing and Retail Stores, NFPA No. 43-1967."

N.F.P.A. Standards for the storage, handling and use of pyroxylin plastic in factories making articles therefrom, "Pyroxlin Plastics in Factories, NFPA No. 42-1967."

N.F.P.A. Standards on the fundamental principles for the prevention of dust explosions in industrial plants, "Prevention of Dust Explosions in Industrial Plants, NFPA No. 63-1971."

N.F.P.A. Code for the prevention of dust explosions in the manufacture of aluminum powder, "Aluminum Powder, NFPA No. 651-1972."

Note: The above standards are published by the National Fire Protection Association in the several volumes of National Fire Codes and may be obtained from them at 60 Batterymarch Street, Boston, Mass.

501.2 — DRY CLEANING, DYEING OR SIMILAR HIGH FIRE HAZARD OCCUPANCY

(a) No building used for dry cleaning or similar hazardous occupancy shall be located within the Fire District, unless only non-flammable liquids are used for cleaning purposes.

(b) Dry cleaning, dyeing, or similar establishments using combustible or flammable liquids or solvents with a flash point of 190° F, or lower (closed cup test), shall be of Type I, or Type II Construction, and shall not exceed 1 story in height or 10,000 square feet in area, without attics, concealed roof spaces, basements or pits. Floors shall not be below grade.

(c) Roofs shall be flat. If, due to local conditions, the Building Official deems it desirable to vent possible explosions upward, the roof may be of light, non-combustible construction.

(d) An approved automatic sprinker system shall be installed throughout each drying room in accordance with Section 901 of Chapter IX.

(e) Partitions shall have not less than 2-hour fire-resistance.

(f) Drying rooms, if under the same roof as the dry cleaning and dry dyeing rooms, shall be separated from such rooms by each wall having a fire-resistance of not less than 4 hours. The entrance to such drying rooms shall be provided with self-closing fire doors.

(g) Except for necessary openings for vents, ducts, piping and shafting, all openings in exterior walls shall be protected with fire doors or windows. Windows shall be of wired glass in metal sash so hung that they will readily swing out in case of explosion.

(h) Exterior walls except those on street fronts, which are located less than 10 feet from adjacent property lines shall have no openings therein and shall have a fire-resistance rating of not less than four hours, or the equivalent, but in no case shall more than two sides of the building have blank walls.

(i) Skylights shall be provided. They shall be constructed with metal frame and sash with plain thin glass and with wire screen provided above the skylight as prescribed in Section 707, or with wired glass arranged to swing outward readily in case of an explosion.

(j) Mechanical systems of ventilation, of explosion-proof type, shall be provided to insure complete and continuous change of air once every 3 minutes in dry-cleaning and dry-dyeing rooms.

(k) All other regulations contained in this Code pertaining to construction, ventilation, storage, heating and lighting, or the like, shall apply as well as any laws of the State regulating the construction and maintenance of dry-cleaning, dyeing, or similar plants.

(l) The installation, ventilation, erection, alteration, maintenance or use of equipment, of buildings or structures for dry-cleaning or dry-dyeing purposes shall be in accordance with the provisions of the Standards of the National Fire Protection Association for dry cleaning and dry dyeing plants, "Dry Cleaning Plants, NFPA No. 32-1972."

501.3 — HANDLING OR STORAGE OF CELLULOSE NITRATE FILM

(a) Construction of Buildings Where Films Are Stored or Processed.

(1). All buildings in which cellulose nitrate films are stored or processed, such as film exchanges, film laboratories, motion picture studios, etc. shall be of Type I or Type II construction and shall be equipped throughout with approved automatic sprinklers in accordance with Section 901 of Chapter IX. Such buildings shall not be located in the Fire District and shall not exceed the maximum height and area limitations specified for Group H in Section 411.5.

(2). The following regulations shall govern the handling and storage of cellulose nitrate film except that they do not apply to the following: Films in original packages in quantities less than 50 pounds, and films stored in motion picture projection booths (See Section 512.16).

(3). Except as otherwise specified herein, the handling and storage of combustible film shall be governed by the Standards of the National Fire Protection Association for storage and handling of cellulose nitrate motion picture film, "Cellulose Nitrate Motion Picture Film, NFPA No. 40-1967."

(4). All rooms in which cellulose nitrate films are stored or handled, except motion picture projection booths and film vaults, shall be enclosed in partitions of non-combustible construction having not less than 2 hours fire-resistance. Openings in such partitions shall be protected by approved fire doors. Floors and ceilings of such rooms shall provide fire-resistance of not less than 2 hours and vents that open automatically in case of fire shall also be provided. Tables and racks used in connection with the handling of film shall be of metal or other non-combustible material and shall be at least 4 inches away from any radiator or heating apparatus. Fire-fighting appliances using water, or water solutions, shall be provided in every room. In rooms where film is stored or handled in quantities greater than 50 lbs., cabinets shall be provided with insulated metal vents. Film storage rooms in which two or more persons work, shall have at least two exits remote from each other.

(5). Cellulose nitrate film in amounts of more than 1,000 lbs. shall be kept in vaults constructed as prescribed in this Section.

(6). Amounts of cellulose nitrate film in excess of 25 lbs. shall be kept in approved metal cabinets of capacity not exceeding 375 lbs. Cabinets having a capacity of over 50 lbs. of film, shall be provided with insulated metal vents of at least 14 square inches per 100 lbs. of film. Cabinets holding over 75 lbs. of film shall be provided with at least one automatic sprinkler, unless so built that each roll is in a separate compartment so constructed that the film will burn out without communicating fire to film in any other compartment.

(7). Unexposed film, when stored in the original shipping cases with each roll in a separate container, shall be stored only in a room provided with an approved automatic sprinkler system. (Section 901 of Chapter IX.)

(b) Film Vaults.

(1). Vaults used for the storage of cellulose nitrate film shall not exceed 750 cubic feet inside and shall not be located near chimneys or other sources of heat.

(2). Walls, floors and roofs of film vaults and their supports shall be of not less than 4-hour fire-resistance construction built without cracks or holes that will permit escape of gases. Drains, or scuppers, to the outside of the building shall be provided. All door openings shall be protected with approved fire doors on each face of the wall; the inner door shall be automatic, the outer door shall be of the self-closing swinging type.

(3). Each vault shall have an independent vent having not less than 140 square inches effective area per 1,000 lbs. of film capacity (equivalent to 70 square inches per 100 standard rolls) but the vent area for a vault of 750 cubic feet shall in no case be less than 1400 square inches. Vents shall be of non-combustible materials and shall be located at least 50 feet from all openings exposed thereto.

(4). Film shall be protected against ignition by rays of the sun and by radiated heat.

(5). Vaults shall have no skylights or glass windows except as specified for vents. Vents may be protected against the weather by a single thickness of glass (1/16" thick) not less than 200 square inches in area, in a sash arranged to open automatically in case of fire, or by equivalent protection.

(6). Vaults shall be protected by an approved system of automatic sprinklers (Chapter IX) with a ratio of one head to each 62½ cubic feet of total vault space. A vault of 750 cubic feet shall have not less than 12 sprinkler heads.

(7). Wire guards shall be provided so that no film could be placed within 12 inches of heating pipes or radiators.

(8). Vault heating shall be automatically controlled so as not to exceed a temperature of 70 degrees F. or a steam pressure of 10 lbs.

(9). All racks and equipment in vaults shall be of metal or other non-combustible material.

501.4 — GRAIN ELEVATORS

(a) Grain elevators, or structures used to store grain, shall not be located within fifty (50) feet of adjoining property lines of other structures, except railway rights of way or adjoining navigable waters, nor shall they be located within the Fire District.

(b) Grain elevators, or structures used to store grain, shall be constructed of steel, concrete, or other non-combustible material or with lumber exterior or interior framing, including plank and laminated walls, when the sizes of the members used conform to the requirements for Type III Construction to meet the approval of the Building Official, and all such structures, buildings, and equipment shall be erected, altered, or installed in accordance with the provisions of Section 501.1 (f).

(c) Where combustible material, other than grain, is present in quantity sufficient to produce a serious fire, fire protection equivalent to Type I Construction shall be provided unless approved automatic sprinkler protection is provided (Section 901 of Chapter IX). In no case, however, shall the requirements for grain elevators, or grain storage buildings, be less restrictive than those applying to Group H occupancies.

SECTION 502 – AIRPLANE HANGARS

(a) Airplane hangars may be of any type of construction. Exterior walls that are located within thirty (30) feet and facing common property, interior lot lines or the opposite side of a public street or thoroughfare shall provide not less than 2 hour fire-resistance.

(b) The floor areas of hangars shall not exceed those permitted for Group F Storage buildings in Section 409.4 (see area exceptions, Section 403).

(c) Where hangars have basements, the floor over the basement shall be of Type I Construction and shall be made tight against seepage of water, oil or vapors. There shall be no opening or communication between basement and hangar. Access to basement shall be from outside only.

(d) Floors shall be graded and drained to prevent water or gasoline from remaining on the floor. Floor drains shall discharge through an oil separator to the sewer or to an outside vented sump.

(e) Heating of hangars shall be from plants located in a detached building.

(f) The process of "doping", involving use of a volatile flammable solvent, or of painting, shall be carried on in a separate detached building equipped with automatic sprinkler equipment in accordance with Section 901 of Chapter IX.

(g) Each hangar of area exceeding 10,000 square feet shall be equipped with approved automatic sprinklers in accordance with Section 901 of Chapter IX.

SECTION 503 – COAL POCKETS

Coal pockets, and other similar structures, shall be constructed of steel, concrete, or other non-combustible material, or of lumber sizes which meet the requirements of Type III Construction.

SECTION 504 – TEMPORARY STRUCTURES

504.1 — PERMIT

A special building permit for a limited time shall be obtained before the erection of Temporary Structures such as construction sheds, seats, canopies, tents and fences used in construction work or for temporary purposes such as reviewing stands. Such structures shall be completely removed upon the expiration of the time limit stated in the permit.

504.2 — TENTS FOR PUBLIC ASSEMBLY

(a) Before a temporary permit is granted, the owner or agent shall file with the Building Official a certificate executed by an acceptable testing laboratory, certifying that the tent, decorative materials and tarpaulins meet the requirements for fire resistance prescribed in the National Fire Protection Association "Standard Methods of Fire Tests for Flame-Resistant Textiles and Films, NFPA No. 701-1969," and that such fire-resistance is effective for the period specified by the permit.

(b) **Tent Exits**—Tent exits, aisles, seating, etc., shall conform with the requirements for places of assembly. All exits shall be kept free and clear of obstructions while the tent is occupied by the public.

(c) Ground within and adjacent to tents shall be cleared of all grass, underbrush or similar fire hazards.

504.3 — TEMPORARY SEATS

A special permit shall not be issued unless all seats, stands and structures conform to the requirements of Chapter XII (Minimum Design Loads). All seats shall be marked allowing a space for each person of not less than eighteen inches in width. Aisles and seating arrangements shall conform to the requirements of Assembly Occupancies (Section 512.7).

SECTION 505 — AUTOMOTIVE SERVICE STATIONS

(a) An automotive service station of Group B occupancy is a place of retail business at which outdoor automotive refueling is carried on using fixed dispensing equipment connected to underground storage tanks by a closed system of piping, and at which goods and services generally required in the operation and maintenance of motor vehicles and fulfilling of motorist needs may also be available. The building consists of a sales office where automotive accessories and packaged automotive supplies may be kept or displayed. It may also include one or more service bays in which vehicle washing, lubrication and minor replacement, adjustment and repair services are rendered. An automotive service station building shall have no cellar or basement, but may have open pits if such pits are continually ventilated. An automotive service station building shall be of Type I, Type II, Type III, Type IV, Type V, or Type VI.

(b) Canopies and their supports over pumps shall be of non-combustible materials, wood of Type III sizes, or of construction providing 1- hour fire resistance. Any such canopy covering fuel islands shall be a minimum height of 12' 6" from ground level at its lowest point.

(c) All equipment likely to cause an explosion, or to be capable of igniting gasoline vapor from heat, sparks, or open flames, shall be located at least 4 feet above the floor, or be completely and tightly enclosed by non-combustible construction, or construction of not less than 1-hour fire-resistance. Any openings to such enclosures shall be from the outside with the sill raised at least 1 foot above the adjoining outside level, and shall be located at least 5 feet from any property line or adjacent building.

SECTION 506 — PRIVATE GARAGES

(a) Garages which are provided for the storage of motor vehicles owned by tenants of buildings on the premises, and with maximum undivided space used for storage of not more than four automobiles, or trucks of one ton or less capacity, but not exceeding 850 square feet, shall be considered private garages. All other garages shall be considered public garages.

(b) Private garages may be of Type I, II, III, IV, V or VI construction, but no private garage shall occupy space above the first floor of Type VI building or shall be erected in the fire district except as provided in Section 304. No private garage shall be located within, or attached to, a building occupied for any other purpose, unless it is separated from such other occupancy by walls, partitions, floors and ceilings that have a fire-resistance rating as specified in Section 412.4 (Mixed Occupancy Separations). Walls, floors, partitions and ceilings that effect such separation shall be continuous and unpierced. A single flush-type solid core wooden door of not less than 1¾ inch nominal thickness, equipped with a self-closing device, may be permitted provided the sill is raised at least 8 inches above the garage floor when the doorway connects directly with any room in which there is any direct-fired heating device or gas fixture. In no case, however, shall a garage have an opening directly into a room used for sleeping purposes.

SECTION 507 — PARKING LOTS AND PUBLIC PARKING DECKS

507.1 — PARKING LOTS

Open sheds or canopies may be erected up to two-thirds (⅔) the area of a lot, provided such construction is not less than required for Type IV Construction, and that all such construction meets the approval of the Building Official.

507.2 — PUBLIC PARKING DECKS

(a) As defined in Section 201.2, Public Parking Decks may be constructed of Types I, II, III, and IV Construction without exterior walls. When such structures are within six (6) feet of common property lines they shall be provided with an enclosure wall along the common property line of not less than two (2) hours fire resistance without openings therein, except that doors opening to buildings adjacent thereto may be permitted provided that such door openings meet the requirements of Section 703.4.

(b) Type III structures shall be limited to a height of four (4) stories and an area limitation of 30,000 square feet per floor with roof parking permitted. Type IV structures shall be limited to a height of eight (8) stories and an area limitation of 30,000 square feet per floor with roof parking permitted except that Type IV structures shall not be limited in area when such structures are four (4) stories or less in height, are surrounded on all sides by a permanent open space of not less than thirty (30) feet, are provided with exterior walls at least 50% open on all sides, and the horizontal distance from any point on any level to an exterior wall does not exceed 200 feet. When of Type I or Type II Construction, the height and area shall not be limited. When of Type III or IV Construction area increases may be allowed in accordance with Section 403.

(c) Each floor of such structure shall have a continuous wheel guard not less than six (6) inches in height above the floor, with a clear passage of four (4) feet between the wheel guard and edge of structure. In such structures without exterior walls there shall be placed in addition to the wheel guard a continuous protective railing not less than three (3) feet six (6) inches above the floor around the entire outside perimeter of the structure.

SECTION 508 — PUBLIC GARAGES

(a) A garage shall be any building or part thereof wherein is kept or stored a motor vehicle having any gasoline or other volatile flammable fuel in its fuel storage tank, or wherein painting, body and fender work, engine overhauling or other major repair of motor vehicles is performed. This occupancy shall not include automotive service stations as defined in Section 505. A garage exceeding 850 square feet in area or used to store more than four automobiles, shall be considered a public garage.

(b) A public garage shall be of Type I, II, III, IV, V, or VI Construction. If of Type V Construction, a public garage shall not exceed one story in height, nor shall it exceed the maximum height and area allowed for Group F storage buildings. Public garages of Type VI Construction may be used only for dead storage and display of automobiles.

(c) No public garage shall be located within, or attached to, a building occupied for any other purpose, unless separated from the other occupancies as prescribed in Section 412, but in no case by walls having fire-resistance less than two hours. Such separation shall be continuous and unpierced, except for doors leading to salesrooms, or offices, operated in connection with such garages, provided such openings are approved by the Building Official as being required or essential, and provided such openings are equipped with self-closing fire doors conforming to the requirements of Section 703.

(d) Unenclosed ramps shall not be considered as providing required exit facilities. Enclosed ramps shall be in accordance with the Exit Requirements of Chapter XI.

(e) Basement and sub-basement garages shall be continuously ventilated by a mechanical system with positive means for both inlet and exhaust of at least 1 cubic foot of air per minute per square foot of floor area, controlled from a location close to the entrance door.

(f) Garage floors shall be of concrete or similar non-combustible and non-absorbent materials. Floors which drain to sewers or storm drains shall be provided with an oil separator or trap.

(g) Sprinkler equipment shall be provided as required in Section 901 of Chapter IX.

(h) Heating equipment, other than direct-fired unit heaters, shall be placed in a room separated by construction equivalent to 8-inch brick walls and 4-inch reinforced concrete floor and ceiling with no openings except as required for heating pipes and ducts. Access shall be open directly to exterior of building.

(i) Connection between garage and any room having a direct-fired heating device, or gas fixture, shall be by means of a doorway with sill raised at least 8 inches above the garage floor level, or through a vestibule providing two doorway separations.

SECTION 509 – GREENHOUSES

Greenhouses more than 35 feet in height shall have a non-combustible structural frame. Greenhouses not over 400 square feet in area, or 15 feet high, shall be considered accessory structures and may be of any construction except that a greenhouse with wood frame construction shall be located not less than 5 feet from an adjoining structure or property line.

SECTION 510 – STADIUMS AND GRANDSTANDS

(a) Stadiums and grandstands may be constructed of steel, iron, reinforced concrete, or wood, designed for live loads and for wind pressures in accordance with the requirements of this Code. They shall not be erected on the roof of any building or structure.

(b) In stands constructed of wood or other combustible materials, the level of the highest seats above the ground (level of ground at immediate front of the stand) shall not exceed 25 feet, and such stands shall not be located within 20 feet of adjoining property lines, or within 50 feet of adjoining Type VI structures.

(c) When the space under a stand is used for any purpose, the space shall be enclosed in construction having not less than 1-hour fire-resistance and shall meet the separation requirements of Section 412.

(d) Aisles not less than 3 feet 6 inches wide shall be provided. Aisles shall be so located that there will be not more than 20 intervening seats between any seat and the nearest aisle. Where backs are provided, seats shall be spaced not less than 30 inches back to back.

(e) A distance of 18 inches along any bench shall constitute one seat in computing the required exit facilities.

SECTION 511 – AMUSEMENT PARK BUILDINGS

(a) Amusement park buildings used as dining rooms, theaters, or for other purposes shall conform to the requirements of this Code governing the particular use or occupancy.

(b) Amusement park buildings over one story in height, or 1200 square feet in floor area, shall have exterior walls, floors and their supports of not less than 1-hour fire resistive construction.

(c) Where amusement park buildings are located within 30 feet of adjacent property lines, buildings or other structures, the exterior walls shall be constructed of non-combustible materials, or shall be protected to provide not less than 1-hour fire-resistance.

(d) Structures of open skeleton frame type shall not be limited in height or area, except that grandstands shall comply with the requirements of Section 510.

(e) Amusement structures shall provide adequate safety for all loads to which they may be subjected and shall be equipped with approved safety devices and safeguards.

SECTION 512 — ASSEMBLY OCCUPANCIES

This Section shall apply to all places of public assembly, except churches or places of worship. Churches or places of worship shall be governed by the regulations as set forth in Section 514.

512.1 — TYPES OF CONSTRUCTION

(a) Buildings of Group E-1 Large Assembly with stage shall be of Type I or of Type II Construction, except that in Auditoriums, ornamental wood, trusses and paneling may be of wood.

(b) Buildings of Group E-1 Large Assembly—without stage shall conform to the limitations of use prescribed in Section 408.6.

(c) Buildings of Group E-2 Small Assembly shall conform to the limitations of use prescribed in Section 408.6 as modified herein.

(d) Gymnasiums and similar occupancies may have running tracks constructed of wood or unprotected metal.

(e) For requirements for stadiums and grandstands see Section 510, for amusement park structures see Section 511.

512.2 — EXCEPTION TO AREA LIMITATIONS

Where there are no balconies or galleries in Group E-2, Small Assembly Places, and the assembly floor is located at, or within, 21 inches of street or grade level and all exits meet the street or grade level by ramps having a slope not exceeding 1 foot in 10 feet, the maximum allowable areas of Type III, IV, and V construction may be increased 50 per cent over those specified for Group E Assembly occupancies in Table 408.6.

512.3 — INTERIOR FINISH AND DECORATIONS

(a) For regulations governing flame spread of ceiling and wall finish materials see Section 704.3. Use of materials which give off smoke that is more dense, or toxic gases that are more toxic than are given off by the burning of untreated wood or untreated paper under comparable exposure to heat or flame shall not be permitted.

(b) In no event shall imitation leather or other material, consisting of, or coated with, a pyroxylin or similarly hazardous base, be used in Group E, Assembly occupancies. The use of combustible materials for decorative purposes in Group E Assembly occupancies, including among others, curtains, cloth, paper, streamers, draperies, vines, leaves, trees, moss, or other interior decorations, shall be prohibited.

512.4 — WALLS AND PARTITIONS

(a) All walls and partitions for enclosing stairs, passageways or corridors (except foyers or waiting space) which are used for exits, or enclosing rooms used for exit purposes in Group E, Assembly occupancies, shall be constructed of not less than 2-hour fire-resistance. Also refer to Section 702.

(b) Where Group E, Assembly occupancy is involved in a building used for any other purpose, separation of occupancies shall be as provided in Section 412.

512.5 — LOCATION OF BUILDING

All buildings of Group E, Assembly occupancy shall front directly upon at least one public street or public place not less than 30 feet wide, in which front shall be located the main entrance and exit of such building.

512.6 — EXITS

Refer to Chapter XI, Means of Egress Requirements, for exits and exit access requirements.

512.7 — AISLES AND SEATING

(a) Every aisle shall lead to an exit door or to a cross aisle running parallel to the seats and leading directly to an exit.

(b) Aisles, cross-aisles, corridors, and passageways shall be of width at least equal to the minimum width required for exits in this Code, but in no case shall the width of an aisle or cross-aisle be less than the width of the widest aisle, passage, cross-aisle or exit which it serves. No aisle shall be less in width than 36 inches, measured at its narrowest point at the end farthest from the foyer, plus an increase of 1½ inches for each 5 feet of length of such aisle from its beginning to an exit, except that aisles with seats on one side only may be 6 inches less in width, and except that when not to exceed 60 seats are served by an aisle, its width may be 30 inches. Where egress is provided at both ends of an aisle, the aisle may have a uniform width not less than the average widths herein specified. No cross-aisle shall be less than 3 feet-6 inches wide. An aisle bordering on a means of entrance shall be not less than 4 feet wide.

(c) In all balconies and galleries having more than 20 rows of seats, there shall be provided a cross-aisle not less than 4 feet wide leading directly to an exit.

(d) There shall be no obstructions of any kind in any aisle. Aisles shall not have a slope of more than one in ten except that the maximum gradient in aisles on the main auditorium floor shall not exceed one in five. Ramps steeper than one in eight shall have non-slip surface.

(e) Rows of seats between aisles shall have not more than 14 seats.

(f) Rows of seats opening on to an aisle at one end only shall have not more than 7 seats. Seats without dividing arms shall have their capacity determined by allowing 18 inches per person.

(g) Exits and aisles shall be so located that the travel distance to an exit door shall not be greater than 150 feet measured along the line of travel.

(h) Steps shall not be used in aisles of the main auditorium floor, or in other aisles, where differences of level can be overcome by gradients not exceeding those permitted herein. Where steps are used in aisles, such steps shall extend across the full width of aisles and shall be illuminated; treads and risers shall be as required elsewhere in the Code for exit stairs. No isolated steps shall be permitted nor shall the aggregate rise of a group of steps exceed 21 inches.

(i) In places of assembly used regularly for theatrical or similar performances, or for the display of motion pictures, the seats shall be securely fastened to the floor. In restaurants, cafeterias, cafetoriums, gymnatoriums and similar multi-purpose places of assembly, the seats shall not be required to be fastened to the floor. All other Group E, Assembly occupancies seating more than 200 persons shall have seats fastened to the floor. All seats in balconies or galleries shall be secured to the floor except that in railed-in enclosures, boxes, or loges, with level floors and having no more than 14 seats, the seats need not be fastened to the floor, or have separating arms.

(j) The spacing of rows of seats from back to back shall be not less than 30 inches, and not less than 27 inches plus the sum of the thickness of the back and the inclination of that back; but in all cases there shall be a space of not less than 12 inches between the back of one seat and the front of the seat immediately behind it as measured between plumb lines.

512.8 — CONTINENTAL SEATING SYSTEMS

(a) Chairs shall be of the fixed type installed in continuous rows across the auditorium, except in areas where aisles are designated.

(b) Chairs shall be floor attached, riser attached, or mounted on bars which are firmly anchored to the floor or riser facing.

(c) No chair shall be less than 20 inches in width as measured across the chair from center lines of arm rests.

(d) Seats of chairs may be an automatic or self-rising type seat, or a non-lifting or non-automatically lifting type seat. The minimum space requirements for walkways shall be measured in the seat-up position for automatic or self-rising seats, all other seats shall be measured in the seat-down position.

(e) Backs of chairs may be stationary or they may be spring operated to encroach on the space between the rows when occupied, provided that such encroachment is limited to a maximum of 5 inches as measured between vertical plumb lines from a point at the center of the upper rear edge of the back before and after depressing the back to its maximum rearward position. The back shall automatically return to a stabilized position of zero encroachment when unoccupied.

(f) Maximum number of chairs per row shall not exceed eighty (80), except as provided in paragraph (g) of this Section.

(g) The spacing of adjacent rows of unoccupied chairs shall be arranged so as to form a minimum spacing of walkways between rows of seats of not less than 21 inches in width and shall be measured between the plumb lines of the front edge of the seats and the top edge of the back of the seats in the row directly ahead. When the row of chairs exceeds eighty (80) chairs per row, the minimum spacing of walkways between rows of seats shall not be less than 24 inches.

(h) The top edges of chair backs shall extend a minimum of 21 inches above the front edge of platform riser treads immediately to the rear. When it is necessary to increase riser heights in order to meet the desired sight-lines to the stage or screen, a safety rail 21 inches high shall be installed along the line of chair backs at the front edge of platform riser treads and shall be anchored to the tread.

(i) Platform riser facing shall be in a true vertical plane and the joint between the tread and facing shall not include a fillet of more than ½ inch radius.

(j) Ventilator hoods or similar devices which are specified for attachment to floors directly under the chair seats shall not exceed 4 inches in height.

(k) Incline of the areas of the floor on which chairs are installed shall not exceed 1 inch per foot forward or .75 inch per foot reverse. Inclines within the same limitations may be used in the tread areas of platform riser systems.

(l) Concrete floors shall have a minimum thickness of 3 inches and a minimum strength of 3000 PSI. Concrete risers for treads shall have a minimum thickness of 3 inches and riser facings a minimum thickness of 5 inches.

(m) Pre-stressed or pre-formed concrete risers and/or steel faced riser systems shall provide permanent anchored locks between tread and facing which are designed to withstand the full amount of tortion forces generated when chairs which are attached to riser facing are occupied. Steel face risers shall have a minimum thickness of ¼ inch.

(n) Wood floors are not approved for auditorium floors in buildings designed for continental seating systems.

(o) Side wall exits and side aisles shall be provided in accordance with one of the following methods:

1. Two exits, one each side of the auditorium, may serve not more than five rows of chairs directly between these exits provided the width of these exits shall not be less than 6'-0" for each exit. They may be so placed that the ends of each fifth row may terminate no less than 18" from the side walls and the ends of the other four rows shall terminate so as to form side aisles of not less than 48-inch width.
2. Two exits, one each side of the auditorium, may serve not more than four rows of chairs directly between these exits provided the width of these exits shall be not less than 5'-0" for each exit. They may be so placed that the ends of each fourth row may terminate no less than 18" from the side walls and the ends of the other three rows shall terminate so as to form side aisles of not less than 42-inch width.
3. Two exits, one each side of the auditorium, may serve not more than three rows of chairs directly between these exits provided the width of these exits shall be not less than 4'-0" for each exit. They may be so placed that the ends of each third row may terminate no less than 18" from the side walls and the ends of the other two rows shall terminate so as to form side aisles of not less than 36-inch width.

(p) Aisles, except as provided herein, shall be not less than 44 inches in width. Where exits are provided on only one side of the auditorium, no aisle shall serve more than 50 seats per row, and exit doors shall be provided along each such aisle at the rate of one pair of exit doors for each 5 rows of seats. Such exit doors shall provide a minimum clear width of 6'-0" for each exit.

(q) Aisles shall provide access to a cross aisle, foyer or exit.

(r) The length of travel to an exit door by any aisle shall not be greater than 150 feet.

(s) No dead-end aisle shall be greater than 20 feet in length. This does not refer to walkways required between rows of seats.

(t) The width of cross aisles, foyers, exit lobbies, or exit passageways shall be of sufficient width to accommodate 50% of the total occupant load served by such aisles or spaces as determined in accordance with Sections 1105.1 and 1105.3.

(u) Main entrance doorways and exit doorways which open directly to the outside of the building shall open directly to a street or open public space and shall be of sufficient width to accommodate

50% of the total occupant load but shall not be less than the sum of the required width of all aisles, exit passageways, exit ways, or exit foyers leading thereto.

(v) Where doors open from the main auditorium seating area, lighted exit signs will not be required at each individual exit doorway. When such doorways open into aisles, corridors, exit passageways, exit foyers, or exit lobbies, lighted directional exit signs shall be provided in such a manner as to be visible from each doorway leading from the main auditorium seating area.

512.9 — RAILINGS

(a) The facia of boxes, balconies and galleries shall have substantial railings not less than 26 inches high above the floor. The railing at the ends of aisles extending to the facia shall be not less than 30 inches high for the width of the aisle, or 36 inches high if at foot of steps.

(b) Cross-aisles, except where the backs of seats on the front of the aisle project 24 inches or more above the floor of the aisles, shall be provided with railings not less than 26 inches high.

(c) In balconies, galleries, or other locations where seats are arranged on platforms or successive tiers, and the height of the rise from one platform to another exceeds 21 inches, a substantial railing of not less than 30 inches high shall be placed at the edge of the platform along the entire row of seats.

512.10 — PLACARD INDICATING CAPACITY

A placard indicating the allowable maximum legal capacity of every Group E, Assembly occupancy, in number of occupants other than employees, shall be displayed in a prominent place. Such signs shall read as follows:

"Occupancy by more than __________ persons is dangerous and unlawful."

______________________________Building Official

512.11 — CONSTRUCTION OF STAGE, PROSCENIUM AND APPURTENANT ROOMS

(a) Any working stage (see definition) shall be enclosed on all sides with walls having a fire-resistance rating of not less than four hours and extending from foundation to a height of four feet above roof.

(b) There shall be no openings in the wall separating a working stage from the auditorium except the stage or proscenium opening, one doorway at each side of the proscenium opening at the stage floor level or the auditorium floor level, at the level of the musicians pit, and where necessary to the organ. Each such doorway shall be not more than 21 square feet in area and shall be protected by a self-closing fire door of 3-hr. fire-resistive rating.

(c) There shall be no windows in such enclosure walls of a working stage within 5 feet of property line other than a street line, and all windows shall be of approved fire-resistive type.

(d) All mouldings and decorations around proscenium opening shall be constructed entirely of non-combustible or fire-resistant materials.

(e) Above the proscenium opening of a working stage shall be a girder or other structural member of adequate strength to support all loads, constructed of non-combustible material and protected to provide not less than 4-hour fire-resistance.

(f) All that portion of the working stage except that used for the working of scenery, traps and other mechanical apparatus for the presentation of a scene, approximately equal to the width of the proscenium opening, shall be of Type I Construction, and appurtenant rooms and compartments shall be of Type I or Type II Construction.

(g) The rigging loft, fly galleries, including pin-rails, shall be of non-combustible materials.

(h) The roof over the working stage shall be of Type I Construction.

(i) Dressing rooms, scene docks, property rooms, workshops, store-rooms, and other rooms or compartments appurtenant to the stage shall be of Type I or Type II Construction and shall be separated from the working stage and other parts of the building by walls having a fire-resistance rating of not less than 3-hours. Such rooms and spaces shall be separated from each other by non-combustible partitions providing not less than 2-hours fire-resistance, except that partitions separating dressing rooms from each other shall have at least 1-hour fire-resistance. In no case shall openings other than the necessary doorways at stage level, protected with self-closing fire doors, connect such rooms with the stage.

(j) Openings through stage floors shall be equipped with tight-fitting trap doors of wood not less than 2 inches thick.

(k) The troughs or frames for footlights and border lights shall be of metal or other non-combustible materials. The suspension lines of border lights shall be of wire for at least 10 feet from the frames.

(l) All electrical equipment shall be protected from falling objects and from contact with stage equipment, and shall conform with the electrical requirements of the authority having jurisdiction.

(m) All woodwork and all scenery, drapes, and sets used upon the stage shall be coated or treated by approved method to make them non-flammable or fire-resistive.

(n) All shelving, closets, etc., property rooms, or storage rooms, shall be constructed of metal or other non-combustible material.

512.12 — VENTILATION OF STAGE

Over the working stage shall be provided one or more ventilators of metal or other non-combustible material, equipped with movable shutters or sash, having an aggregate clear area of not less than one-eighth the area of the stage, constructed to open automatically and instantly by approved heat-actuated devices. Suitable means for manual operation shall be provided in addition. If glass is used in the construction, only wired glass shall be used in such parts where the breaking of glass would cause it to fall on the stage.

512.13 — PROSCENIUM CURTAIN

(a) Every proscenium opening shall be provided with a curtain of metal or other non-combustible material, so designed and constructed that for at least thirty minutes it will prevent all passage of flame and withstand without failure a temperature of not less than 1700 degrees F. and an air pressure normal to its surface of not less than 10 lbs. per square foot. When closed, proscenium curtain shall be reasonably tight against the passage of smoke. The Building Official may require a fire test or other satisfactory evidence of its sufficiency in respect to these requirements. Curtain shall be subjected to operating tests and be approved by the Building Official before initial performance shall be held and shall be lowered after every performance.

(b) Every proscenium curtain shall overlap the proscenium opening by at least 2 feet at the top and 18 inches at each side, and shall slide vertically at each side within iron or steel grooves which shall have a minimum depth of 12 inches. Every such curtain shall be so arranged and maintained that, in case of fire, it would be released automatically and instantly by an approved heat-actuated device, and will descend slowly and safely by its own weight to completely close the proscenium opening within 30 seconds, taking not over 5 seconds for the bottom 5 feet. It shall also be equipped with effective devices to permit prompt and immediate closing of the proscenium opening by manual means.

(c) No part of any proscenium curtain shall be supported by or fastened to combustible material.

(d) For requirements pertaining to non-working stage, see definition.

512.14 — SPRINKLERS

(a) Every theater classified as a Group E-1, Large Assembly Place, shall have an approved system of automatic sprinklers conforming with Section 901 over the stage, in toilet rooms, lounges, smoking rooms, and all other parts including basements, cellars, property rooms, dressing rooms, storerooms, workshops, and all portions of stage and rooms under the stage floor level except as noted below.

(b) Sprinklers may be omitted in auditoriums, foyers and lobbies; and over dynamos or switchboards.

(c) Sprinklers shall not be placed in the immediate vicinity of automatic stage ventilators.

512.15 — STANDPIPES

In Group E-1, Large Assembly occupancies, a standpipe outlet with hose attached shall be provided on each side of the rear of each balcony and gallery, on each side of the stage, on each tier of dressing rooms, and within 50 feet of all property rooms, store rooms and work rooms. Such outlets shall connect with a standpipe which shall conform to the requirements of Section 902.1, but which shall have a diameter of not less than 4 inches except that standpipes on each side of the stage shall be of diameter not less than 2½ inches.

512.16 — MOTION PICTURE PROJECTION ROOMS

(a) The provisions of this Section shall apply where ribbon-type cellulose acetate or other safety film is used in conjunction with electric arc, Xenon or other light source projection equipment which develops hazardous gases, dust or radiation. Where cellulose nitrate film is used, projection rooms shall be in compliance with the provisions of N.F.P.A. Standard No. 40, Article 29.

(b) Every motion picture machine projecting film as mentioned within the scope of this Section shall be enclosed in a projection room. Appurtenant electrical equipment, such as rheostats, transformers and generators, may be within the projection room or in an adjacent room of equivalent construction.

(c) There shall be posted on the outside of each projection room door and within the projection room itself a conspicuous sign with one-inch block letters stating: "SAFETY FILM ONLY PERMITTED IN THIS ROOM."

(d) Every projection room shall be of permanent construction consistent with the construction requirements for the type of building in which the projection room is located. Openings need not be protected.

(e) The room shall have a floor area of not less than 80 square feet for a single machine, and at least 40 square feet for each additional machine. Each motion picture projector, floodlight, spotlight or similar piece of equipment shall have a clear working space not less than thirty inches by thirty inches on each side and at the rear thereof, but only one such space shall be required between two adjacent projectors.

(f) The projection room and the rooms appurtenant thereto shall have a ceiling height of not less than seven feet six inches.

(g) The projection room shall be provided with not less than one exit having a minimum opening of not less than 30 inches wide and 80 inches high.

(h) The aggregate of openings for projection equipment shall not exceed twenty-five percent of the area of the wall between the projection room and the auditorium or assemblage area. All such openings shall be provided with glass or other approved material so as to completely close the opening.

(i) The projection room itself shall be provided with two or more separate fresh air inlet ducts with screened opening terminating within twelve inches of the floor and they shall be located at opposite ends of the room. Such air inlets shall be of sufficient size to permit an air change every three minutes. Fresh air may be supplied from the general building air conditioning system—but when this is done, it shall be so arranged that the projection room will continue to receive one change of air every three minutes, regardless of the status of the general air conditioning system. Each projection room shall be provided with one or more exhaust air outlets which may be manifolded into a single duct outside the room. Such outlets shall be so located as to insure circulation throughout the room. Projection room exhaust air systems shall be independent of any other air systems in the building. Exhaust air ducts shall terminate at the exterior of the building in such a location that the exhaust air cannot be readily recirculated into the supply air system. The exhaust system shall be mechanically operated and be of such capacity as to provide a minimum of one change of air every three minutes. The blower motor shall be outside the duct system. The projection room ventilation system may also serve appurtenant rooms such as the generator room and the rewinding room.

(j) Each projection machine shall be provided with an exhaust duct which will draw air from each lamp and exhaust it directly to the outside of the building in such a fashion that it will not be picked up by supply inlets. Such a duct shall be of rigid materials, except for a continuous flexible connector approved for that purpose. The lamp exhaust system shall not be interconnected with any other system. For electric arc projection equipment the exhaust capacity shall be 200 cfm for each lamp connected to the lamp exhaust system, or as recommended by the equipment manufacturer. Auxiliary air may be introduced into the system through a screened opening to stabilize the arc. When Xenon projection equipment is used the lamp exhaust system shall exhaust not less than 300 cfm per lamp nor less than that exhaust volume required or recommended by the equipment manufacturer, whichever is greater. The external temperature of the lamp housing shall not exceed 130° F when operating.

(k) Each projection room shall be provided with rewind and film storage facilities.

(l) A maximum of four containers for flammable liquids not greater than sixteen ounce capacity and of a non-breakable type may be permitted in each projection room.

512.17 — SUPPLEMENTARY LIGHTING SYSTEM

There shall be installed in every Group E-1 Large Assembly a supplementary lighting system in addition to the regular system by local electric power. Such supplementary or emergency lighting shall be automatically actuated in case of power failure. Every supplementary lighting system shall be maintained in good working order and shall be tested at least once every ten days.

SECTION 513 — BOWLING ALLEYS

513.1 — GENERAL

(a) Bowling Alleys shall comply with Section 405, Group "B"—Business and all provisions related thereto and with the provisions of this Section.

(b) Where bowling pin finishing or refinishing operations are carried on, such a separate building, or a separate room, constructed as specified herein, shall be provided.

(c) Such a room shall be located at or above street level and shall have one or more windows opening to the outside of the building.

(d) Walls and ceiling of such rooms shall have not less than one-hour fire-resistance. Floors shall be of concrete at least two inches thick or of equivalent non-combustible protective material.

(e) Door openings shall be provided with non-combustible sills, raised six inches above floor level and protected with approved fire doors.

(f) Shelving, containers, and all furnishings shall be of non-combustible material. Machinery shall be effectively grounded. See Section 501.1 (f).

(g) Ventilation sufficient to effect complete change of air at least once every three minutes shall be provided.

SECTION 514 — CHURCHES

514.1 — SCOPE

This section shall apply to churches or places of worship. All other places of public assembly shall be governed by the regulations as set forth in Section 512.

514.2 — TYPES OF CONSTRUCTION

For types of construction permitted and other limitations related thereto, see Section 408.6.

514.3 — EXITS

Refer to Chapter XI, Means of Egress Requirements, for exits and exit access requirements.

514.4 — INTERIOR FINISH AND DECORATIONS

All interior finishes and decorations shall conform with Section 512.3 except that nothing in this Section shall prevent the use of wood for ornamental purposes, trusses, paneling or chancel furnishing.

514.5 — AISLES AND SEATING

(a) Every aisle shall lead to an exit door or to a cross aisle running parallel to the seats and leading directly to an exit. No aisle shall be less in width than 36 inches plus an increase of 1½ inches for each five feet of such aisle from its beginning to an exit, except that aisles with seats on one side may be six inches less in width; where egress is provided at both ends of an aisle, the aisle may have a uniform width of not less than specified herein. No cross aisle shall be less than 3 feet 6 inches. An aisle bordering on a means of entrance shall be not less than 4 feet wide.

(b) There shall be no obstructions of any kind in an aisle. Aisles shall not exceed a gradient of more than one in eight. No steps shall be used in any aisle where differences of level can be overcome by gradients. Where it is necessary in balconies to use steps, they shall extend the full width of aisles and risers shall not exceed six and one-half inches.

(c) Rows of seats between aisles shall have not more than 20 seats. Rows of seats opening onto an aisle at one end shall have not more than 7 seats. Seats without dividing arms shall have their capacity determined by allowing 18 inches per person.

(d) The spacing of rows of seats from back to back shall be not less than 30 inches. In every case there shall be a clear space of not less than 12 inches between the back of one seat and the front of one seat immediately behind it, measured at the seat line.

514.6 —AREA INCREASE

Where there are no balconies or galleries in churches and places of worship and the assembly floor is located at, or within, 21 inches of the street or grade level and all exits meet the street or grade level by ramps having a slope not exceeding 1 foot in 10 feet, the maximum allowable areas of Type III, IV, and V construction may be increased 50 per cent over those specified for Group E Assembly occupancies in Table 408.6.

SECTION 515 – FARM BUILDINGS

515.1 — GENERAL

Farm Buildings shall include those structures other than residences and structures appurtenant thereto, for on-farm use (barns, sheds, poultry houses, etc.). Maximum allowable deflection for structural members of such farm buildings shall not exceed 1/180 of span. Design limitations based on deflection as prescribed elsewhere in this code shall not be applicable.

SECTION 516 — COVERED MALLS, WALKWAYS AND TUNNELS

516.1 — SCOPE

This section shall apply to connections between buildings such as covered malls and walkways or tunnels, located at, above or below grade level, that are used as a means of travel by persons.

516.2 — DEFINITIONS

(a) Covered mall: is a covered or roofed interior area having a minimum horizontal dimension of 30 feet used as a pedestrian publicway and connecting buildings and/or group of buildings housing individual or multiple tenants.

(b) Covered walkway: is a roofed, unobstructed walkway, where the least horizontal dimension is less than thirty (30) feet, connecting buildings and used as a means of travel by persons and where less than fifty (50) percent of the perimeter is enclosed.

(c) Enclosed walkway: is a roofed, unobstructed walkway, where the least horizontal dimension is less than thirty (30) feet, connecting buildings and used as a means of travel by persons and where fifty (50) percent or more of the perimeter is enclosed.

(d) Tunneled walkway: is an unobstructed underground walkway connecting buildings and used as a means of travel by persons.

516.3 — CONSTRUCTION

(a) Covered Malls: The roof construction and supporting members of a covered mall shall be required to be of a type of construction permitted for the buildings connected and where Type IV, V, or VI construction is used it shall provide not less than 1 hour fire resistance. All unprotected walls and openings separating a tenant area from the mall area shall be provided with a water curtain unless the tenant area is provided with a complete automatic sprinkler system. Where there is an occupied area above the mall, the occupancy separation provisions of this code shall apply.

Concealed spaces in a mall roof assembly shall be separated from adjoining buildings by not less than 1 hour fire resistive construction.

Except where an approved automatic sprinkler system is provided, Class 1 hose cabinets shall be provided for each 200 feet of mall length.

(b) Covered walkway: A covered walkway shall be of any type of construction permitted by this code, provided the walls and openings at the point of connection to the building shall be protected so as to reasonably prevent the spread of fire from one building into the other.

(c) Enclosed walkway: An enclosed walkway shall be required to be of a type of construction permitted for the buildings connected. Separation between the enclosed walkway and the building to which it is connected, except when used as an exit outlet, shall be of not less than 1 hour fire resistant construction, and openings therein shall be protected in accordance with Section 703.

(d) Tunneled walkway: A tunneled walkway shall be of a type of of construction suitable for underground location. Separation between the tunneled walkway and the building to which it is connected shall be not less than 2 hour fire resistant construction and openings therein shall be protected in accordance with Section 703.

516.4 — ALLOWABLE AREAS

When complying with the provisions of this code, covered malls of Types I and II construction may be unlimited in area. For all other types of construction the basic allowable area for covered malls shall be 12,000 square feet. The area of covered malls may be increased:

(a) 200 percent when the covered mall is provided with a complete automatic sprinkler system and

(b) at the rate of 25 percent for each side of the building provided with at least thirty (30) foot width of mall leading to a public place or street not less than thirty (30) feet in width, but not to exceed 100 percent.

516.5 — EXITS

See Chapter XI.

516.6 — VENTILATION

Smoke and heat venting shall be provided for covered malls and enclosed walkways and tunneled walkways. Such venting systems shall be in accordance with "Guide for Smoke and Heat Venting, NFPA 204-68," or other accepted engineering practice.

SECTION 517 – TIRE RECAPPING

517.1 — GENERAL

Tire recapping facilities shall be classified as Group "H"—Special Hazardous occupancy if they exceed the square footage shown below:

Type Construction	Square Feet
Type I	No Limit
Type II	No Limit
Type III	12,000
Type IV	9,000
Type V	9,000
Type VI	Not Permitted

CHAPTER VI

CLASSIFICATION OF BUILDINGS BY CONSTRUCTION

SECTION 601 — CLASSIFICATION BY TYPE OF CONSTRUCTION

601.1 — TYPES

All buildings shall be classified into six general types according to the character of materials employed and their method of assembly, as follows:

TYPE I

TYPE II

TYPE III

TYPE IV

TYPE V

TYPE VI

601.2 — FIRE-RESISTANCE REQUIREMENTS

All fire-resistance requirements are expressed in terms of the number of hours of satisfactory performance in accordance with the "Standard Methods of Fire Tests of Building Construction and Materials, ASTM E119-71."

601.3 — MATERIALS AND CONSTRUCTION APPROVED FOR FIRE PROTECTION

(a) The degree of fire resistance and the materials, assemblies, and constructions providing such resistance shall be as defined in Chapter X of this Code, except that other materials, assemblies, and constructions shall be approved, provided test data of a recognized engineering or testing laboratory are submitted, establishing that they develop the required fire-resistance ratings under tests made in accordance with the "Standard Methods of Fire Tests of Building Construction and Materials, ASTM E119-71."

(b) Where structural requirements necessitate assemblies providing greater fire resistance than specified in this Chapter, such structural requirements shall govern.

SECTION 602 — TYPE I

602.1 — GENERAL

Type I Construction, is that in which all exterior walls are of masonry or reinforced concrete, or of other approved materials or combination of materials, and in which all the structural members are of non-combustible materials, and provide fire-resistance not less than stipulated in this Section.

602.2 — FIRE DISTRICT — SECTION 301

602.3 — ALLOWABLE HEIGHT — SECTIONS 404 to 411, inclusive

602.4 — ALLOWABLE AREA — SECTIONS 404 to 411, inclusive

602.5 — FIRE PROTECTIVE REQUIREMENTS — TYPE I

TABLE 602.5 — FIRE PROTECTIVE REQUIREMENTS, TYPE I

STRUCTURAL MEMBERS	Required Fire Resistance (Hours)	
WALLS		
Party Walls	4	Party and fire walls shall extend not less than three (3) feet above the roof, except that fire walls need not extend above the roof where the roof is of non-combustible construction for the area within forty (40) feet of each side of the wall.
Fire Walls	4	
Exterior Bearing	4	All walls except:
	3	Where wall faces on street or public place thirty (30) feet or more in width.
	2	Where wall faces on street or public place fifty (50) feet or more in width.
Exterior Non-Bearing (See Section 608.3) (See Section 507.2 for Public Parking Decks).	2	All walls except:
	1	Where protection of wall openings is not required by Section 703.
Inner Court Walls	3	All walls except:
Penthouse Walls	2	Where penthouse walls set back five (5) feet or more from exterior walls. Where set back is

TABLE 602.5 — FIRE PROTECTIVE REQUIREMENTS
TYPE I—(Continued)

STRUCTURAL MEMBERS	Required Fire Resistance (Hours	
Inner Court Penthouse (Continued)		less than five (5) feet, penthouse walls shall conform to fire-resistance ratings for exterior walls.
PARTITIONS		Enclosures and vertical openings—Sect. 701. Buildings with mixed occupancies — Sect. 412. All other partitions—Sect. 702.
Interior Bearing	4	
Interior Non-Bearing	Sec. 702.2	
COLUMNS		
Supporting Masonry or Bearing Walls	4	
Supporting Roofs only	3	
Other Columns	4	
TRUSSES		
Supporting Masonry or Bearing Walls, Columns, Girders, Trusses	4	
Supporting Roofs only	2	
	1*	See Footnote
Other Trusses	2½	
GIRDERS		
Supporting Masonry or Bearing Walls, Columns, Girders, Trusses	4	
Supporting Roofs only	2	
	1*	See Footnote
Other Girders	2½	
BEAMS		
Supporting Masonry or Bearing Walls, Columns, Girders, Trusses	4	
Supporting Roofs only	1½	
	1*	See Footnote
Other Beams	2½	

TABLE 602.5 — FIRE PROTECTIVE REQUIREMENTS, TYPE I—(Continued)

STRUCTURAL MEMBERS	Required Fire Resistance (Hours)	
FLOORS		
Deck Construction	2½	
ROOFS		
Deck Construction	1½*	See Footnote

NOTE: *In two (2) story buildings approved Fire Retardant Treated Wood may be used.

In buildings of Group C and E (School and Assembly) occupancies where structural members supporting a roof only are not less than 18' clear above any floor or balcony, one hour fire resistance shall be provided; where the clear distance is 25' or more, fire protection of structural members supporting roof construction may be omitted.

PROTECTION OF WALL OPENINGS .... Section 703
FIRESTOPPING .... Section 705
STAIRWAY CONSTRUCTION .... Section 1115
ROOF COVERINGS .... Sections 301 and 706

REGULATIONS GOVERNING EXTERIOR USE OF COMBUSTIBLE MATERIALS:

a. Gutters and Leaders .... Section 712

b. Dormer Windows .... Section 709

c. Towers, Spires and Cupolas .... Section 713

d. Cooling Towers .... Section 715

e. Tanks .... Section 714

f. Skylights .... Section 707

REGULATIONS GOVERNING INTERIOR USE OF COMBUSTIBLE MATERIALS:

a. Floor Finish .... Section 704.2

b. Ceilings and other interior finishes .... Section 704.

SECTION 602.6 — STRUCTURAL AND ENGINEERING REQUIREMENTS

a. Minimum Design Loads ....Chapter XII
b. Foundations ....Chapter XIII
c. Steel ....Chapter XV
d. Concrete ....Chapter XVI
e. Masonry ....Chapter XIV
f. Aluminum ....Chapter XXVIII
g. Lathing, Plastering and Gypsum Wallboard..........................Chapter XVIII
h. Stair Construction ....Section 1115
i. Elevators and Escalators ....Chapter XXIV
j. Safeguards During Construction ....Chapter XXI

SECTION 603 – TYPE II

603.1 — GENERAL

Type II Construction is that in which all exterior walls are of masonry or reinforced concrete, or of other approved materials or combinations of materials and in which all the structural members are of non-combustible materials, and provide fire-resistance not less than stipulated in this section. Fire Retardant Treated Wood may be used as specified in Table 603.5.

603.2 — FIRE DISTRICT — SECTION 301

603.3 — ALLOWABLE HEIGHT — SECTIONS 404 to 411, inclusive

603.4 — ALLOWABLE AREA — SECTIONS 404 to 411, inclusive

603.5 — FIRE PROTECTIVE REQUIREMENTS — TYPE II

TABLE 603.5 — FIRE PROTECTIVE REQUIREMENTS, TYPE II

STRUCTURAL MEMBERS	Required Fire Resistance (Hours)	
WALLS		Party and fire walls shall extend not less than three (3) feet above the roof, except that firewalls need not extend above the roof where the roof is of non-combustible construction for the area within forty (40) feet of each side of the wall.
Party Walls	4	
Fire Walls	4	
Exterior Bearing	3**	See Footnote
	2	Where wall faces on street or public place thirty (30) feet or more in width.
Exterior Non-Bearing	2	All walls except:
(See Section 608.3) (See Section 507.2 for Public Parking Decks)	1	Where protection of wall openings is not required by Section 703.
Inner Court Walls	2	
Penthouse Walls	2	Where penthouse walls set back five (5) feet or more from exterior walls. Where set back is less than five (5) feet, penthouse walls shall conform to fire-resistive ratings for exterior walls.

TABLE 603.5 — FIRE PROTECTIVE REQUIREMENTS, TYPE II—(Continued)

STRUCTURAL MEMBERS	Required Fire Resistance (Hours)	
PARTITIONS		Enclosures and vertical openings—Sect. 701. Buildings with mixed occupancies — Sect. 412. All other partitions—Sect. 702.
Interior Bearing	3	
Interior Non-Bearing	Sec. 702.2	
COLUMNS		
Supporting Masonry or Bearing Walls	3	
Supporting Roofs only	2	
Other Columns	2	
TRUSSES		
Supporting Masonry or Bearing Walls, Columns, Girders, Trusses	3	
Supporting Roofs only	1*	See Footnote
Other Trusses	1½	
GIRDERS		
Supporting Masonry or Bearing Walls, Columns, Girders, Trusses	3	
Supporting Roofs only	1*	See Footnote
Other Girders	1½	
BEAMS		
Supporting Masonry or Bearing Walls, Columns, Girders, Trusses	3	
Supporting Roofs only	1*	See Footnote
Other Beams	1½	
FLOORS		
Deck Construction	1½	
ROOFS		
Deck Construction	1*	See Footnote

TABLE 603.5 — FIRE PROTECTIVE REQUIREMENTS, TYPE II—(Continued)

Note: *In two (2) story buildings approved Fire Retardant Treated Wood may be used.

In buildings of Group C and E (School and Assembly) occupancies, fireproofing may be omitted where structural members supporting a roof only and are twenty (20) feet or more clear above any floor or balcony.

In one (1) story buildings structural members of heavy timber sizes may be used as an alternate to unprotected structural roof members.

**Exterior walls of Type II buildings not over three (3) stories in height of Group A, B, C, D or E occupancy may be constructed of framed wall assemblies, that have fire resistance against outside exposure, as specified in Table 603.5, and that has not less than 1½ hours fire-resistance inside the building.

PROTECTION OF WALL OPENINGS .... Section 703
FIRESTOPPING .... Section 705
STAIRWAY CONSTRUCTION .... Section 1115
ROOF COVERINGS .... Sections 301 and 706

REGULATIONS GOVERNING EXTERIOR USE OF COMBUSTIBLE MATERIALS:

a. Gutters and Leaders .... Section 712
b. Dormer Windows .... Section 709
c. Towers, Spires and Cupolas .... Section 713
d. Cooling Towers .... Section 715
e. Tanks .... Section 714
f. Skylights .... Section 707

REGULATIONS GOVERNING INTERIOR USE OF COMBUSTIBLE MATERIALS:

a. Floor finish .... Section 704.2
b. Ceilings and other interior finishes .... Section 704

SECTION 603.6 — STRUCTURAL AND ENGINEERING REQUIREMENTS

a. Minimum Design Loads .... Chapter XII
b. Foundations .... Chapter XIII
c. Steel .... Chapter XV
d. Concrete .... Chapter XVI
e. Masonry .... Chapter XIV
f. Lathing, Plastering and Gypsum Wallboard .... Chapter XVIII
g. Stair Construction .... Section 1115
h. Elevators and Escalators .... Chapter XXIV
i. Safeguards During Construction .... Chapter XXI

SECTION 604 – TYPE III

604.1 — GENERAL

Type III Construction is that type in which fire-resistance is attained by the sizes of heavy timber members (sawn or glue-laminated) being not less than indicated in this Section or by providing fire-resistance not less than one-hour where materials other than wood are used; by the avoidance of concealed spaces under floors and roofs; by the use of approved fastenings, construction details, and adhesives for structural members; and by providing the required degree of fire-resistance in exterior and interior walls.

604.2 — FIRE DISTRICT — SECTION 301

604.3 — ALLOWABLE HEIGHT — SECTIONS 404 to 411, INCLUSIVE

604.4 — ALLOWABLE AREA — SECTIONS 404 to 411, INCLUSIVE

604.5 — FIRE PROTECTIVE REQUIREMENTS — TYPE III

TABLE 604.5 — FIRE PROTECTIVE REQUIREMENTS
TYPE III

STRUCTURAL MEMBERS	Required Fire-Res. (H. T. Sizes or Hours)	
WALLS		See Note (1) for parapet requirements.
See Note (1)	4	
Party Walls	4	
Fire Walls	4	
Exterior Bearing Walls	3	All walls except:
	2	Where a horizontal separation of more than three (3) feet is provided.
Exterior Non-Bearing Walls	3	All walls except:
	2	Where a horizontal separation of more than three (3) feet but less than twenty (20) feet is provided.
	1	Where a horizontal separation of twenty (20) feet but less than thirty (30) feet is provided.

TABLE 604.5 — FIRE PROTECTIVE REQUIREMENTS—
TYPE III—(Continued)

STRUCTURAL MEMBERS	Required Fire Resistance (H. T. Sizes or Hours)	
Exterior Non-Bearing Walls (Cont'd.)	0	Where a horizontal separation of thirty (30) feet or more is provided. (See also Section 608.3.)
Inner Court Walls	3	
Penthouse Walls	2	Where penthouse walls set back five (5) feet or more from exterior walls. Where set back is less than five (5) feet, penthouse walls shall conform to fire-resistive rating of exterior walls.
PARTITIONS		
Interior Bearing	3	
Interior Non-Bearing	Sec. 702.2	Buildings with mixed occupancies — Sec. 412. All other partitions — Sec. 702.
Vertical Openings		Enclosure for vertical openings —Section 701.
COLUMNS		See Note (5).
Supporting Masonry Bearing Walls	2	Columns shall be continuous or superimposed by means of properly designed reinforced concrete or metal caps, or by timber splice plates affixed to the columns by means of timber connectors or by other approved methods.
Supporting Roof only	6 x 8 or 1 Hr.	
Supporting Floors	8 x 8 or 1 Hr.	
TRUSSES		See Note (4).
Supporting Masonry Bearing Walls	2	
Supporting Roof only	4 x 6 or 1 Hr.	
Supporting Floors	8 x 8 or 1 Hr.	

TABLE 604.5 — FIRE PROTECTIVE REQUIREMENTS— TYPE III—(Continued)

STRUCTURAL MEMBERS	Required Fire Resistance (H. T. sizes or Hours)	
ARCHES Supporting Roof only	6 x 8 for lower half 6 x 6 for upper half	See Note (4) and Note (5). Arches springing from floor line or grade.
	4 x 6 or 1 Hr.	Arches springing from top of wall or wall abutments.
Supporting Floors	8 x 8 or 1 Hr.	
BEAMS AND GIRDERS		See Note (5).
Supporting masonry bearing walls	2	Girders and beams entering masonry shall have ½ inch air space on all sides and bear on self-releasing wall plate boxes or approved hangers. Where the wall is required to have a 2-hour or more fire rating, it shall have not less than 4 inches of solid masonry between the beam and the outside or adjacent beams.
Supporting Roof only	4 x 6 or 1 Hr.	
Supporting Floors	6 x 10 or 1 Hr.	
FLOORS Deck Construction		See Note (2).
ROOFS Deck Construction		For roof anchorage see Section 1205.3. See Notes (3) and (5).

NOTE (1) PARTY AND FIRE WALLS shall extend not less than three (3) feet above the roof.

EXTERIOR WALLS shall extend not less than eighteen (18) inches above the roof, except that parapet walls need not be constructed on buildings where the roof slopes more than four (4) inches vertical to twelve (12) inches horizontal from the back of the exterior wall of such building, or where the exterior of such building is located thirty (30) feet or more distance from the property line or other building on the same property, or faces on an alley or public way thirty (30) feet or more in width.

NOTE (2) FLOORS shall be of sawn or glued-laminated plank, splined, or tongued and grooved, of not less than 3 inches, nominal, in thickness or of planks not less than 4 inches, nominal, in width set on edge and well spiked together. The planks shall be laid so that no continuous line of joints will occur except at points of support.

Planks shall be covered with 1 inch, nominal, tongued and grooved flooring laid crosswise or diagonally. Planks and flooring shall not extend closer than ½ inch to walls to provide an expansion joint, and the joint shall be covered at top and bottom.

NOTE (3) ROOF DECKS shall be sawn or glue-laminated, splined or tongue and grooved plank, not less than 2 inches, nominal, in thickness, one and one-eighth (1⅛) inches thick interior plywood (exterior glue), or of planks not less than 3 inches, nominal, in width, set on edge and spiked together, as required for floors. Other types of roof decking may be used that provide equivalent fire resistance, is not more combustible than two (2) inch nominal wood sheathing, and is used within spans which have been proved structurally safe by approved tests.

NOTE (4) ROOF FRAMING: Spaced members may be composed of two or more pieces not less than 3 inches, nominal, in thickness when blocked solidly throughout their intervening spaces or when such spaces are tightly closed by a continuous wood cover plate of not less than two (2) inches, nominal, in thickness, secured to the underside of the members. Splice plates shall be of no less than 3 inches, nominal, in thickness. When protected by approved automatic sprinklers under the roof deck, framing members shall be not less than 3 inches, nominal, in width.

NOTE (5) Where a horizontal separation of 20 feet or more is provided, wood columns, arches, beams and roof decks, conforming to heavy timber sizes may be used externally.

TABLE 604.5 — FIRE PROTECTIVE REQUIREMENTS—
TYPE III—(Continued)

REGULATIONS GOVERNING EXTERIOR USE OF COMBUSTIBLE MATERIALS:

REGULATIONS GOVERNING INTERIOR USE OF COMBUSTIBLE MATERIALS:

a. In Group H, Special Hazardous Occupancies, only non-combustible finishes shall be used.

b. For Group E, Assembly Occupancies, see Section 512.3.

c. For other occupancies, see Section 704.

SECTION 604.6 — STRUCTURAL AND ENGINEERING REQUIREMENTS:

SECTION 605 — TYPE IV

605.1 — GENERAL

Type IV Construction, is that in which all structural members, including wall framing, floors, roofs and their supports, shall be of steel, iron or other metal, or of other non-combustible materials, and in which the exterior surface of the building is of steel, iron or other metal, or of asbestos, masonry, reinforced concrete, or other non-combustible materials, and that are fire-protected only where specified in this Section.

605.2 — FIRE DISTRICT — SECTION 301

605.3 — ALLOWABLE HEIGHT — SECTIONS 404 to 411, inclusive

605.4 — ALLOWABLE AREA — SECTIONS 401 to 411, inclusive

605.5 — FIRE-PROTECTIVE REQUIREMENTS — TYPE IV

TABLE 605.5 — FIRE-PROTECTIVE REQUIREMENTS— TYPE IV

STRUCTURAL MEMBERS	Required Fire Resistance (Hours)	
WALLS		Party and fire walls shall extend not less than three (3) feet above the roof, except that fire walls need not extend above the roof where the roof is of non-combustible construction of the area within forty (40) feet of each side of the wall.
Party Walls	4	
Fire Walls	4	
		FIRE DISTRICT
Exterior Walls (See Section 608.3)	2***	All walls except:
(See Section 507.2 for Public Parking Decks)	N.C.	Walls of one story buildings not exceeding 2000 sq. ft. in area facing and more than 15 feet from a common property or interior lot line.

TABLE 605.5 — FIRE-PROTECTIVE REQUIREMENTS — TYPE IV (Continued)

		OUTSIDE OF FIRE DISTRICT
Exterior Walls (See Section 608.3)	1***	Walls facing and within 8 feet of common property or interior lot lines.
	N.C.*	Other walls.
	N.C.	Except as prescribed above and except where fire protection is required by footnote* all exterior walls may be of unprotected non-combustible construction.
Inner Court Walls	N.C.*	Same as Exterior Walls.
Penthouse Walls	N.C.	
PARTITIONS		For fire-resistance requirements see Footnote*
Interior Bearing	N.C.*	
Interior Non-Bearing	Sec. 702.2	Enclosure for vertical openings —Sect. 701. Buildings with mixed occupancies — Sect. 412. All other partitions — Sect. 702.
COLUMNS		
Supporting Masonry or Bearing Walls	2**	
Supporting Roofs only	N.C.	
Other Columns	N.C.*	
TRUSSES		
GIRDERS		
BEAMS		
Supporting Masonry or Bearing Walls, Columns, Girders, Trusses	2**	
Other Trusses	N.C.*	
Other Girders	N.C.*	
Other Beams	N.C.*	

TABLE 605.5 — FIRE-PROTECTIVE REQUIREMENTS — TYPE IV
(Continued)

FLOORS		
Deck Construction	N.C.*	For fire-resistance requirements, see Footnote.*
ROOFS		
Deck Construction	N.C.	
High above Floor	N.C.	

ABBREVIATIONS: "N.C." means Non-Combustible.

NOTE: *In buildings three (3) stories or more in height, (unless they are Public Parking Decks or sprinklered) all walls, partitions, floors, roofs and their supporting structural members shall provide not less than one (1) hour fire resistance within the building except that roofs of such buildings, need not be protected.

In every Group D (Institutional) building; and in Group C (Schools) two or more stories in height; at least one-hour interior fire-resistance shall be provided throughout the building.

**This requirement applies only to structural members supporting masonry walls, except that this does not apply in one (1) story buildings or where the only masonry supported is a masonry veneer.

***Fire-resistance against outside fire exposure.

PROTECTION OF WALL OPENINGS ....Section 703

FIRESTOPPING ....Section 705

STAIRWAY CONSTRUCTION ....Section 1115

ROOF COVERINGS ....Sections 301 and 706

REGULATIONS GOVERNING EXTERIOR USE OF COMBUSTIBLE MATERIALS:

a. Gutters and Leaders ....Section 712

b. Dormer Windows ....Section 709

c. Towers, Spires and Cupolas ....Section 713

d. Cooling Towers ....Section 715

e. Tanks ....Section 714

f. Skylights ....Section 707

REGULATIONS GOVERNING INTERIOR USE OF COMBUSTIBLE MATERIALS:

a. In Group H, Special Hazardous Occupancy, only non-combustible finishes shall be used.

b. For Group E, Assembly Occupancies, See Section 512.3.

c. For other occupancies, See Section 704.

SECTION 605.6 — STRUCTURAL AND ENGINEERING REQUIREMENTS:

a. Minimum Design Loads .... Chapter XII

b. Foundations .... Chapter XIII

c. Steel .... Chapter XV

d. Concrete .... Chapter XVI

e. Masonry .... Chapter XIV

f. Aluminum .... Chapter XXVIII

g. Lathing, Plastering and Gypsum Wallboard .... Chapter XVIII

h. Stair Construction .... Section 1115

i. Elevators and Escalators .... Chapter XXIV

j. Safeguards During Construction........ Chapter XXI

SECTION 606 – TYPE V

606.1 — GENERAL

Type V Construction—is that construction not meeting the requirements of Type III, but in which the exterior walls are of masonry or reinforced concrete or of approved materials or assembly of materials that provide fire-resistance as required in this Section, and in which the interior framing is partially or wholly of unprotected wood, or of unprotected iron or steel, except that fire protection shall be provided, as required by this Section.

606.2 — FIRE DISTRICT — SECTION 301

606.3 — ALLOWABLE HEIGHTS — SECTIONS 404 to 411, inclusive

606.4 — ALLOWABLE AREA — SECTIONS 404 to 411, inclusive

606.5 — FIRE-PROTECTIVE REQUIREMENTS — TYPE V

TABLE 606.5 — FIRE PROTECTIVE REQUIREMENTS — TYPE V

STRUCTURAL MEMBERS	Required Fire Resistance (Hours)	
WALLS		
Party Walls	4	Party, and fire walls shall extend not less than three (3) feet above the roof, except that fire walls need not extend above the roof where the roof is of non-combustible construction for the area within forty (40) feet of each side of the wall.
Fire Walls	4	
Exterior Bearing Walls	3	
	2	For walls facing on a street or public place thirty (30) feet or more in width.***
Exterior Non-Bearing Walls (See Section 608.3)	2	
	2	For buildings located in the Fire District***.
	N.C.*	For buildings located outside the Fire District.***

TABLE 606.5 — FIRE PROTECTIVE REQUIREMENTS — TYPE V (Continued)

PARTITIONS		
Interior Bearing	*	The use of combustible construction for interior bearing partitions shall be limited to the support of not more than two (2) floors and a roof.
Interior Non-Bearing	* Sec. 702.2	Enclosure for vertical openings—Sect. 701. Buildings with mixed occupancies—Sect. 412. All other partitions—Sect. 702.
COLUMNS		
Supporting Masonry or Bearing Walls		Shall be same rating as required for the wall it supports.
Supporting Roof only	*	
Other Columns	*	
TRUSSES		
GIRDERS		
BEAMS		
Supporting Masonry or Bearing Walls, Columns, Girders, Trusses	2**	
Supporting Roof only	*	
Other Trusses	*	
Other Girders	*	See Footnote for fire-resistance requirements.
Other Beams	*	See Footnote for fire-resistance requirements.
FLOORS		
Deck Construction	*	See Footnote for fire-resistance requirements.
ROOFS		
Deck Construction	*	

ABBREVIATIONS: "N.C." means Non-Combustible.

NOTE (1) Buildings of Type V Construction two or more stories in height, except one and two family dwellings, floors located immediately above usable spaces in basements or cellars and above furnaces shall have one hour fire protection except where basement or cellar is equipped with an approved automatic sprinkler system.

*In buildings three or more stories in height, (unless

sprinklered) all walls, partitions, floors and their supporting structural members shall provide not less than one (1) hour fire-resistance within the building and the ceiling underneath the roof shall be the same as required for the floors.

In all Group D (Institutional) occupancies; and in Group C (Schools) two or more stories in height; at least one (1) hour interior fire-resistance shall be used throughout.

**This requirement applies only to structural members supporting masonry walls except that this does not apply in one story buildings or where the only masonry supported is a masonry veneer.

***Exterior walls shall extend not less than eighteen (18) inches above the roof, except that parapet walls need not be constructed on buildings where the roof slopes more than four (4) inches vertical to twelve (12) inches horizontal from the back of the exterior wall of such buildings or where the exterior wall of such building is located fifteen (15) feet or more distance from the property line or is located on one alley or public way of fifteen feet or more in width.

PROTECTION OF WALL OPENINGS ....Section 703

FIRESTOPPING ....Section 705

STAIRWAY CONSTRUCTION ....Section 1115

ROOF COVERING ....Sections 301 and 706

REGULATIONS GOVERNING EXTERIOR USE OF COMBUSTIBLE MATERIALS:

a. Gutters and Leaders ....Section 712

b. Dormer Windows ....Section 709

c. Towers, Spires and Cupolas ....Section 713

d. Cooling Towers ....Section 715

e. Tanks ....Section 714

f. Skylights ....Section 707

REGULATIONS GOVERNING INTERIOR USE OF COMBUSTIBLE MATERIALS:

a. In Group H, Special Hazardous Occupancy, only non-combustible finishes shall be used.

b. For Group E, Assembly Occupancies, See Section 512.3.

c. For other occupancies, See Section 704.

SECTION 606.6 — STRUCTURAL AND ENGINEERING REQUIREMENTS

a. Minimum Design Loads .... Chapter XII

b. Foundations .... Chapter XIII

c. Wood .... Chapter XVII

d. Steel .... Chapter XV

e. Concrete .... Chapter XVI

f. Masonry .... Chapter XIV

g. Aluminum .... Chapter XXVIII

h. Lathing, Plastering and Gypsum Wallboard .... Chapter XVIII

i. Stair Construction .... Section 1115

j. Elevators and Escalators .... Chapter XXIV

k. Safeguards During Construction .... Chapter XXI

SECTION 607 — TYPE VI

607.1 — GENERAL

Type VI Construction—is that in which the enclosing walls are of wood or other combustible materials, including construction having exterior masonry veneer, stucco, or metal, which is dependent upon wood for support, stability or rigidity, and in which interior framing is of wood or other combustible materials.

607.2 — LOCATION OF PROPERTY

All exterior walls of Type VI buildings located less than three (3) feet from property lines shall provide not less than one hour fire-resistance protection.

607.3 — BUILDINGS LOCATED ON THE SAME LOT

Where two or more buildings of Type VI construction are on the same lot, there shall be a clear space of not less than six (6) feet between the buildings unless one of the exposed walls has at least one hour fire-resistance.

607.4 — FIRE DISTRICT

Type VI Construction is prohibited in the Fire District except as provided in Section 304.

607.5 — ALLOWABLE HEIGHT — SECTIONS 404 to 411, inclusive

607.6 — ALLOWABLE AREA — SECTIONS 404 to 411, inclusive

607.7 — FIRE PROTECTIVE REQUIREMENTS — TYPE VI

TABLE 607.7 — FIRE PROTECTIVE REQUIREMENTS — TYPE VI

STRUCTURAL MEMBERS	Required Fire Resistance (Hours)	
WALLS		
Party Walls	4	Party and fire walls shall extend not less than three (3) feet above the roof, except that fire walls need not extend above the roof where the roof is of non-combustible construction for the area within forty (40) feet of each side of the wall.
Fire Walls	4	

TABLE 607.7 — FIRE PROTECTIVE REQUIREMENTS — TYPE VI—(Continued)

PARTITIONS Interior	Sec. 702.2	Enclosure for vertical openings — Sect. 701. Buildings with mixed occupancies — Sect. 412. All other partitions—Sec. 702.
COLUMNS **TRUSSES** **GIRDERS** **BEAMS** Supporting Masonry or Bearing Walls, Columns, Girders, Trusses	1	This requirement applies only to structural members supporting masonry walls except that this does not apply in one (1) story buildings or where the only masonry supported is a masonry veneer.

FIRESTOPPING .... Section 705
PRIVATE GARAGES .... Section 506
ROOF COVERING .... Sections 301 and 706

SECTION 607.8 — STRUCTURAL AND ENGINEERING REQUIREMENTS

a. Wood .... Chapter XVII
b. Minimum Design Loads .... Chapter XII
c. Foundations .... Chapter XIII
d. Concrete .... Chapter XVI
e. Masonry .... Chapter XIV
f. Aluminum .... Chapter XXVIII
g. Lathing, Plastering and Gypsum Wallboard .... Chapter XVIII
h. Safeguards During Construction .... Chapter XXI

SECTION 608 — EXCEPTIONS TO FIRE PROTECTION

608.1 — ELEVATOR FRAMES

Structural members of frames for elevators will not be required to have the fire protection required for structural steel, provided such members are erected within an enclosure of the prescribed fire-resistance rating. Section 701—Enclosure of Vertical Openings.

608.2 — LINTELS

Lintels over openings in walls shall be protected to provide a fire-resistance rating at least equal to that required for beams, except that when such lintels are used over openings less than four (4) feet wide, such protection may be omitted. The outer member of an assembled steel lintel, which supports face masonry that is securely bonded to backing need not be protected, provided that the load carrying member of such lintel is protected as herein required.

608.3 — UNPROTECTED EXTERIOR WALLS OR PANELS

Unprotected walls or panels may be permitted in exterior non-bearing walls under the following conditions:

(1) Provided such walls are of non-combustible material or of exterior grade Fire Retardant Treated Wood.

(2) Provided such walls face a street or permanent open space of 30 feet or more in width.

(3) Provided that in buildings three stories or more in height, exterior openings located in a story above a Group D, Institutional; Group F, Storage; Group G, Industrial; Group H, Hazardous Occupany are separated from such an occupancy by a 2-hour fire-resistive wall construction not less than 3 feet in height.

608.4 — WOOD VENEERS ON EXTERIOR WALL PANELS

Wood veneers of not less than one (1) inch nominal thickness or three-eighths (⅜) inch exterior type plywood or particleboard may be used on exterior walls under the following conditions:

(1) The wall to which the veneer is attached faces a street or permanent open space of thirty (30) feet or more in width.

(2) The veneer does not exceed two stories in height, measured from grade.

(3) The veneer is attached to or furred from a non-combustible backing of the fire resistance required by other provisions of this Chapter.

(4) Where open or spaced wood veneers are used, they shall not project more than twenty-four (24) inches from the building wall.

Where the wood veneer is furred from the wall and forms a solid surface, the distance between the back of the veneer and the wall shall not exceed one and five-eighths (1⅝) inches and the space thereby created shall be firestopped in accordance with Section 1703 and arranged so that there will be no open space exceeding one hundred (100) sq. ft. Where wood furring strips are used, they shall be of approved wood of natural decay-resistance or pressure treated wood.

608.5 — UNUSABLE SPACE

In one hour fire resistant construction the ceiling may be omitted over unusable crawl space and flooring may be omitted when unusuable attic space occurs above.

SECTION 609 – MIXED TYPES OF CONSTRUCTION

609.1

When two or more types of construction not separated by fire walls occur in the same building not classified as a mixed occupancy, the entire building shall then be subject to the occupancy restrictions of the least fire resistive type of construction used in the building.

609.2

Where a building is constructed of more than one type of construction, the following limitations shall be observed:

TYPE I construction shall not be supported by any other type.

TYPE II construction shall not be supported by construction other than Type I or Type II.

TYPE III construction shall not be supported by construction other than Type I, Type II or Type III.

TYPE IV construction shall not be supported by Type V, or Type VI.

TYPE V construction shall not be supported by Type VI.

SECTION 610 – BUILDINGS LOCATED ON THE SAME LOT

Where the exterior walls of two or more buildings located on the same lot face one another, and one of the walls is not constructed as required for a fire wall, a common-property line shall be assumed between them. The fire-resistance requirements for such facing walls and for the protection of openings therein shall be the same as required by this code for walls and openings facing common-property lines, except as otherwise prescribed in Section 607.3 for buildings of Type VI Construction.

CHAPTER VII

FIRE PROTECTION REQUIREMENTS

SECTION 701

PROTECTION OF VERTICAL OPENINGS, STAIRS AND ELEVATORS

701.1 — GENERAL REQUIREMENTS FOR ENCLOSURE OF VERTICAL OPENINGS OR SHAFTS

(a) Every series of openings in floors or roofs, except one and two family dwellings, shall be enclosed to prevent spread of fire from story to story, as herein specified, unless otherwise specifically provided in this Code.

(b) In buildings more than two (2) stories in height, except one and two family dwellings, all vertical shafts extending through more than one story shall be enclosed in all stories with materials of not less than 2-hour fire-resistive construction, unless otherwise prescribed in Sections 701.2, 701.3 and 701.4. In Group A Residential buildings not over three (3) stories in height, such construction shall provide at least 1-hour fire resistance. A shaft that does not extend through the roof shall have its top enclosed with construction having fire-resistance at least equal to that of the enclosing walls.

(c) For bearing partition requirements see Types of Construction, Chapter VI.

(d) Parapet walls at least thirty-six (36) inches in height above the roof shall be provided around all open shaft enclosures that extend through a roof except that where the roof is of non-combustible construction a non-combustible handrail at least thirty (30) inches high may be used around openings instead of a wall.

(e) All shaftway enclosures extending above the roof of a building, except those open at the top, shall have a skylight with metal frame glazed with plain glass and with protective wire mesh screen below, conforming with Section 707; such skylight shall have an area equal to at least seventy five (75) per cent of the area of the shaftway at the top story, or other equivalent ventilation shall be provided. A window of plain glass of equivalent area, located in the side of the shaft may be used instead of a skylight provided it has its sill not less than two (2) feet above the roof and does not front a property line within ten (10) feet.

(f) In every building where the enclosing walls or shafts are open to the outer air at the top, such wall shall be constructed to provide fire resistance equivalent to that specified in Chapter VI, for the inner court walls of such a building.

(g) Openings in all shaft enclosures shall be limited to those absolutely necessary for the purposes of the shaft and shall be protected with approved fire doors, fire shutters, or fire windows (see Section 703.6).

701.2 — STAIRWAY AND EXIT ENCLOSURES

Refer to Chapter XI for requirements of stairway and exit enclosures.

701.3 — ELEVATOR ENCLOSURES

(a) The number of elevators to be placed in one shaft enclosure shall be as specified in Paragraph 100.1d of "American Safety Code for Elevators, Dumbwaiters, Escalators and Moving Walks, ANSI A17.1-1971." Such shaft enclosure shall have at least 2-hour fire resistance, except that in residential buildings of Group A, or any Group B building, not over three (3) stories in height, enclosures may be of construction having 1-hour fire resistance.

(b) In shaftways containing only one elevator, there shall be at least one (1) door in every 36 feet of height. All openings shall be protected as required in Section 703.6.

(c) Elevators counted as exits shall not be in a common enclosing shaft with a stairway, and the path of travel from one flight of stairs to the next shall not pass directly in front of elevator doors.

(d) An elevator shaft that does not extend to the bottom of a building shall be enclosed at its lower point to provide fire-resistance of not less than that required for the lowest floor through which the elevator passes, but in no case less than is required for Type II construction.

(e) Partitions between fire resistive hoistways and machine rooms having fire resistive enclosures and which are located at a side or beneath the hoistway, may be of unperforated non-combustible, at least equal to, #16 Manufacturer's Standard gauge sheet steel in strength and stiffness with openings therein essential for ropes, drums, sheaves and other elevator equipment.

Where machine room is placed over the hoistway at the top, floor over the hoistway shall be non-combustible in accordance with "American Standard Safety Code for Elevators, Dumbwaiters, Escalators and Moving Walks, ANSI A17.1-1971, Section 100.3."

701.4 — ESCALATOR ENCLOSURES

Protection of floor openings shall be as required by "American Safety Code for Elevators, Dumbwaiters, Escalators and Moving Walks, ANSI A17.1-1971, Section 800, Rules 800.1, 800.2 and 800.3 and Section 902, Rule 902.1."

SECTION 702 — PARTITIONS

702.1 — GENERAL

(a) This section shall apply to partition requirements for the various occupancies and types of construction.

(b) All partitions required to have a fire resistance rating shall extend from the top of the floor below to the ceiling above and shall be securely attached thereto. Where said ceiling is not a part of an assembly having a fire resistance rating at least equal to that required for the partition, the partition shall be constructed tight against the floor deck above. Where said fire resistant partitions surround vertical shafts, stairways and exitways and are offset at intermediate floors, the offset and floor construction shall be of construction having a fire resistance of not less than that required for the enclosing partitions.

(c) Where a greater degree of fire resistance is required by other Sections of this Code, the provisions of Section 702.2 and 702.3 shall not apply.

(d) Masonry partitions shall conform to the provisions of Chapter XIV.

(e) Plaster and wallboard partitions shall conform to the provisions of Chapter XVIII.

(f) Wood frame partitions shall conform to the provisions of Chapter XVII.

(g) For fire resistance rating of partitions, see Chapter X and Appendix B.

702.2 — PARTITION REQUIREMENTS BY TYPE OF CONSTRUCTION

All non-bearing partitions shall conform to the requirements of this Section, except as specified elsewhere in this Code. See Interior finish requirements of Section 704.

TYPE I AND TYPE II—Partitions shall provide not less than one-hour fire resistance and be constructed of noncombustible materials or Fire Retardant Treated Wood.

TYPE III Partitions shall be of one-hour fire-resistant construction or may be of wood construction formed of two (2) layers of one (1) inch nominal matched boards, or of solid wood laminated construction not less than three and five-eighths (3⅝) inches thick.

TYPE IV (Non-combustible) and,

TYPE V—In buildings three (3) stories or more in height, unless sprinklered, all permanent partitions shall be of not less than one-hour fire-resistant construction.

702.3 — PARTITION REQUIREMENTS BY OCCUPANCY

GROUP A—RESIDENTIAL—All partitions (except in one and two family dwellings) along public hallways or partitions that separate apartments, or that separate apartments from other occupancies, shall be of not less than one-hour fire-resistant construction. Non fire rated partitions may be permitted within individual dwelling units.

GROUP B—BUSINESS BUILDINGS—In buildings more than one-story in height, partitions along public hallways shall be of one-hour fire resistant construction. Regardless of Type of Construction, temporary non-fire resistant partitions may be constructed within rooms or spaces not exceeding three thousand (3,000) square feet in area. Such room or space shall be enclosed with permanent partitions having not less than one-hour fire-resistive construction. The area occupied by each tenant shall be separated from adjacent tenants by a partition of not less than one-hour fire-resistant construction. (See Section 412 for mixed occupancy and separation requirements.)

GROUP C—SCHOOLS—Partitions in buildings over one-story in height shall be of not less than one-hour fire-resistive construction.

GROUP D—INSTITUTIONAL—All partitions shall be of not less than one-hour fire-resistive construction.

GROUP E—THEATERS AND PUBLIC ASSEMBLY—See Section 512.4 and Section 1106 for partition requirements.

GROUP F—STORAGE,

GROUP G—INDUSTRIAL, and

GROUP H—SPECIAL OCCUPANCIES—Partitions in Group F and Group G buildings four (4) stories or more in height, other than Public Parking Decks, and all partitions in Group H buildings shall be of not less than one-hour fire resistant construction, except where greater fire-resistance is required by this Code.

SECTION 703 — PROTECTION OF WALL OPENINGS

703.1 (a) — WHERE PROTECTION IS REQUIRED

For the purpose of this Code, when a building is divided by fire walls into two or more sections, each section shall be regarded as a separate building.

(a) Every building except one and two family dwellings which are not over two (2) stories in height, churches, buildings of Type VI Construction, and public parking decks as defined in Section 507.2, shall have approved fire windows, fire doors or other approved protectives in every opening in the exterior walls as follows:

1. In buildings three stories or more in height where the opening is fifteen feet or less from a common property line. In buildings less

than three stories in height, protection shall be required only where the opening is eight feet or less from a common property line.

2. In buildings where such opening is above and is less than fifteen (15) feet from any part of a neighboring roof.

Exceptions: Such protection shall not be required for a show window fronting on a street or public space which does not extend over the second story above grade nor shall such protection be required when the openings are located in walls in the same plane of direction. All required opening protection shall be of an approved type as defined elsewhere in this section.

703.1 — EXTERIOR WALLS WITHOUT ACCESS OPENINGS

(b) Automatic sprinklers shall be installed throughout every building which does not have access to each story above grade on at least one accessible side of the building up to a height of eighty-five feet. Such access shall be openings (for fire department use) through the wall at each story, a minimum of 32 inches wide and 48 inches high and with the bottom of the opening not more than 32 inches above the floor. Openings shall be so spaced that there will be one opening in each 50 feet of exterior wall on the accessible side of the building. Buildings equipped with an automatic sprinkler system throughout shall have access panels as set forth above for each 200 feet of wall.

(c) An access door or covered opening with suitable hardware and identifying marking shall be installed in the veneered facing on each floor above the main floor, with free access into the building.

(1) Opening shall be of sufficient size to permit ready access of fire fighting personnel and hand carried fire fighting equipment.

(2) Exterior of the opening shall have distinctive markings for purpose of ease in locating panels.

(3) Such access opening shall open into a fire aisle within the building and no shelving, loose or fixed, no containers or equipment of any description, nor any loose merchandise shall be placed so as to block aisleway.

703.2 — APPROVED TYPES OF FIRE WINDOWS, FIRE DOORS, AND FIRE SHUTTERS

(a) Requirements

Wall openings required to be protected by Section 703.1 (a), 703.3 and 703.4, shall be protected by fire doors, windows and shutters and their accompanying hardware in accordance with the requirements of NFPA standard for the installation of "Fire Doors and Windows, NFPA 80-1970," except as otherwise specified in this Code.

(b) Classification and Protection of Wall Openings

Openings shall be classified as A, B, C, D, E and F in accordance with the character and location of the wall in which they are located. In each of the following classes, the minimum fire protection ratings are shown:

Fire Protection ratings for products meeting the requirements of this section shall be as determined and reported by a national recognized testing agency in accordance with "Standard Methods of Fire Tests of Door Assemblies, NFPA 252-72;" UL 10(b)-1970; ASTM E 152-66; and ANSI A 2.2. All such products shall bear the label or identification of an approved testing agency showing the classification.

Class A openings are in walls separating buildings or dividing a single building into fire areas. Doors for the protection of Class A openings shall have a fire protection rating of 3 hours. Such doors shall be without glazing.

Class B openings are in enclosures of vertical communication through buildings, (stairs, elevators, etc.). Doors for the protection of Class B openings shall have a fire protection rating of a least 1 hour. Doors used in stairway enclosures in buildings four or more stories in height shall have a fire protection rating of 1½ hours and shall have an average temperature rise of not more than 450° F at the end of 30 minutes of standard fire test exposure.

Class C openings are in corridor and room partitions. Doors for the protection of Class C openings shall have a fire protection rating of ¾ hour, and may have not more than 1296 square inches of ¼ inch wired glass. (For exception see Section 703.4 (d))

Class D openings are in exterior walls which are subject to severe fire exposure from outside of the building. Doors and shutters for the protection of Class D openings shall have a fire protection rating of 1½ hours. Such doors and shutters shall be without glazing.

Class E and F openings are in exterior walls which are subject to moderate or light fire exposure respectively from outside the building. Doors, shutters or windows (Sections 13, 14, and 15—NFPA 80-1970) for the protection of Class E and F openings shall have a fire protection rating of ¾ hour.

703.3 — VERTICAL SEPARATION OF OPENINGS

In buildings, exterior openings that are located vertically above one another and which are not protected by approved types of fire windows or fire doors shall have a space of not less than three (3) feet between the top of one opening and the bottom of the next above, or such opening above the lower opening shall be protected against fire by an approved protective device. Such wall space shall meet the requirements for exterior walls of the type of construction used as prescribed in Chapter VI of this Code.

703.4 — PROTECTION OF DOOR OPENINGS IN WALLS AND PARTITIONS

(a) Wherever protection of door openings is required and in all walls and partitions which are required to have 2-hour or greater fire-resistance, door openings shall be protected with approved fire door assemblies meeting the requirements of Section 703.2. In addition, approved fire doors may be required by the Building Official for the protection of exits or of adjoining property.

(b) In 4-hour and 3-hour fire-resistive walls or partitions, no opening shall exceed one hundred twenty (120) Sq. Ft. in area with no dimension greater than twelve (12) Ft., and the aggregate width of all openings at any level shall not exceed twenty-five percent of the length of such wall or partition.

Every door opening in such wall or partition shall be an approved automatic fire door; provided that when such wall or partition serves also as a horizontal exit, it shall have no openings other than exit doors. Such horizontal exit doors may be self-closing and so maintained or shall be automatic closing when activated by a device which will operate upon the detection of invisible products of combustion generated by a fire or when smoke from that fire reduces the intensity of a one foot long beam of white light by 4%. All such automatic door closing equipment shall operate on the fail safe principle.

(c) In 2-hour fire resistive walls or partitions, doors meeting a fire protection rating of 1½ hours shall be used. No single door opening shall exceed one hundred eighty (180) sq. ft. area. The aggregate width of all openings in such walls or partitions at any level shall not exceed that permitted for three hour walls.

Every door opening shall be protected with an approved automatic or self-closing fire door, provided that when such wall serves also as a horizontal exit, doors shall be maintained self-closing or shall be made automatic closing in a method as described in Section 703.4(b).

(d) In 1-hour fire-resistive walls or partitions, unless otherwise specified, all door openings shall be protected with approved metal or metal covered doors or wooden doors of the solid core flush type of nominal thickness of at least one and three quarters (1¾) inch in all parts.

(e) See Section 703.2 for fire-resistance requirements of fire doors.

(f) When proof satisfactory to the Building Official is furnished that a larger size of opening than prescribed herein is necessary, the area may be increased if such opening is provided with protective devices that meet the approval of the Building Official.

703.5 — FIRE SHUTTERS

(a) When equipped with fire shutters of the swinging type, at least one in every three openings facing a street in each story shall have such shutters arranged to be readily opened from the outside. Distinguishing marks shall be provided on such shutters.

(b) Fire shutters of the rolling type shall be carefully counterbalanced and so arranged that they can be readily opened from the outside.

703.6 — OPENINGS IN STAIRWAYS OR SHAFTWAYS

Shaft walls or enclosures of vertical openings shall have no openings other than such as are necessary for the purpose of the shaftway; all openings in shafts shall be protected with approved fire doors, approved fire shutters or approved fire windows.

703.7 — OPENINGS IN MIXED OCCUPANCY SEPARATIONS

See Section 703.4 for requirements governing door openings in walls and partitions required to be of fire-resistive construction.

703.8 — FIRE DAMPERS

Except when proper fire tests have shown that fire dampers are not necessary to maintain the required fire endurance of construction (see exceptions below), fire dampers bearing a U-L label shall be installed in the following locations:

a. Where a duct passes through a partition which serves to restrict the spread of fire and is required to have a standard fire resistance rating of not less than 2 hours.

b. At each opening in a required enclosure of a vertical shaft.

c. Where duct systems serve two or more floors:

1. At each direct air outlet or air inlet in the enclosures for a main vertical duct, or at each point where such vertical duct pierces a floor it serves.

AND

2. At the point where each branch duct pierces the enclosure for a main vertical duct.

d. Where an aluminum duct or Class I duct regardless of sizes passes through a fire resistive floor, unless encased as specified in NFPA Pamphlet 90A-1971.

e. At fresh air intakes except where permission to omit them, because of light exposure, is granted by the authority having jurisdiction.

f. As an alternate to enclosure of vertical ducts which extend through only one floor, dampers to be located at each point where the floor is pierced.

Exceptions:

1. In branch ducts not of aluminum or Class I duct, having a cross-sectional area of less than 20 square inches which supply only air conditioning units discharging air at not over 4 feet above the floor.

2. Where a duct less than 20 square inches in cross-sectional area and not made of aluminum or Class I duct pierces the floor at one place only and supplies air conditioning units in one story only that discharge air at not over 4 feet above the floor.

3. In small buildings with unprotected floor openings.

4. In duct systems serving only one floor and used only for exhaust of air to the outside and not penetrating a fire wall or partition which serves to restrict the spread of fire and is required to have a standard fire resistance rating of not less than two hours, or passing entirely through the enclosure for a vertical shaft.

5. Where branch ducts connect to return risers in which the air flow is upward and subducts at least 22 inches in length are carried up inside the riser from each inlet.

Access

Suitable hand hole opening with tightly fitted cover shall be provided to make all fire dampers in ducts accessible for inspection and service.

Construction

Fire dampers, as well as having the U-L label showing compliance with all U-L requirements for 1½ hour rating, shall be of bolted or welded construction. Same shall include a factory installed sleeve of appropriate length for the wall thickness wherein it is to be installed. Included shall be a U-L approved fusible link so located to be readily affected by an abnormal rise of temperature in the duct. Such link shall have a temperature rating approximately 50° above the maximum temperature normally encountered with the system in operation or shut down. Selection of this temperature rating shall recognize the prevailing ambient temperature at the site.

All fire dampers for horizontal air flow less than 10′ x 10′ in size shall be a single assembly with blade stack spanning the full width. Multiple assemblies incorporating mullions or meeting rails may be used only for horizontal air flow sizes where either width or height exceeds 10′ x 10′.

Installation

Installation shall be made in accordance with the instructions of NFPA Pamphlet #90A-1971 as required by U-L test ratings. The manufacturer of fire dampers shall furnish a copy of these approved installation instructions with each order for fire dampers supplied. Same shall include a table of U-L tested sleeve gauge requirements, sleeve projection requirements, retaining angle sizes, and permissible tolerances for opening sizes.

Design

The designer of an air duct system shall show on the plans the location of all fire dampers, as required by NFPA Pamphlet #90A-1971.

SECTION 704 – RESTRICTIONS ON INTERIOR USE OF COMBUSTIBLE MATERIALS

704.1 — GENERAL

Combustible materials may be used for ceilings, floor finish or other interior finish of buildings as provided in this Section. Show windows in the first story of buildings may be of wood or of unprotected metal framing.

704.2 — FLOOR FINISH

(a) In buildings of Type I Construction, or Type II Construction, floor finish, if of combustible material, shall be applied directly upon the floor construction, except that a floor finish of wood, linoleum, rubber, tile or cork may be secured to a sub-floor of wood. Where wood sleepers are used for laying wood floors or sub-floors in such buildings, they shall be firestopped so that there will be no open space extending under any permanent partition. Where wood sleepers are used and the space between the floor slab and the underside of the floor or subfloor is more than 2½ inches, such space shall be filled with non-combustible material so that such space is not more than 2½ inches.

(b) Combustible insulating boards may be used for sound deadening, or insulating of floors, except that in buildings required to be of Type I Construction or of Type II Construction, such insulating board shall not be more than one-half (½) inch thick and cemented directly to the floor slab or secured to wood sleepers fire stopped as called for above and covered with approved finish flooring.

704.3 — CEILINGS AND INTERIOR WALL FINISH*

In every building except one and two family dwellings, and sprinklered buildings, flame spread ratings for walls and ceilings using "Methods of Test for Surface Burning Characteristics of Building Materials, ASTM E84-68," shall not exceed the following:

(1) For vertical enclosures in buildings four (4) or more stories in height, interior finish shall not have a flame spread rating higher than 75. For required horizontal enclosures, the interior finish shall not have a flame spread rating higher than 200, including grade lobbies in hotels and apartment buildings.

(2) For rooms or spaces in institutional buildings where occupants are restrained, interior finish shall not have a flame spread rating higher than 75. In other rooms or spaces in institutional buildings, the rating shall not exceed 200.

(3) For assembly, school, business, mercantile, apartment and hotel occupancies, interior finish shall not have a flame spread rating of over 200. The use of a surface finish of paper or of material of no greater fire hazard than paper shall not be prohibited provided

such finish does not exceed 0.025 of an inch in thickness, and is applied directly to a non-combustible base.

*For interior finish and decoration for Group E, Assembly Occupancy, see Section 512.3.

SECTION 705 — FIRESTOPPING

(a) Firestopping shall be provided in all walls and partitions to cut off all concealed draft openings both horizontal and vertical, and to form an effectual fire barrier between stories and between the upper story and the roof space.

(b) Walls, including masonry walls furred with combustible material, and stud partitions shall be effectively firestopped with non-combustible material at floors, ceilings, and roofs, except in those parts of a building which are framed with wood, the firestopping may be of wood not less than two (2) inches in nominal thickness. See Section 1703.

(c) All openings around exposed pipes or power shafting shall be filled with approved non-combustible material, or shall be closed off by close-fitting metal caps at the ceiling and floor line, and on each side of a wall or partition.

(d) All openings for belts and conveyors shall be provided with approved slotted doors, or be otherwise closed off. Belts shall not pass through fire-walls.

(e) No firestopping shall be covered or concealed until inspected by the Building Official.

(f) In combustible roof construction, where ceilings or concealed spaces occur, such spaces shall be divided into horizontal areas of not more than three thousand (3,000) square feet (except one and two family dwellings) with tight partitions of non-combustible material or of approved wood construction consisting of one-half inch exterior plywood or of not less than two thicknesses of one (1) inch nominal lumber with joints broken.

(g) All openings through these partitions shall be protected by self-closing doors of approved construction meeting the partition requirements.

(h) Except in 1 and 2-family dwellings, when stairs are of wood or of combustible construction, the space between stair stringers shall be firestopped at top and bottom, and firestopping shall also be provided between studs, along and in line with run of stair adjoining such partition.

(i) Floors and roof constructed of combustible materials shall be firestopped at walls and partitions where openings occur. When wood joists run parallel to a wall, the space between the wall and the nearest joist shall be not less than two and one-half ($2\frac{1}{2}$) inches and shall be solidly filled with non-combustible material.

(j) Joists in all types of construction shall be firestopped at the ends and over supports for the full depth of the joists.

(k) Spaces between chimneys and wood framing shall be solidly firestopped with mortar, concrete, or other non-combustible material, except as provided for in Section 802 — Chimneys. (See Section 1703.)

(l) In firestopping, any of the following materials may be used: brick, concrete, gypsum, steel, iron, asbestos, metal lath and cement or gypsum plaster, mineral wool, rock wool, or other approved non-combustible materials, securely fastened in place to cut off drafts and provide an effective fire stop.

SECTION 706 – ROOF COVERINGS

706.1 — GENERAL

Roof coverings shall be divided into the classes defined below, whose use within the Fire District shall be governed by the requirements of Sec. 301.3(d).

706.2 — TYPE A ROOF COVERINGS

Type A Roof Coverings shall include the following: Brick, Concrete, Slate, Tile, Corrugated Asbestos Cement and Built-up or Prepared Roof Coverings and/or assemblies listed and identified as Class "A" by an approved testing agency having a re-examination service.

706.3 — TYPE B ROOF COVERINGS

Type B Roof Coverings shall include the following: Corrugated iron sheets, galvanized iron sheets, galvanized iron shingles, sheet copper, galvanized iron and Built-up or Prepared Roof Coverings and/or assemblies listed and identified as Class "B" by an approved testing agency having a re-examination service.

706.4 — TYPE C ROOF COVERINGS

(a) Type C Roof Coverings shall include the following: Asphalt rag-felt smooth surfaced roll roofing laid in single thickness with 2" or more and side laps, Aluminum .019 inches in thickness and Built-up or Prepared Roof Coverings and/or assemblies listed and identified as Class "C" by an approved testing agency having a re-examination service.

(b) Red Cedar or Redwood Shingles and Shakes may be used as provided in Section 706.6.

706.5 — REQUIREMENTS FOR ROOFS OUTSIDE FIRE DISTRICT

Roofs on buildings outside of the Fire District, as established in Chapter III, may have Type A, Type B, or Type C roof coverings, as specified herein. Wood Shingles or Shakes may be used as provided in this section.

706.6 — WOOD SHINGLES AND SHAKES

Outside the Fire District, buildings not over three stories in height and/or 9,000 sq. ft. (See Section 403 for Allowable Area Increases) in area and located not less than six (6) feet from the property line may be roofed with grade-labeled No. 1 or No. 2 Wood Shingles or Shakes with a minimum butt thickness of 1/2 inch, securely fastened with rust resistant nails. Shingles and Shakes shall be applied in accordance with the recommendations of the Red Cedar Shingle and Handsplit Shake Bureau, and with the following exposure table.

TABLE—WEATHER EXPOSURE AND ROOF PITCH

	Roof Pitch		3/12	4/12	5/12 & Up
	16-inch	No. 1	3¾"	4½"	5 "
		No. 2	3¾"	4 "	4 "
Shingles	18-inch	No. 1	4¼"	5 "	5½"
		No. 2	4¼"	4½"	4½"
	24-inch	No. 1	5¾"	6¾"	7½"
		No. 2	5¾"	6 "	6 "
	18-inch			7½"	7½"
Shakes	24-inch			10 "	10 "

706.7 — ROOF INSULATION

The use of cork, fiberboard or other approved insulation shall be permitted in all types of construction provided it is covered with approved roof coverings applied directly thereto.

*In addition to those listed, any roof covering that is equivalent to listing of Underwriters' Laboratories, Inc. may be accepted.

SECTION 707 — SKYLIGHTS

(a) Except in Type VI buildings, sashes and frames of skylights shall be constructed of steel or other approved metal; except that in foundries or buildings where acid fumes, which may be injurious to those metals, are present as an incident to the occupancy of such buildings, sashes and frames may be constructed of wood when specifically approved by the Building Official.

(b) Skylights shall be glazed with wired glass; except that skylights placed over shaftways, vent shafts, and stair enclosures shall be glazed with plain glass not over one-eighth (1/8) inch in thickness. No single pane of wired glass shall exceed seven hundred twenty (720) square inches in area, or forty-eight (48) inches in any dimension. Skylights of approved plastics may be used where glass of one-eighth (⅛) inch thickness is required.

(c) Every skylight in which plain glass is used shall be protected by a substantial wire screen, having a mesh not less than three-quarter by three-quarter (¾ x ¾) inch nor coarser than one by one (1 x 1) inch and made of wire not smaller than No. 12 B. and S. gauge, located at a distance not less than four (4) inches nor more than ten (10) inches above the glazed portion of such skylight at all points, and extending beyond such glazed portion on all sides, a distance not less than the height of the screen above the glass. A similar screen shall be placed below such skylight in such position as to serve as a protection from falling glass.

(d) Skylights of approved plastics may be used in accordance with provisions of Chapter XXVI.

(e) The above provisions shall not apply to skylights used in or as the roofs of greenhouses.

SECTION 708 — MANSARD OR SLANTING ROOFS

Every mansard or other slanting roof having a pitch of more than sixty degrees (60 degrees) on any building over fifty (50) feet in height, shall be of non-combustible construction providing not less than 1-hour fire resistance, except that where such building exceeds eighty (80) feet in height, roofs shall be of construction providing not less than 1½ hour fire resistance.

SECTION 709 — DORMER WINDOWS

Dormer windows shall be of the same type of construction as the roof on which they are placed, or of the side walls of the building. The top and sides shall be covered with roofing materials conforming with the requirements governing the roofing of the building.

SECTION 710 — CORNICES, BALCONIES, BAY WINDOWS

(a) All cornices, including those on show windows, on the exterior of buildings within the Fire District or on buildings over forty (40) feet in height located outside the Fire District shall be of non-combustible materials. The exterior of cornices on buildings forty (40) feet or less in height located outside of Fire District, except 1 and 2-family dwellings, and buildings of Type VI Construction, shall be of non-combustible material, or shall be covered with non-combustible material, when located within five (5) feet of a lot line of another building.

(b) Continuous exterior cornices of wood, or of wood frames, shall be firestopped at intervals not exceeding twenty (20) feet.

(c) Balconies not used as required exits, and bay windows, shall conform to the type of construction required for the building to which they are attached, except that on buildings four (4) or more stories in height, all exterior balconies attached to, or supported by, walls of material other than wood, shall have brackets or beams of steel, concrete, or other noncombustible materials.

SECTION 711 – CANOPIES ON EXTERIOR WALLS

Canopies extending over public property shall comply with the requirements of Chapter XXII.

Other permanent canopies may extend over adjacent open spaces and be of any material permitted by this Code provided:

(a) that when located in the Fire District or less than thirty (30) feet from an interior lot line or other structure the canopy and its supports shall be of non-combustible material, fire retardant treated wood, wood of Type III sizes, or of one hour fire-resistant construction;

(b) the canopy shall have one long side open;

(c) the maximum horizontal width of the canopy shall not exceed 15 feet;

(d) the fire resistance of exterior walls shall not be reduced.

SECTION 712 – GUTTERS AND LEADERS

Gutters and leaders hereafter placed on buildings other than one or two-family dwellings, private garages, or buildings of Type VI Construction, shall be of non-combustible material.

(See section 1406 for Parapet Wall relief opening requirements.)

SECTION 713 – TOWERS, SPIRES, CUPOLAS, AERIAL SUPPORTS, POLES, ETC.

(a) Any tower, spire, dome or cupola shall be of a type of construction not less in fire-resistance rating than required for the building to which it is attached except that any such tower, spire, dome or cupola which exceeds sixty (60) feet in height above grade, and all construction upon which it is supported, shall be of Type I or Type II Construction when the area at any horizontal section of such tower, spire, dome, or cupola exceeds two hundred (200) Sq. Ft. or when it is used for any purpose other than a belfry or an architectural embellishment.

(b) Any tower, spire, dome or cupola which exceeds twenty-five (25) feet in height above the highest point at which it comes in contact with the roof or which exceeds two hundred (200) Sq. Ft. in area at any horizontal section or which is intended to be used for any purpose other than a belfry or architectural embellishment, shall be entirely constructed of and supported by non-combustible materials. Such structures shall be separated from the building below by construction having a fire-resistance rating of not less than 1½ hours and, if access doors are provided, such doors shall be of approved fire-resistive type.

(c) All structures, except aerial supports not over twelve (12) feet high, flag poles, water tanks and cooling towers, placed above the roof of any building within the Fire District, or above the roof of any building more than fifty (50) feet in height, wherever located, shall be of non-combustible material, and shall be supported by construction of non-combustible material.

SECTION 714 — TANKS

(a) Tanks of more than five hundred (500) gallons capacity placed in or on a building shall be supported on masonry, reinforced concrete or steel construction, except that portion of the supporting structure which is above the roof of the building may be of heavy timbers; provided that when such construction is within the building it shall be as required for Type I Construction.

(b) Such tanks shall have in the bottom or on the side near the bottom, a pipe or outlet, fitted with a suitable quick opening valve for discharging the contents in an emergency through an adequate drain.

(c) Such tanks shall not be placed over nor near a line of stairs or an elevator shaft, unless there is a solid roof or floor underneath the tank.

(d) All unenclosed roof tanks shall have covers sloping toward the outer edges.

(e) When hoops are used in the construction of tanks, they shall be of metal, and provision shall be made to guard against corrosion.

SECTION 715 — COOLING TOWERS

Cooling towers in excess of two hundred and fifty (250) square feet in base area or in excess of fifteen (15) feet in height when located on buildings more than fifty (50) feet in height in or out of the Fire District, shall be of non-combustible construction; except that drip boards may be of wood not less than one (1) inch nominal thickness and the enclosing frame work may be of wood, if covered on the exterior of the tower with non-combustible material. Cooling towers shall not exceed one-third of the supporting roof area.

SECTION 716 — DRYING ROOMS

(a) Drying rooms or dry kilns located within a building shall be constructed entirely of non-combustible materials where used or intended to be used at temperatures exceeding one hundred twenty five degrees (125 degrees) fahrenheit; if enclosure is of metal, it shall be insulated from all combustible material by not less than a twelve (12) inch air space, one-quarter (¼) inch asbestos or other approved insulation.

(b) All drying rooms shall have approved ventilation.

(c) Heating pipes, not located overhead, shall be shielded to maintain not less than two (2) inch clearance between them and the contents.

CHAPTER VIII

CHIMNEYS, FIREPLACES AND HEATING EQUIPMENT

SECTION 801 — GENERAL

801.1 — DEFINITIONS

AUXILIARY EQUIPMENT—Auxiliary equipment shall mean that equipment used in connection with combustion equipment such as stokers, burners, draft regulators, etc.

BARBECUE—A portable or stationary open hearth or brazier used for cooking. It may be of a type in which fuel is burned, or it may be one provided with electric elements.

BUILDING HEATING APPLIANCE—Fuel burning and electric boilers operating at not over 50 psig pressure, central furnaces, and heaters intended primarily for heating spaces having volume exceeding 25,000 cubic feet.

CENTRAL HEATING BOILERS AND FURNACES—Heating furnaces and boilers shall include warm air furnaces, floor mounted direct-fired unit heaters, hot water boilers, and steam boilers operating at not in excess of 15 pounds gauge pressure, used for heating of buildings or structures.

Mechanical warm air furnaces are here defined as warm air furnaces which are equipped with a fan to circulate the air.

Devices referred to as "such that flame from gases do not come in contact with the base" include conventional type heating furnaces and boilers having an ash pit or similar space beneath the burning fuel (whether solid, liquid or gas) and like devices in which the base is not directly exposed to flame or products of combustion.

CHIMNEY—One or more passageways, vertical or nearly so, for conveying flue gases to the outer air.

CHIMNEY, FACTORY-BUILT—Is a listed chimney.

CHIMNEY, MASONRY—Is a chimney of solid masonry units, bricks, stones, listed hollow unit masonry units or reinforced concrete.

CHIMNEY, METAL—Is a chimney constructed of metal.

CHIMNEY CLASSIFICATIONS:

CHIMNEY, RESIDENTIAL APPLIANCE TYPE is a factory-built or masonry chimney suitable for removing products of combustion from residential type appliances producing combustion gases not in excess of 1000° F. measured at the appliance flue outlet.

CHIMNEY, LOW-HEAT APPLIANCE TYPE is a factory-type, masonry or metal chimney suitable for removing the products of combustion from fuel-burning low-heat appliances producing combustion gases not in excess of 1000° F. under normal operating conditions but capable of producing combustion gases of 1400° F. during intermittent forced firing for periods up to one hour. All temperatures are measured at the appliance flue outlet.

CHIMNEY, MEDIUM-HEAT APPLIANCE TYPE is a factory-built, masonry or metal chimney suitable for removing the products of combustion from fuel-burning medium heat appliances producing combustion gases, not in excess of 2000° F. measured at the appliance flue outlet.

CHIMNEY, HIGH-HEAT APPLIANCE TYPE is a factory-built masonry or metal chimney suitable for removing the products of combustion from fuel-burning high-heat appliances producing combustion gases in excess of 2000° F. measured at the appliance flue outlet.

CHIMNEY CONNECTOR—The pipe which connects a fuel burning appliance to a chimney.

DRAFT HOOD—Is a device placed in, and made part of the flue pipe from an appliance, or in the appliance itself, which is designed to (1) insure the ready escape of the products of combustion in the event of no draft, back draft or stoppage beyond the draft hood; (2) prevent a backdraft from entering the appliance; and (3) neutralize the effect of stack action of the chimney flue upon the operation of the appliance.

EXHAUSTER, FLUEGAS (Draft Booster)—A fan installed in or on a chimney, vent, or chimney or vent connector, to induce a draft at the appliance.

FIREPLACE—A hearth and fire chamber or similarly prepared place in which a fire may be made and which is built in conjunction with a chimney.

(a) FACTORY-BUILT FIREPLACE. A fireplace composed of listed factory-built fire chamber and chimney sections designed to be readily joined one to the other without requiring field construction, to form the completed fireplace.

(b) MASONRY FIREPLACE. A field-constructed fire chamber and chimney of solid masonry units such as bricks, stones, listed masonry units, or reinforced concrete; lined with suitable flue liners.

FIREPLACE STOVE—A chimney-connected, solid fuel burning stove having part of its fire chamber open to the room.

FLUE COLLAR—That portion of an appliance designed for attachment of the chimney or vent connector.

FLUE GASES—Combustion products and excess air.

FURNACE, CENTRAL, WARM-AIR—A self-contained indirect-fired appliance designed to supply heated air through ducts to spaces remote from or adjacent to the appliance location.

(a) FORCED-AIR TYPE. A central furnace equipped with a fan or blower which provides the primary means for circulation of air.

(b) GRAVITY TYPE. A central furnace depending primarily on circulation of air by gravity.

FURNACE, DOWNFLOW—A forced-air type central furnace designed with air flow through the furnace essentially in a vertical path, discharging air at or near the bottom of the furnace.

FURNACE, DUCT—A central furnace designed for installation in a duct of an air-distribution system to supply warm air for heating and which depends for air circulation on a blower not furnished as part of the furnace.

FLOOR FURNACE—A completely self-contained unit furnace suspended from the floor of the space being heated, taking air for combustion from outside this space, and with means for observing flames and lighting the appliance from such space.

(a) GRAVITY TYPE FLOOR FURNACE. A floor furnace depending primarily upon circulation of air by gravity. This classification shall also include floor furnaces equipped with booster type fans which do not materially restrict free circulation of air by gravity flow when such fans are not in operation.

(b) FAN TYPE FLOOR FURNACE. A floor furnace equipped with a fan which provides the primary means for circulating air.

FURNACE, HORIZONTAL—A forced-air type central furnace designed with air flow through the furnace essentially in a horizontal path.

FURNACE, UPFLOW—A central furnace designed with air flow thru the furnace essentially in a vertical path, discharging air at or near the top of the furnace.

HEAT PUMP—A refrigeration system arranged to accomplish either heating or cooling.

HOOD—As applied to a heating device, is a canopy or similar device placed over a stove, range or other heating installation connected to a ventilating duct.

INCINERATOR, COMMERCIAL AND INDUSTRIAL—An incinerator intended for the reduction of rubbish, garbage by-product wastes and other types of waste matter resulting from or incidental to any class of occupancy.

INCINERATOR, RESIDENTIAL—A direct-fed appliance designed primarily for use in one and two family dwellings for the burning of ordinary combustible waste materials and garbage incidental to domestic occupancy and having a fire box or charging compartment of not over 5 cubic feet in capacity. Such incinerator may be either of the type manufactured and installed as a complete unit or of the type designed to be encased in brick or installed in a masonry wall or chimney.

INDIRECT-FIRED APPLIANCE—An appliance in which combustion products (flue gases) are not mixed in the appliance with the medium (e.g., air) being heated.

INDUSTRIAL APPLIANCES—

(a) LOW-HEAT INDUSTRIAL APPLIANCE—An industrial appliance such as a commercial cooking range, presing machine boiler at any pressure, bake oven, candy furnace, stereotype furnace drying and curing appliance, and other process appliances in which materials are heated or melted at temperatures (excluding flue-gas temperatures) not exceeding 600° F. Appliances otherwise classed as medium-heat appliances may be considered as low-heat appliances if not larger than 100 cubic feet in size excluding any burner equipment and blower compartment.

(b) MEDIUM-HEAT INDUSTRIAL APPLIANCE — An industrial appliance such as an annealing furnace (glass or metal), charcoal furnace, galvanizing furnace, gas producer, commercial or industrial incinerator, and steam boiler operating at over 50 psig pressure when such appliance is larger than 100 cubic feet in size, and other furnaces classified as medium-heat appliances in accordance with nationally recognized good practice. Appliances otherwise classed as medium-heat appliances may be considered as low-heat appliances if not larger than 100 cubic feet in size excluding any burner equipment and blower compartment.

(c) HIGH-HEAT INDUSTRIAL APPLIANCE. An industrial appliance such as a billet and bloom furnace, blast furnace, brass melter, cupola, glass furnace, open-hearth furnace, and ceramic kiln and vitreous enameling oven (ferrous metals) when such appliances are larger than 100 cubic feet in size, and other furnaces classified as high-heat appliances in accordance with nationally recognized good practice.

LISTED—Refers to products which are shown in a list published by an approved testing agency, qualified and equipped for experimental testing, and maintaining an adequate periodic inspection of current production of listed models and whose listing states that the product complies with nationally recognized safety requirements.

METAL PIPE, SINGLE WALL—Is an unlisted sheet metal hollow cylinder used for conveying products of combustion.

PLENUM CHAMBER—A compartment or chamber to which one or more ducts are connected and which form a part of either the supply or return air system.

PLENUM FLOOR SYSTEM—The use of space between the floor structural system, earth (separated by non-porous vapor barrier and sand or concrete), and foundation walls in a one-story building for the distribution of conditioned air.

RESIDENTIAL-TYPE APPLIANCE—Fuel-burning and electric heating appliances, except high pressure steam boilers, for heating building spaces having a volume of not more than 25,000 cubic feet and other heat producing appliances of the type mainly used in residences, but which may be used in other buildings, such as cooking stoves and ranges, clothes dryers, fireplace stoves, domestic incinerators, laundry stoves, water heaters, and heat pumps.

ROOM OR SPACE HEATER—Space heaters are above the floor devices for direct heating of the space in and adjacent to that which the device is located without heating pipes or ducts.

UNDERFLOOR HORIZONTAL FURNACES—Means horizontal flow forced warm air devices located under the floor of the heated structure and connected to a duct system.

UNIT HEATERS, CEILING-TYPE—Unit heaters are here defined as appliances consisting of a combination of heating element and fan, having a common enclosure and placed within or adjacent to the space to be heated.

VENT PIPE—As applied to heating, means a pipe for removing products of combustion from appliances.

VENTING SYSTEM—A continuous open passageway from the flue collar or draft hood of a fuel-burning appliance to the outside atmosphere for the purpose of removing products of combustion.

KINDS OF VENTING SYSTEMS:

(a) Type B and Type B-W; a gas venting system consisting of vent piping and fittings listed for use with listed gas appliances with draft hoods and other gas appliances listed for use with Type B gas vents.

(b) Type L; a venting system consisting of listed vent piping and fittings for use with oil burning appliances listed for use with Type L or with unlisted gas appliances.

WALL FURNACE—A self-contained, vented appliance complete with grilles or equivalent, designed for incorporation in or permanent attachment to the structure of a building, mobile home or travel trailer, and furnishing heated air circulated by gravity or by a fan directly into the space to be heated through openings in the casing. Such appliances shall not be provided with duct extensions beyond the vertical and horizontal limits of the casing proper, except that boots not to exceed 10 inches beyond the horizontal limits of the casing for extension through walls of nominal thickness may be permitted. When such boots are provided, they shall be supplied by the manufacturer as an integral part of the appliance. This definition excludes floor furnaces, unit heaters, sealed combustion system wall furnaces and central furnaces.

(a) GRAVITY-TYPE WALL FURNACE. A wall furnace depending on circulation of air by gravity.

(b) FAN-TYPE WALL FURNACE. A wall furnace equipped with a fan.

WATER HEATER—A device for the heating and storage of water to be used for other than heating or industrial purposes.

SECTION 802 – CHIMNEYS

802.1 — GENERAL

(a) Chimneys shall be required for all heating or heat producing appliances except electrical heating and appliances listed for use with venting systems.

Every chimney shall be constructed and every venting system shall be installed in accordance with the applicable requirements of this Chapter.

(b) Draft. Every chimney, vent or venting system shall be capable of producing a draft at the appliance not less than that required for the safe operation of the appliance connected thereto.

A power exhauster may be used, except with incinerators, to increase low draft. When an exhauster is used, provision shall be made to shut off the fuel supply to the appliance in the event of failure of the exhauster.

(c) Nothing in this Code shall prohibit the joining of two or more connectors into a combined connector, provided that all pipes are of sufficient size to serve all of the appliances thus connected, and provided that all pipes are constructed to comply with the severest requirements for any of those connected.

(d) The connector of a fuel burning appliance shall not be connected into the chimney flue of an incinerator which has the rubbish chute identical with the chimney flue.

802.1 B — CHIMNEY CONSTRUCTION — General Requirements.

(a) No chimney shall carry any load other than its own dead weight.

(b) Chimney crickets of metal or other roofing materials shall be laid or installed on solid roof decking consistent with the construction of the roof.

(c) All chimneys which are or become unsafe or dangerous shall be made safe or taken down.

(d) Chimneys shall extend at least three (3) feet above the highest point where they pass through the roof of the building and at least two (2) feet higher than any ridge within ten (10) feet of such chimney. (See Figure 2.)

(e) An appliance equipped with a forced or induced draft system which may result in positive pressure in the venting system shall be connected to a pressure-tight venting system.

(f) RAISING ADJOINING CHIMNEYS

1. Whenever a building is hereafter erected, enlarged or raised, the owner of such building shall, at his own expense, carry up, either independently or on his own building, all chimneys, smokestacks, and smoke flues of an adjoining building which are within ten (10) feet of any portion of the wall extending above such chimney or flue.

2. The construction of such chimney shall conform to all requirements of this Code. Such a chimney shall be carried up simultaneously with the walls.

3. It shall be the duty of the owner of the building to be erected, enlarged or raised to notify, in writing, at least ten (10) days before such work is to begin, the owner of the chimney affected, of his intention to carry up such chimney.

802.2 — FACTORY-BUILT CHIMNEYS

Factory-built chimneys shall be listed and shall be installed in accordance with the conditions of the listing, and the manufacturer's instructions.

(a) Nothing contained in this code shall be construed as prohibiting the use of insulated suspended factory-built chimneys provided such assemblies have been tested and approved by a recognized laboratory, for the use intended and are installed in accordance with their approval.

(b) Factory-Built chimneys may be installed with zero clearance on wood structural members, such as framing, roof rafters, floor and ceiling joists and other component structural members, when it has been determined by test reports of recognized and approved testing laboratories that the unit does not transmit heat to the supporting combustible members of more than 90 degrees F. above room temperature.

(c) Supports for such chimneys attached to ceiling or floor joists shall be permitted provided the joists are of adequate size to support additional load.

(d) All chimney installations shall conform with the height requirements of this code.

SECTION 803 – MASONRY CHIMNEYS

803.1 — GENERAL REQUIREMENTS

(a) Support. Masonry chimneys shall be supported on properly designed foundations of masonry or reinforced concrete. Noncombustible material having a fire resistance rating of not less than 3 hours may be used to support masonry chimneys where such supports are independent of the floor construction and the load transferred to the ground.

(b) Corbeling. Masonry chimneys shall not be corbeled from a wall more than 6 inches, nor shall a chimney be corbeled from a wall which is less than 12 inches in thickness unless it projects equally on each side of the wall provided that in the second story of 2-story dwellings corbeling of chimneys on the exterior of the enclosing walls may equal the wall thickness. Corbeling shall not exceed one inch projection for each course of brick projected.

(c) Change in size or shape at roof not permitted. A change in size or shape of a chimney flue where the chimney passes through the roof, shall not be made within a distance of 6 inches above or below the roof joists or rafters.

(d) Cleanout openings. Cleanout openings provided in chimneys shall be equipped with ferrous metal doors and frames arranged to remain tightly closed when not in use. Adequate clearance between cleanout doors and combustible material shall be provided.

(e) Firestopping. All spaces between chimneys and floors and ceilings through which chimneys may pass shall be firestopped with noncombustible material. The firestopping of spaces between chimneys and wood joists, beams, or headers shall be to a depth of one inch only placed on strips of metal or metal lath laid across the spaces between combustible material and the chimney.

(f) Smoke test. Masonry chimneys shall be proved tight by a smoke test after erection and before being put into use.

803.2 — MASONRY CHIMNEYS FOR RESIDENTIAL TYPE APPLIANCES — (See Table 805-A)

(a) Construction

1. Masonry chimneys shall be constructed of solid masonry units or reinforced concrete with walls not less than 4 inches thick or rubble stone masonry not less than 12 inches thick.
2. Masonry chimneys for residential type appliances shall be lined with approved fire clay tile flue liners not less than ⅝ of an inch thick, or with other approved liner of material that will resist corrosion, softening or cracking from flue gases at temperatures up to 1800 degrees F.
3. Fire clay tile liners shall be installed ahead of the construction of the chimney as it is carried up, carefully bedded one on the other in fire clay mortar or as listed in Section 1402.12 (Table 3) with close fitting joints left smooth on the inside.
4. Liners shall be separate from the chimney wall and the space between the liner and masonry shall not be filled; only enough mortar shall be used to make a good joint and hold the liners in position.
5. Flue liners shall start from a point not less than 8 inches below the intake. They shall extend, as nearly vertically as possible, for the entire height of the chimney.

6. Where two adjoining flues in the same chimney are separated only by flue liners the joints of the adjacent flue liners shall be staggered at least 7 inches.

7. Where more than two flues are located in the same chimney, masonry wythes (partitions) at least 4 inches wide and bonded into the masonry walls of the chimney shall be built at such points between adjacent flue linings that there are not more than two flues in any group of adjoining flues without such wythe separation.

(b) Termination. Masonry chimneys for residential type appliances shall extend at least 3 feet above the highest point where they pass through the roof of a building and at least 2 feet higher than any portion of a building within 10 feet. (See Figure 2.)

(c) Clearance from combustible material.

1. All wood beams, joists and studs shall be trimmed away from chimneys. Headers, beams, joists and studs shall be not less than 2 inches from the outside face of a chimney or from masonry enclosing a flue. Ends of wood girders may be supported on corbeled shelf of a masonry chimney provided there is not less than 8 inches of solid masonry between the ends and the flue liner.

2. No combustible lathing, furring or plaster grounds shall be placed against a chimney at any point more than 1½ inches from the corner of the chimney, but this shall not prevent plastering directly on the masonry or on metal lath and metal furring, nor shall it prevent placing chimneys for residential type appliances entirely on the exterior of a building against the sheathing.

803.3 — MASONRY CHIMNEYS FOR LOW HEAT APPLIANCES

(a) Construction.

1. Masonry chimneys shall be constructed of solid masonry units or reinforced concrete with walls not less than 8 inches thick, except that rubble stone masonry shall be not less than 12 inches thick.

2. Masonry chimneys for low heat appliances shall be lined with approved fire clay tile flue liners not less than ⅝ of an inch thick, or with other approved liner of material that will resist corrosion, softening or cracking from flue gases at temperatures up to 1800 degrees F.

3. Fire clay tile flue liners shall be installed ahead of the construction of the chimney as it is carried up, carefully bedded one on the other in fire clay mortar or as listed in Section 1402.12 (Table 3) with close fitting joints left smooth on the inside.

4. Flue liners shall start from a point not less than 8 inches below the intake. They shall extend, as nearly vertically as possible, for the entire height of the chimney.

5. Where two adjoining flues in the same chimney are separated only by flue liners, the joints of the adjacent flue liners shall be staggered at least 7 inches.

6. Where more than two flues are located in the same chimney, masonry wythes (partitions) at least 4 inches wide and bonded into the masonry walls of the chimney shall be built at such points between adjacent flue linings that there are not more than two flues in any group of adjoining flues without such wythes separation.

(b) Termination. Masonry chimneys for low-heat appliances shall extend at least 3 feet above the highest point where they pass through the roof of a building and at least 2 feet higher than any portion of a building within 10 feet. (See Figure 2.)

(c) Clearance from combustible material.

1. All wood beams, joists and studs shall be trimmed away from chimneys. Headers, beams, joists, and studs shall be not less than 2 inches from the outside face of a chimney or from masonry enclosing a flue.

2. Combustible lathing, furring or plaster grounds shall not be placed against a chimney at any point more than 1½" from the corner of the chimney, but this shall not prevent plastering directly on the masonry or on metal lath and metal furring, nor shall it prevent placing chimneys for low heat appliances entirely on the exterior of a building against the sheathing.

803.4 — MASONRY CHIMNEYS FOR MEDIUM HEAT APPLIANCES

(a) Construction. Masonry chimneys for medium heat appliances shall be constructed of solid masonry units or of reinforced concrete not less than 8 inches thick, except that stone masonry shall be not less than 12 inches thick; and in addition, shall be lined with not less than 4½ inches of fire brick laid on the 4½ inch bed in fire clay mortar, starting not less than 2 feet below the chimney connector entrance and extending for a distance of at least 25 feet above the chimney connector entrance. Chimneys extending 25 feet or less above the chimney connector shall be lined to the top.

(b) Termination. Masonry chimneys for medium heat appliances shall extend not less than 10 feet higher than any portion of any building within 25 feet.

(c) Clearance from combustible material. Masonry chimneys for medium heat appliances shall have a clearance of 4 inches from buildings and structures.

803.5 — MASONRY CHIMNEYS FOR HIGH HEAT APPLIANCES

(a) Construction. Masonry chimneys for high heat appliances shall be constructed with double walls of solid masonry units or of reinforced concrete each not less than 8 inches in thickness, with an air space of not less than 2 inches between them. The inside of the interior walls shall be of fire brick not less than 4½ inches in thickness laid on the 4½ inch bed in fire clay mortar.

(b) Termination. Masonry chimneys for high heat appliances shall extend not less than 20 feet higher than any portion of any building within 50 feet.

(c) Clearance from combustible material. Masonry chimneys for high heat appliances shall have sufficient clearance from buildings and structures to avoid overheating combustible material, to permit inspection, maintenance operations on the chimney and to avoid danger of burns to persons. Clearances shall be based on good engineering practice and acceptable to the authority having jurisdiction.

803.6 — MASONRY CHIMNEYS FOR INCINERATORS

A. RESIDENTIAL. Masonry chimneys for residential type incinerators shall be constructed in accordance with the requirements for Chimneys for Residential Type Appliances, Section 803.2.

B. FUEL-FED INCINERATORS (apartment house type).

(a) Construction.

1. The flue of fuel-fed incinerators shall serve the incinerator only and be used for no other purpose.

2. The flue liner shall be straight and plumb and shall be smooth on the inside.

3. The size of incinerator flues shall be in accordance with the following:

   Where not more than one service opening is provided, the size of flue shall be not less than 14 by 14 inches or 196 square inches, inside measurements, except that in one family dwellings the size shall be not less than 12 by 12 inches or 144 square inches.

   Where two to six service openings are provided, the size of flue shall be not less than 18 by 18 inches or 324 square inches, inside measurements.

   Where seven or more service openings are provided, the size of flue shall be not less than 22 by 22 inches or 484 square inches, inside measurements.

4. A masonry chimney serving an incinerator with a combustion chamber having a horizontal combined hearth and grate area of 7 square feet or less shall have walls of clay or shale brickwork not less than 4 inches thick with a lining of 4½ inches of fire brick for a distance of not less than 10 feet above the roof of the combustion chamber; beyond this point chimney walls shall consist of not less than 8 inches of clay or shale brickwork with a standard fire clay tile flue liner not less than ⅝ inch in thickness extending from the top of the fire brick lining to the top of the chimney.

5. A masonry chimney serving an incinerator with a combustion chamber having a horizontal combined hearth and grate area exceeding 7 square feet shall have walls of clay or shale brickwork not less than 4 inches thick with a lining of 4½ inches of fire brick for a distance of not less than 40 feet above the roof of the combustion chamber; beyond this point, chimney walls shall consist of not less than 8 inches of clay or shale brickwork with a standard fire clay tile flue liner not less than ⅝ inch in thickness extending from the top of the fire brick lining to the top of the chimney.

6. Other constructions may be used if equivalent to the construction outlined in the preceding paragraphs, in structural strength, insulating value and ability to withstand thermal expansion and flame impingement.

7. Fire brick shall be laid in high temperature cement or fire clay mortar.

8. A chimney flue that is divided into two channels, one for feeding refuse and the other for the discharge of combustion gases, shall be constructed as specified in this section.

9. Masonry chimneys for fuel-fed incinerators shall be supported on properly designed foundations of masonry or reinforced concrete. Noncombustible material having a fire resistance rating of not less than 3 hours may be used to support masonry chimneys where such supports are independent of the floor construction and the load is transferred to the ground. They shall be so constructed as not to place excessive stress upon the roof of the combustion chamber. Masonry chimneys may be supported on incinerator walls.

10. All flues shall terminate in a substantially constructed spark arrester with openings not greater than ¾ inch, or be provided with other suitable means of avoiding discharge of fly particles. Expansion chambers used as a secondary combustion chamber shall be constructed equivalent to that of the incinerator combustion chamber. Those used only for settling shall be of construction equivalent to that of the upper portion of incinerator.

(b) Termination. Masonry chimneys of fuel-fed incinerators shall extend at least 4 feet above sloping roofs measured from the highest point at which the chimney passes through the roof and at least 8 feet above flat roofs. In either case, the chimney shall extend at least 2 feet higher than any portion of a building within 20 feet.

(c) Clearances.

1. A clearance of not less than 2 inches shall be provided between the exterior surface of chimneys for fuel-fed apartment house type incinerators and combustible material.

2. Combustible lathing, furring or plaster grounds shall not be placed against a chimney at any point more than 1½ inches

from the corner of the chimney but this shall not prevent plastering directly on the masonry or on metal lath and metal furring; nor shall it prevent placing chimneys entirely on the exterior of a building against the sheathing.

(d) Service openings to incinerators. All service openings into an incinerator flue shall be provided with a hopper or other charging device constructed of metal of sufficient thickness and durability to prevent cracking, breakage, or deformation in normal usage. Such hopper or other charging device shall be firmly built into the masonry and shall be so designed and installed that no part will project into the flue and that the opening to the flue interior will be closed off while the service opening (hopper) door is fully open. The hopper or other device shall be counter-weighted or otherwise devised so that it will close automatically upon release and be so constructed as to be tightly fitted when in the closed position.

C. COMMERCIAL AND INDUSTRIAL TYPE INCINERATORS.

(a) Construction.

1. Masonry chimneys of commercial and industrial type incinerators, except as provided in the following paragraphs b, and c, shall be not less than 8 inches of clay or shale brickwork or reinforced concrete lined with fire brick not less than 4½ inches thick for the full height of the chimney.

2. Subject to approval by the authority having jurisdiction, commercial and industrial type incinerators may be connected to chimneys constructed of 8 inches of clay or shale brickwork or reinforced concrete lined with fire clay flue liner, where the incinerator is specially constructed to produce low flue gas temperatures.

3. Other constructions may be used if equivalent to the construction outlined in the preceding paragraphs, in structural strength, insulating value and ability to withstand thermal expansion and flame impingement.

4. Fire brick and other refractory lining shall be laid in high temperature cement or fire clay mortar.

5. Masonry chimneys for commercial and industrial incinerators shall be supported on properly designed foundations of masonry or reinforced concrete. Noncombustible material having a fire resistance rating of not less than 3 hours may be used to support masonry chimneys where such supports are independent of the floor construction and the load is transferred to the ground. They shall be so constructed as not to place excessive stress upon the roof of the combustion chamber. Masonry chimneys may be supported on incinerator walls.

6. Incinerators used for the burning of rubbish or other readily combustible solid waste material shall include effective means for arresting sparks and fly particles, such as an expansion chamber, baffle walls, or other effective arrangement, or the

flues of such incinerators shall be provided with an approved spark arrester having opening not greater than ½ inch.

(b) Termination. Masonry chimneys for commercial and industrial type incinerators shall extend at least 4 feet above sloping roofs measured from the highest point at which the chimney passes through the roof and at least 8 feet above flat roofs. In either case, the chimney shall extend at least 2 feet above any ridge, parapet, cornice, penthouse or other obstruction within 20 feet.

(c) Clearances. A clearance of not less than 4 inches shall be provided between the exterior surface of masonry chimneys for commercial and industrial type incinerators and combustible material.

SECTION 804 – METAL CHIMNEYS (Smokestacks)

804.1 — GENERAL REQUIREMENTS

(a) Single metal wall chimneys shall not be used inside 1 and 2-family dwellings.

(b) Metal chimneys shall be of adequate thickness based on good engineering practice, properly riveted or welded, and securely supported.

NOTE: When selecting the gage of metal chimneys consideration should be given to factor such as location, maintenance, use, etc., as well as engineering design factors. As a guide the following are gages of uncoated sheet steel for given cross-sectional areas:

Mfgrs. Std. Gage No.	Area (Sq. In.)
16	Up to 154
14	154 to 201
12	201 to 254
10	Larger than 254

(c) Metal chimneys shall not be carried up inside of ventilating ducts unless such ducts are constructed and installed as required by this standard for chimneys and the ventilating ducts are used solely for exhaust of air from the room or space in which the appliance served by the metal chimney is located.

(d) Metal chimneys shall have sufficient clearance from buildings and structures to avoid overheating combustible material, to permit inspection and maintenance operations on the chimneys, and to avoid danger of burns to persons.

804.2 — METAL CHIMNEYS FOR RESIDENTIAL TYPE OR LOW HEAT APPLIANCES

(a) Termination.

1. Metal chimneys for residential or low heat appliances except as provided in 804.2(a)2 shall extend at least 3 feet above the highest point where they pass through the roof of a building and at least 2 feet higher than any portion of a building within 10 feet. (See Figure 2)

2. The outlet of a metal chimney for residential type and low heat appliances equipped with an exhauster may terminate at a location not less than 3 feet from an adjacent building or building opening and at least 10 feet above grade or walkways. In any case, the outlet shall be so arranged that the flue gases are not directed so as to jeopardize people, overheat combustible structures, or enter building openings in the vicinity of the outlet.

(b) Clearances.

1. Exterior metal chimneys used only for residential type or low heat appliances as defined in Table 805A shall have a clearance of not less than 6 inches from a wall of wood frame construction and from any combustible material.

2. Exterior metal chimneys over 18 inches in diameter shall have a clearance of not less than 4 inches, and those 18 inches or less in diameter a clearance of not less than 2 inches from a building wall of other than wood frame construction.

3. An exterior metal chimney shall not be nearer than 24 inches to any door or window or to any walkway, unless insulated or shielded in an approved manner to avoid burning a person who might touch the chimney.

(c) Interior.

1. Where a metal chimney extends through any story above that in which the appliances connected to the chimney are located, it shall be enclosed in such upper stories with walls of noncombustible construction having a fire resistance rating of not less than one hour.

2. The enclosure shall provide a space on all sides of the chimney sufficient to permit inspection and repair but in no case shall it be less than 12 inches.

3. The enclosing walls shall be without openings, except doorways equipped with approved self-closing fire doors at various floor levels for inspection purposes.

4. Where a metal chimney is used for residential type or low heat appliances as defined in Table 805A is located in the same story of a building as that in which the appliances connected thereto are located, it shall have a clearance of not less than 18 inches from a wall of wood frame construction and from any combustible material.

5. Where a metal chimney serving only low heat appliances as defined in Table 805A passes through a roof constructed of combustible material, it shall be guarded by a ventilating thimble of noncombustible material, of galvanized iron or approved corrosion resistant metal, extended not less than 9 inches below and 9 inches above the roof construction, and of a size to provide not less than 6 inches clearance on all

sides of the chimney; or the combustible material in the roof construction shall be cut away so as to provide not less than 18 inches clearance on all sides of the chimney, with any material used to close up such opening entirely noncombustible.

804.3 — METAL CHIMNEYS FOR MEDIUM HEAT APPLIANCES

(a) Termination. Metal chimneys for medium heat appliances shall extend not less than 10 feet higher than any portion of any building within 25 feet.

(b) Clearances.

(c) Exterior.

1. Exterior metal chimneys used for medium heat appliances as defined in Table 805A shall have a clearance of not less than 24 inches from a wall of wood frame construction and from any combustible material.
2. Exterior metal chimneys over 18 inches in diameter shall have a clearance of not less than 4 inches, and those 18 inches or less in diameter a clearance of not less than 2 inches from a building wall of other than wood frame construction.
3. No portion of an exterior metal chimney shall be nearer than 24 inches to any door or window or to any walkway, unless insulated or shielded in an approved manner to avoid burning a person who might touch the chimney.

(d) Interior.

1. Where a metal chimney extends through any story of a building above that in which the appliances connected to the chimney are located, it shall be enclosed in such upper stories with walls of noncombustible construction having a fire resistance rating of not less than one hour if the building is less than 4 stories in height, and not less than 2 hours if the building is 4 stories or more in height.
2. The enclosing walls shall provide a space on all sides of the chimney to permit inspection and repair, but in no case shall it be less than 12 inches.
3. The enclosing walls shall be without openings, except doorways equipped with approved self-closing fire doors at various floor levels for inspection purposes.
4. Where a metal chimney serving a medium heat appliance as defined in Table 805A passes through a roof constructed of combustible material, it shall be guarded by a ventilating thimble of galvanized iron or approved corrosion resistant metal, extending not less than 9 inches below and 9 inches above the roof construction and of a size to provide not less than 18 inches clearance on all sides of the chimney.
5. Where a metal chimney used for medium heat appliances as defined in Table 805A is located in the same story of a build-

ing as that in which the appliances connected are located, it shall have a clearance of not less than 36 inches from a wall of wood frame construction and from any combustible material.

804.4 — METAL CHIMNEYS FOR HIGH HEAT APPLIANCES

(a) Construction. Metal chimneys used for high heat appliances as defined in Table 805A shall be lined with not less than $4\frac{1}{2}$ inches of fire brick laid in fire clay mortar extending not less than 25 feet above the chimney connector entrance. Chimneys extending 25 feet or less above the chimney connector shall be lined to the top.

(b) Termination. Metal chimneys for high heat appliances shall extend not less than 20 feet higher than any portion of any building within 50 feet.

(c) Clearance from combustible material. Metal chimneys for high heat appliances shall have sufficient clearance from buildings and structures to avoid overheating combustible material, to permit inspection, maintenance operations on the chimney and to avoid danger of burns to persons. Clearances shall be based on good engineering practice and acceptable to the authority having jurisdiction.

804.5 — METAL CHIMNEYS FOR INCINERATORS

A. RESIDENTIAL TYPE INCINERATORS. Metal pipe not less No. 20 galvanized sheet gage number or other equivalent noncombustible corrosion resistant material may be used for residential type (See Figure 2) incinerators installed in locations such as sheds, breezeways or carports provided the metal pipe is exposed and readily examinable for its full length and clearances not less than 18 inches are maintained from combustible material. The metal pipe shall extend at least 3 feet above the highest point where it passes through the roof and at least 2 feet higher than any portion of the building within 10 feet. Where the metal pipe passes through a roof constructed of combustible material clearances shall conform to the requirements for interior metal chimneys for Low Heat Appliances, 804.2(b) 4, 5, 6, 7 and 8.

B. COMMERCIAL AND INDUSTRIAL TYPE INCINERATORS.

(a) Construction.

1. Metal chimneys of commercial and industrial type incinerators, shall be lined with firebrick not less than $4\frac{1}{2}$ inches thick for the full height of the chimney.
2. Firebrick shall be laid in high temperature cement or fire clay mortar.
3. Subject to approval by the authority having jurisdiction, commercial and industrial type incinerators may be connected to

a metal chimney without firebrick flue liner provided the incinerator is specially constructed to produce low flue gas temperatures.

(b) Termination. Metal chimneys of commercial and industrial type incinerators shall extend at least 4 feet above sloping roofs measured from the highest point at which the metal chimney passes through the roof and at least 8 feet above flat roofs. In either case, the chimney shall extend at least 2 feet higher than any portion of a building within 20 feet.

(c) Clearance. Metal chimneys for commercial and industrial incinerators shall be installed to provide clearances as specified in 804.3, 5.

FIGURE 2 — TYPICAL TERMINATION LOCATIONS FOR CHIMNEYS

A. TERMINATION Less THAN 10 FEET FROM RIDGE, WALL OR PARAPET

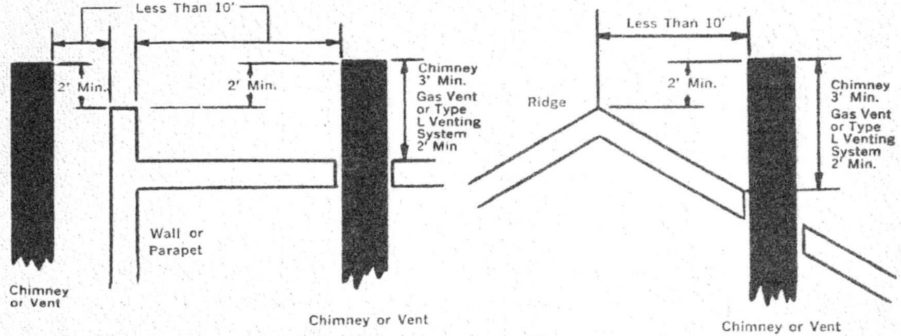

B. TERMINATION More THAN 10 FEET FROM RIDGE, WALL OR PARAPET

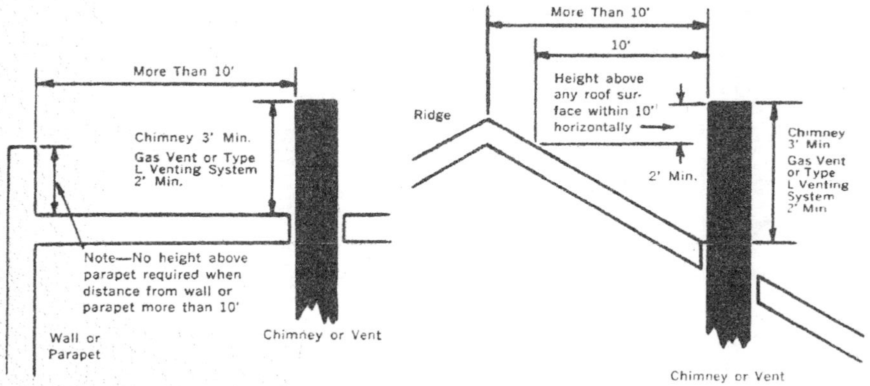

Typical Termination Locations for Gas Vents and Chimneys

TABLE 805-A — CHIMNEY SELECTION CHART

Chimneys for Residential Type Appliances	Chimneys for LOW HEAT Appliances		Chimneys for MEDIUM HEAT Appliances[1]	Chimneys for HIGH HEAT Appliances[1]
	Building Heating Appliances	Industrial Type Low Heat Appliances		
1. Factory built (low heat) 2. Masonry (residential).[4]	1. Factory built (low heat). 2. Masonry (low heat type).[4] 3. Metal (smokestack).[5]	1. Factory Built (industrial low heat type). 2. Masonry (low heat type) 3. Metal (smokestack).[5]	1. Factory built (medium heat type). 2. Masonry (medium heat type).[4] 3. Metal (smokestack).[5]	1. Masonry (high heat type).[4] 2. Metal (smokestacks).[5]
TYPES OF APPLIANCES TO BE USED WITH EACH TYPE CHIMNEY				
Column I	**Column II**	**Column III**	**Column IV**	**Column V**
A. Residential type appliances, such as: 1. Ranges. 2. Warm air furnaces. 3. Water heaters. 4. Hot water heating boilers. 5. Low pressure steam heating boilers (not over 15 psig). 6. Domestic incinerators. 7. Floor furnaces. 8. Wall furnaces. 9. Room heaters. 10. Fireplace stoves. B. Fireplaces.	A. All appliances shown in Column I. B. Nonresidential type building heating appliances for heating a total volume of space exceeding 25,000 cubic feet.[5] C. Steam boilers operating at not over 50 lb. per sq. in. gage pressure; pressing machine boilers.	All appliances shown in Columns I and II, and appliances such as: 1. Annealing baths for hard glass (fats, paraffine, salts, or metals). 2. Bake ovens (in bakeries). 3. Boiling vats, for wood fibre, straw, lignin, etc. 4. Candy furnaces. 5. Coffee roasting ovens. 6. Core ovens. 7. Cruller furnaces. 8. Feed drying ovens. 9. Fertilizer drying ovens. 10. Fireplaces, other than residential type. 11. Forge furnaces (solid fuel). 12. Gypsum kilns.	All appliances shown in Columns I, II and III, and appliances such as: 1. Alabaster gypsum kilns. 2. Annealing furnaces (glass or metal). 3. Charcoal furnaces. 4. Cold stirring furnaces. 5. Feed driers (direct fire heated). 6. Fertilizer driers (direct fire heated). 7. Galvanizing furnaces. 8. Gas producers. 9. Hardening furnaces (cherry to pale red). 10. *Incinerators, commercial and industrial type.* 11. Lehrs and glory holes. 12. Lime kilns.	All appliances shown in Columns I, II, III, and IV and appliances[2] such as 1. Bessemer retorts. 2. Billet and bloom furnaces. 3. Blast furnaces. 4. Bone calcining furnaces. 5. Brass furnaces. 6. Carbon point furnaces. 7. Cement, brick and tile kilns. 8. Ceramic kilns. 9. Coal and water gas retorts. 10. Cupolas. 11. Earthenware kilns. 12. Glass blow furnaces. 13. Glass furnaces (smelting).

TABLE 805-A — CHIMNEY SELECTION CHART—(Continued)

Chimneys for Residential Type Appliances	Chimneys for LOW HEAT Appliances		Chimneys for MEDIUM HEAT Appliances[1]	Chimneys for HIGH HEAT Appliances[2]
		13. Hardening furnaces (below dark red). 14. Hot air engine furnaces. 15. Ladle drying furnaces. 16. Lead melting furnaces. 17. Nickel plate (drying) furnaces. 18. Paraffine furnaces. 19. Recuperative furnaces (spent materials). 20. Rendering furnaces. 21. Restaurant type cooking appliances using solid or liquid fuel. 22. Rosin melting furnaces. 23. Stereotype furnaces. 24. Sulphur furnaces. 25. Tripoli kilns (clay, coke and gypsum). 26. Type foundry furnaces. 27. Wood drying furnaces. 28. Wood impregnating furnaces. 29. Zinc amalgamating furnaces.	13. Linseed oil boiling furnaces. 14. Porcelain biscuit kilns. 15. Pulp driers (direct fire heated). 16. Steam boilers operating at over 50 lb. per sq. in. gage pressure except pressing machine boilers. 17. Water-glass kiln. 18. Wood-distilling furnaces. 19. Wood gas retorts.	14. Glass kilns. 15. Open hearth furnaces. 16. Ore roasting furnaces. 17. Porcelain baking and glazing kilns. 18. Pot-arches. 19. Puddling furnaces. 20. Regenerative furnaces. 21. Reverberatory furnaces. 22. Stocks, carburetor or superheating furnaces (in water gas works). 23. Vitreous enameling ovens (ferrous metals). 24. Wood carbonizing furnaces.

*Nonresidential type building heating appliances for heating a total volume of space not to exceed 25,000 cubic feet may be connected to chimneys for residential type appliances.

Note 1: Appliances otherwise classed as high heat appliances may be considered as medium heat appliances if not larger than 100 cubic feet in size.

Note 2: When such appliances are larger than 100 cubic feet in size, and other furnaces classified as high heat appliances in accordance with nationally recognized good practice.

Note 3: Continuous operating equipment of the counter current type may not require the type of flue indicated by general types of appliances.

Note 4: For construction and other provisions for masonry chimney installation see Section 802.

Note 5: For construction and other provisions for metal chimney installation see Section 804.

TABLE 805-B

Venting System Selection Chart

TYPE OF VENTING SYSTEM

Type B — Gas[1] (Round or Oval)	Type BW — Gas[1]	Type L — Oil	Metal Pipe[1]
Column I	**Column II**	**Column III**	**Column IV**
All listed gas appliances with draft hoods such as: 1. Central furnaces. 2. Duct furnaces. 3. Floor furnaces. 4. Heating boilers. 5. Ranges. 6. Built-in ovens. 7. Vented wall furnaces listed for use with Type B vents. 8. Room heaters. 9. Water heaters. 10. Horizontal furnaces. 11. Unit heaters.	1. Vented wall furnaces listed for use with Type BW vents only.	1. Low temperature flue gas appliances listed for use with Type L venting systems. 2. Gas appliances shown in Column I.	1. Incinerators used outdoors, such as in open sheds, breezeways or carports as provided in 804.5A. 2. Gas appliances shown in Column I. 3. Listed residential and low heat gas appliances without draft hoods and unlisted residential and low heat gas appliances with or without draft hoods.

Note 1: Installation requirements in Section 811.

SECTION 806 – CHIMNEY CONNECTORS AND VENT CONNECTORS

806.1 — CONNECTORS REQUIRED

Connectors shall be used to connect appliances to the vertical chimney or vent unless the chimney or vent is attached directly to the appliance.

806.2 — MATERIALS

(a) Connectors shall be made of noncombustible corrosion resistant material such as steel or refractory masonry capable of withstanding the flue gas temperatures produced by the appliances and of sufficient thickness to withstand physical damage. Connectors for appliances installed in attics shall be of Type B or Type L vent material for listed gas appliances with draft hoods or of Type L vent material for oil appliances listed as suitable for Type L vents. For other appliances allowed in attics, a chimney shall be attached directly to the appliance.

NOTE: When selecting the gage of metal for connectors of residential type and low heat appliances not installed in attics, consideration should be given to factors such as location, maintenance, use, etc., as well as engineering design factor. As a guide the following are gages of galvanized steel for given diameter connectors, except as provided in paragraphs (b), (c) and (d):

Galvanized Sheet Gage No.	Diameter of Connector, Inches
24	10 or less
22	10 to 12
20	14 to 16
16	16

(b) Metal connectors for medium heat appliances and commercial and industrial incinerators shall be constructed of steel not lighter than that designated for metal chimneys in 806.2(a). In addition, they shall be lined with fire brick (ASTM C106 Type F) not less than 2¼ inches thick, when they are more than 12 inches but not in excess of 18 inches in diameter or greatest dimension, and with fire brick not less than 4½ inches thick when they are over 18 inches in diameter or greatest dimension. Fire brick shall be laid in high temperature cement or fire-clay mortar (ASTM C105 medium duty).

Chimney connectors for appliances specially constructed to produce low flue gas temperatures, and chimney connectors not over 10 inches in diameter and not over 8 feet long, may be of flue tile properly supported and insulated, or of other suitable construction without fire brick lining where located entirely within the appliance room.

(c) Metal connectors for high heat appliances shall be constructed of steel not lighter than that designated for metal chimneys in 806.2(a).

In addition, they shall be lined with fire brick (ASTM C106, Type A) or the equivalent having a thickness not less than 4½ inches thick, laid in high temperature cement or fire-clay mortar (ASTM C105, high duty).

(d) Masonry connectors or breechings shall be made of refractory material equivalent in resistance to heat and corrosion to high-duty regular type (ASTM C106, Type A) fire-clay brick 4½ inches thick.

(e) Connectors used for gas appliances having draft hoods and for listed conversion-burner-equipped appliances having draft hoods may be constructed of materials having resistance to corrosion and heat not less than that of No. 28 gage galvanized steel, or they may be of Type B or Type L vent material.

(f) Connectors made of Type L vent material may be used with gas, oil and solid fuel-burning residential type appliances including residential-type incinerators.

806.3 — LENGTH

A connector shall be as short and straight as possible. The appliance shall be located as close as practicable to the chimney, gas vent, or venting system. The horizontal run of an uninsulated connector to a natural draft chimney, or vent, serving a single liquid or solid fuel burning appliance, shall not be more than 75 percent of the height of the vertical portion of the chimney or vent above the connector, unless part of an engineered venting system.

(b) The horizontal run of an insulated connector to a natural draft chimney, or vent, serving a single gas fuel fired appliance shall be not more than 100 percent of the height of the vertical portion of the chimney or vent above the connector, unless part of an engineered venting system. The horizontal length, design, and construction of combined connectors, or connectors to a manifold joining two or more appliances to a chimney or vent shall be determined in accordance with approved engineering methods.

806.4 — SIZE

The connector, for its entire length, shall be not smaller than the flue collar of the appliance unless otherwise recommended by the appliance, chimney, or vent manufacturer.

806.5 — CLEARANCE

Clearances from combustible material shall be in accordance with Table 811-A or 811-B.

806.6 — LOCATION

When the connector used for a gas appliance having a draft hood must be located in or pass through a crawl space or other cold area, that portion of the connector shall be of listed Type B or Type L vent material or be provided with equivalent means of insulation.

806.7 — INSTALLATION

(a) A connector to a masonry chimney shall extend through the wall to the inner face or liner but not beyond, and shall be firmly cemented to masonry. A thimble may be used to facilitate removal of the chimney connector for cleaning, in which case the thimble shall be permanently cemented in place with high-temperature cement.

(b) A chimney connector or a vent connector shall not pass through any floor or ceiling, nor through a fire wall or fire partition.

(c) Connectors for listed residential and low heat gas appliances with draft hoods except incinerators may pass through walls or partitions constructed of combustible material if:

1. Made of listed Type B or Type L material and installed with not less than listed clearances to combustible material.
2. Made of single wall metal pipe and guarded by a ventilated metal thimble not less than 4 inches larger in diameter than the vent connector.

(d) Connectors of low heat appliances except listed residential and low heat gas appliances with draft hoods (Table 805-B, Column 1) shall not pass through walls or partitions constructed of combustible material unless they are guarded at the point of passage by:

1. Metal ventilated thimbles not less than 12 inches larger in diameter than the connector.
2. Metal or burned fire clay thimbles built in brickwork or other approved fireproofing materials extending not less than 8 inches beyond all sides of the thimble.

(e) In lieu of thimbles all combustible material in the wall or partition shall be cut away from the connector a sufficient distance to provide the clearance required from such connector. Any material used to close up such openings shall be noncombustible insulating material.

(f) A connector of any medium or high heat appliance classified in Table 805-A shall not pass through any wall or partition constructed of combustible material.

(g) Connectors shall maintain a pitch of rise of at least ¼ inch to the foot (horizontal length of pipe) from the appliance to the chimney.

(h) Connectors shall be installed so as to avoid sharp turns or other construction features which would create excessive resistance to the flow of flue gases. A device which will obstruct the free flow of flue gases shall not be installed in a connector, chimney, or vent. This shall not be construed to prohibit the use of devices specifically listed for installation in a connector, such as heat reclaimers, draft regulators, and safety controls.

(i) Connectors shall be securely supported and joints fastened with sheet-metal screws, rivets, or other approved means.

(j) The entire length of a connector shall be readily accessible for inspection, cleaning, and replacement, unless listed materials are used and previous approval has been obtained from the authority having jurisdiction.

(k) A vent connector shall not be connected to a chimney flue serving a fireplace unless the fireplace opening is sealed or the chimney flue which vents the fireplace is permanently sealed below the connection.

806.8 — INTERCONNECTION

(a) Connectors shall not be connected to a chimney, vent, or venting system served by a power exhauster unless the connection is made on the negative pressure side of the exhauster.

(b) Two or more fuel-burning appliances may be connected to a single chimney or vent provided sufficient draft is available for safe combustion in each appliance and removal of all the products of combustion safely to the outdoors. Gas and oil appliances so connected shall be equipped with primary safety controls.

806.9 — DAMPERS

(a) Manually operated dampers shall not be placed in chimneys, vents or connectors of stoker fired, liquid or gas-burning appliances. Fixed baffles on the appliance side of draft hoods and draft regulators shall not be classified as dampers.

(b) Automatically operated dampers shall be of approved type designed to maintain a safe damper opening at all times and arranged to prevent the initiation or increase of firing unless the damper is opened to a safe position.

806.10 — DRAFT HOODS

For information concerning the use and installation of draft hoods, refer to Southern Standard Gas Code.

SECTION 807 – DRAFT REGULATORS

(a) Gas appliances connected to chimeys, other than those required by Southern Standard Gas Code, to be installed with draft hoods, may be installed with draft regulators if in accordance with the appliance manufacturer's instruction.

(b) For information concerning the use and installation of draft regulators with oil-burning appliances, refer to the "Standard for Oil-Burning Equipment, NFPA 31, 1972."

(c) Solid fuel-burning appliances may be installed with draft regulators to reduce draft intensity. Such regulators shall be installed and set in accordance with the instructions furnished with the appliance or the draft regulator.

(d) A barometric draft regulator, if used, shall be installed in the same room or enclosure as the appliance in such a manner that no difference in pressure between the air in the vicinity of the regulator and the combustion air supply will be permitted.

SECTION 808 — FIREPLACES

808.1 — FACTORY-BUILT FIREPLACES

Factory-built fireplaces shall be listed and shall be installed in accordance with the conditions of the listing. Hearth extensions shall be provided in accordance with Section 808.3(f).

808.2 — FACTORY-BUILT FIREPLACE STOVES

Factory-built fireplace stoves shall be listed and shall be installed in accordance with the conditions of the listing.

808.3 — MASONRY FIREPLACES

(a) Fireplaces shall be constructed of solid masonry or of reinforced concrete with back and sides of the thickness specified in this paragraph, except as provided in Section 808.1. Where a lining of firebrick at least 2 inches thick or other approved lining is provided, the total thickness of back and sides, including the lining, shall be not less than 8 inches. Where no such lining is provided, the thickness of back and sides shall be not less than 12 inches.

(b) Steel fireplace units incorporating a firebox liner of not less than ¼-inch thick steel and an air chamber may be installed with masonry to provide a total thickness at the back and sides of not less than 8 inches, not less than 4 inches of which shall be solid masonry.

(c) Warm air ducts employed with steel fireplace units of the circulating air type shall be constructed of metal or masonry.

(d) Fireplace hearth extensions shall be provided of approved noncombustible material for all fireplaces. Where the fireplace opening is less than 6 sq. ft., the hearth extension shall extend at least 16 in. in front of, and at least 8 in. beyond each side of the fireplace opening. Where the fireplace opening is 6 sq. ft. or larger, the hearth extension shall extend at least 20 in. in front of, and at least 12 in. beyond each side of the fireplace opening. Where a fireplace is elevated above or overhangs a floor, the hearth extension shall also extend over the area under the fireplace.

(e) Fireplaces constructed of masonry or reinforced concrete shall have hearth extensions of brick, concrete, stone, tile or other approved noncombustible material properly supported and with no combustible material against the underside thereof. Wooden forms or centers used during the construction of hearth and hearth extension shall be removed when the construction is completed.

(f) Hearth extensions of approved factory built fireplaces and fireplace stoves shall be not less than ⅜-in. thick of abestos, concrete, hollow metal, stone, tile or other approved noncombustible material. Such hearth extensions may be placed on the sub or finish flooring whether the flooring is combustible or not. The hearth extension shall be readily distinguishable from the surrounding floor.

(g) All wood beams, joists and studs shall be trimmed away from fireplaces. Headers supporting trimmer arches at fireplaces shall be not less than 20 inches from the face of the chimney breast. Trimmers shall be not less than 6 inches from the inside face of the nearest flue lining.

(h) Woodwork shall not be placed within 4 inches of the back of a fireplace, but this shall not prevent plastering directly on the masonry or on metal lath and metal furring.

(i) Woodwork shall not be placed within 6 inches of a fireplace opening. Woodwork above and projecting more than 1½ inches from a fireplace opening shall not be placed less than 12 inches from the top of a fireplace opening.

808.4 — FALSE FIREPLACES

False fireplaces may be used in connection with listed gas or electric heaters, provided such fireplaces are constructed of noncombustible materials.

SECTION 809 – INCINERATORS

809.1 — RESIDENTIAL TYPE INCINERATORS

(A) Design and Construction

1. Self-contained, factory-built, residential-type incinerators shall be approved. The design shall be such that in service they will not crack, warp, or otherwise fail structurally so as to permit flame passage or emission of combustion gases or sparks into the building.
2. The enclosing walls for combustion chambers of built-in masonry incinerators, except as provided in Item 3 below shall be constructed of clay or shale brickwork not less than 4 inches thick with a lining of 4½ inches of firebrick.
3. Other construction may be used if equivalent to the constructions outlined in the preceding paragraphs, in structural strength, insulating value, and ability to withstand thermal expansion and flame impingement.
4. Firebrick shall be laid in high-temperature cement or fire-clay mortar. All common brickwork shall be laid with full, push filled, cross, and bed mortar joints.

(B) Mounting

Except where approved specifically for mounting on combustible floors, incinerators shall be mounted on the ground or on floors of fire-resistive construction with noncombustible flooring or surface finish and with no combustible material against the underside thereof, or on fire-resistive slabs or arches having no combustible material against the underside thereof. Such construction shall extend not less than 12 inches beyond the incinerator base on all sides except at the front or side where ashes are removed where it shall extend not less than 18 inches beyond the incinerator.

(C) Chimney Connections

Residential type incinerators shall be connected to chimneys as specified in sections 802.2; 803.6 or 804.5. Connectors for residential type incinerators shall comply with section 806.

809.2 — COMMERCIAL AND INDUSTRIAL TYPE INCINERATORS

(A) Design and Construction

(a) Self-contained, commercial and industrial type incinerators shall be of the approved type. The design shall be such that in service they will not crack, warp or otherwise fail structurally so as to permit flame passage or emission of combustion gases or sparks into the building.

(b) Field-erected incinerators shall be built in accordance with the following:

1. Incinerators shall have the enclosing walls of combustion chambers constructed of clay or shale brick not less than 8 inches in thickness with a lining of fire brick not less than 4½ inches thick. Thicker fire brick may be needed for proper support of large walls, or to allow continuous operation at high temperatures, or to provide for flame impingement, spalling, or mechanical injury from operating tools, etc. Provision shall be made to allow for expansion and contraction of the fire brick. Other constructions may be used if equivalent in structural strengtn, insulating value and ability to withstand thermal expansion and flame impingement.

2. Enclosing walls of incinerators shall be strongly braced and stayed with structural steel shapes designed to withstand interior thrusts and support door and appurtenant assemblies, except that intermittent duty incinerators not over 85 cubic feet in capacity and burning not over 165 pounds per hour need not have a steel frame if otherwise constructed to conform to Paragraph (a) above.

3. Fire brick shall be laid in high temperature cement or fire clay mortar. All common brickwork shall be laid with full, push-filled, cross and bed mortar joints.

4. Metal stays, lintels or other supports shall not be exposed to the interior of the combustion area.

5. Incinerators used for the burning of rubbish or other readily combustible solid waste material shall include effective means for arresting sparks and fly particles, such as an expansion chamber, baffle walls, or other effective arrangement, or the flues or chimneys of such incinerators shall be provided with an approved spark arrestor having openings not greater than ½ inch.

(B) Mounting

Incinerators shall be set on proper foundations on the ground or on fire resistive floors with no combustible material on the underside thereof.

(C) Chimney Connections

Commercial and Industrial Type incinerators shall be connected to chimneys as specified in sections 803.6, C or 804.5, B. Chimney Connectors (Breechings) shall be constructed and installed in accordance with the applicable requirements of of Section 806.

809.3 — CLEARANCES, SERVICE OPENINGS, ETC.

For additional requirements see NFPA Standard No. 82 "Standard on Incinerators—1972."

SECTION 810 – BARBECUES

810 — BARBECUES

(a) Factory built barbecues shall be approved and installed in accordance with the manufacturers instructions.

(b) Masonry built barbecues shall meet the applicable requirements of Section 808.3.

SECTION 811 – VENTING OF GAS APPLIANCES

Gas appliances shall be vented in accordance with the Southern Standard Gas Code.

TABLE 811-A — STANDARD INSTALLATION CLEARANCES, INCHES, FOR HEAT PRODUCING APPLIANCES

TABLE 811-A—STANDARD INSTALLATION CLEARANCES, INCHES, FOR HEAT PRODUCING APPLIANCES (See Note 1.)

These clearances apply unless otherwise shown on listed appliances. Appliances should not be installed in alcoves or closets unless so listed. For installation on combustible floors, see note 2.

Residential Type Appliances For Installation In Rooms Which Are Large (See Note 3)		APPLIANCE				
		Above Top of Casing of Appliance	From Top and Sides of Warm-Air Bonnet or Plenum	From Front See Note 4	From Back	From Sides
Boilers and Water Heaters Steam Boilers—15 psi Water Boilers—250° F Water Heaters—200° F All Water Walled or Jacketed	Automatic Oil or Comb. Gas-Oil	6	—	24	6	6
	Solid	6	—	48	6	6
Furnaces—Central Gravity, Upflow, Downflow, Horizontal and Ducts. Warm-Air 250° F Max.	Automatic Oil or Comb. Gas-Oil	6[5]	6[5]	24	6	6
	Solid	18[6]	18[6]	48	18	18
	Electric	6[5]	6[5]	18	6	6
Furnaces—Floor For Mounting in Combustible Floors	Automatic Oil or Comb. Gas-Oil	36	—	12	12	12
Heat Exchanger Steam—15 psi Max. Hot Water—250° F Max.	—	1	1	1	1	1

Table 811-A — Standard Installation Clearances, Inches, for Heat Producing Appliances (Cont.)

Residential Type Appliances For Installation In Rooms Which Are Large (See Note 3)		APPLIANCE: Above Top of Casing of Appliance	From Top and Sides of Warm-Air Bonnet or Plenum	From Front See Note 4	From Back	From Sides	
Room Heaters Circulating Type Vented or Unvented	Oil or Solid	36	—	24	12	12	
Radiant or Other Type Vented or Unvented	Oil or Solid	36	—	36	36	36	
Ranges—Cooking Stoves		See Note 7				Firing Side	Opp. Side
Vented or Unvented	Oil	30	—	—	9	24	18
	Solid—Clay lined Firepot	30	—	—	24	24	18
	Solid—Unlined Firepot	30	—	—	36	36	18
	Electric	30	—	—	6	6	
Clothes Dryers Listed Types	Electric	6	—	24	0	0	
Incinerators Domestic Types	—	See Note 10 36	—	48	36	36	

Table 811-A — Standard Installation Clearances, Inches, for Heat Producing Appliances (Cont.)

Commercial-Industrial Type Low-Heat Appliances Any and All Physical Sizes Except As Noted		APPLIANCE				
		Above Top of Casing or Appliance See Note 8	From Top and Sides of Warm-Air Bonnet or Plenum	From Front	From Back See Note 8	From Sides See Note 8
Boilers and Water Heaters						
100 cu. ft. or less Any psi Steam	All Fuels	18	—	48	18	18
50 psi or less Any Size	All Fuels	18	—	48	18	18
Unit Heaters						
Floor Mounted or Suspended—Any Size	Steam or Hot Water	1	—	—	1	1
Suspended— 100 cu. ft. or less	Oil or Comb. Gas-Oil	6	—	24	18	18
Suspended— 100 cu. ft. or less						
Suspended— Over 100 cu. ft.	All Fuels	18	—	48	18	18
Floor Mounted Any Size	All Fuels	18	—	48	18	18
Ranges—Restaurant Type						
Floor Mounted	All Fuels	48	—	48	18	18
Other Low-Heat Industrial Appliances						
Floor Mounted or Suspended	All Fuels	18	18	48	18	18

Table 811-A — Standard Installation Clearances, Inches, for Heat Producing Appliances (Cont.)

Commercial-Industrial Type Medium-Heat Appliances		APPLIANCE				
		Above Top of Casing or Appliance See Note 9	From Top and Sides of Warm-Air Bonnet or Plenum	From Front	From Back See Note 9	From Sides See Note 9
Boilers and Water Heaters Over 50 psi Over 100 cu. ft.	All Fuels	48	—	96	36	36
Other Med.-Heat Industrial Appliances All Sizes	All Fuels	48	36	96	36	36
Incinerators All Sizes	—	48	—	96	36	36
Industrial Type High-Heat Appliances						
High-Heat Industrial Appliances All Sizes	All Fuels	180	—	360	120	120

NOTES TO TABLE 811-A

1. Standard clearances may be reduced by affording protection to combustible material in accordance with Table 811-B.

2. An appliance may be mounted on a combustible floor if the appliance is listed for installation on a combustible floor or if the floor is protected in an approved manner. For details of protection reference may be made to the Code for the installation of Heat Producing Appliances, obtainable from the National Board of Fire Underwriters, 85 John Street, New York, N. Y. 10038, or Part 6 of the National Building Code of Canada published by the National Research Council, Ottawa, Canada.

3. Rooms which are large in comparison to the size of the appliance are those having a volume equal to at least 12 times the total volume of a furnace and at least 16 times the total volume of a boiler. If the actual ceiling height of a room is greater than 8 ft., the volume of a room shall be figured on the basis of a ceiling height of 8 ft.

4. The minimum dimension should be that necessary for servicing the appliance including access for cleaning and normal care, tube removal, etc.

5. For a listed oil, combination gas-oil, gas, or electric furnace this dimension may be 2 in. if the furnace limit control cannot be set higher than 250° F or this dimension may be 1 in. if the limit control cannot be set higher than 200° F.

6. The dimension may be 6 in. for an automatically stoker-fired forced warm-air furnace equipped with 250° F limit control and with barometric draft-control operated by draft intensity and permanently set to limit draft to a maximum intensity of 0.13 in. water gauge.

7. To combustible material or metal cabinets. If the underside of such combustible material or metal cabinet is protected with asbestos millboard at least 1/4 in. thick covered with sheet metal of not less than No. 28 gauge the distance may be not less than 24 in.

8. If the appliance is encased in brick, the 18 in. clearance above and at sides and rear may be reduced to not less than 12 in.

9. If the appliance is encased in brick the clearance above may be not less than 36 in. and at sides and rear may be not less than 18 in.

10. Clearance above charging door should be not less than 48 in.

Table 811-B. Clearances, Inches, with Specified Forms of Protection*

Type of Protection	Where the required Clearance with no protection is:											
Applied to the combustible material unless otherwise specified and covering all surfaces within the distance specified as the required clearance with no protection (see Fig. 1). Thicknesses are minimum.	36 inches			18 inches			12 inches		9 inches	6 inches		
	Above	Sides & Rear	Chimney or Vent Con-nector	Above	Sides & Rear	Chimney or Vent Con-nector	Above	Sides & Rear	Chimney or Vent Con-nector	Above	Sides & Rear	Chimney or Vent Con-nector
(a) ¼ in. asbestos millboard spaced out 1 in.†	30	18	30	15	9	12	9	6	6	3	2	3
(b) 28 gage sheet metal on ¼ in. asbestos millboard	24	18	24	12	9	12	9	6	4	3	2	2
(c) 28 gage sheet metal spaced out 1 in.†	18	12	18	9	6	9	6	4	4	2	2	2
(d) 28 gage sheet metal on ⅛ in. asbestos millboard spaced out 1 in.†	18	12	18	9	6	9	6	4	4	2	2	2
(e) 1½ in. asbestos cement covering on heating appliance	18	12	36	9	6	18	6	4	9	2	1	6
(f) ¼ in. asbestos millboard on 1 in. mineral fiber bats reinforced with wire mesh or equivalent	18	12	18	6	6	6	4	4	4	2	2	2
(g) 22 gage sheet metal on 1 in. *mineral fiber bats reinforced with* wire or equivalent	18	12	12	4	3	3	2	2	2	2	2	2
(h) ¼ in. asbestos cement board or ¼ in. asbestos millboard	36	36	36	18	18	18	12	12	9	4	4	4
(i) ¼ in. cellular asbestos	36	36	36	18	18	18	12	12	9	3	3	3

* Except for the protection described in (e), all clearances should be measured from the outer surface of the appliance to the combustible material disregarding any intervening protection applied to the combustible material.

† Spacers should be of noncombustible material.

Fig. 1. Extent of Protection Required to Reduce Clearances

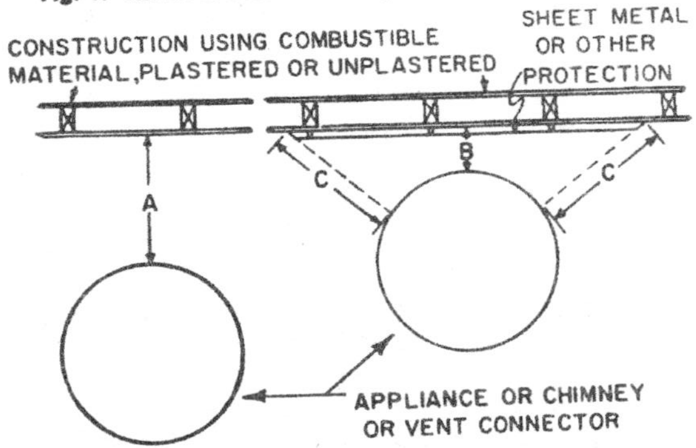

A equals the required clearance with no protection, specified in Table 811-A.

B equals the reduced clearance permitted in accordance with Table 811-B. The protection applied to the construction using combustible material should extend far enough in each direction to make C equal to A.

TABLE 811-C

Chimney Connector and Vent Connector Clearances from Combustible Materials

Description of Appliance	Minimum Clearance, Inches (See Note 1)
RESIDENTIAL TYPE APPLIANCES	
Column 1, Table 805-A	
Single-Wall, Metal Pipe Connectors	
Gas Appliances Without Draft Hoods	18
Electric, Gas, and Oil Incinerators	18
Oil and Solid-Fuel Appliances	18
Unlisted Gas Appliances With Draft Hoods	9
Boilers and Furnaces Equipped With Listed Gas Burners and With Draft Hoods	9
Oil Appliances Listed as Suitable For Use With Type L Venting Systems, but only when connected to chimneys	9
Listed Gas Appliances With Draft Hoods. See Note 3.	6
Type L Venting System Piping Connectors	
Gas Appliances Without Draft Hoods	9
Electric, Gas, and Oil Incinerators	9
Oil and Solid-Fuel Appliances	9
Unlisted Gas Appliances With Draft Hoods	6
Boilers and Furnaces Equipped With Listed Gas Burner and With Draft Hoods	6
Oil Appliances Listed as Suitable For Use With Type L Venting Systems	(See Note 2)
Listed Gas Appliances With Draft Hoods	(See Note 3)
Column 1, Table 805-B	
Type B Gas Vent Piping Connectors	
Listed Gas Appliances With Draft Hoods	(See Note 3)

TABLE 811-C (Cont.)

Description of Appliance	Minimum Clearance, Inches (See Note 1)
COMMERCIAL-INDUSTRIAL TYPE APPLIANCES	
Low-Heat Appliances Column 2, Table 805-A	
Single-Wall Metal Pipe Connectors	
Gas, Oil, and Solid Fuel Boilers, Furnaces, and Water Heaters	18
Ranges, Restaurant Type	18
Oil Unit Heaters	18
Unlisted Gas Unit Heaters	18
Listed Gas Unit Heaters With Draft Hoods	6
Other Low-Heat Industrial Appliances	18
Medium-Heat Appliances Column 3, Table 805-A	
Single-Wall Metal Pipe Connectors	
All Gas, Oil, and Solid-Fuel Appliances	36

NOTE 1: These clearances apply except if the listings of an appliance specifies different clearance, in which case the listed clearance takes precedence.

NOTE 2: If listed Type L venting system piping is used, the clearance may be in accordance with the venting system listing.

NOTE 3: If listed Type B or Type L venting system piping is used, the clearance may be in accordance with the venting system listing.

The clearances from connectors to combustible materials may be reduced if the combustible material is protected in accordance with Table 811-B.

SECTION 812 – FURNACE AND BOILER ROOM

(a) Every Central Heating Boiler and/or Furnace with an input capacity of 250,000 B.T.U. (British Thermal Units) or over, installed in any building, other than a one or two-family dwelling, shall be enclosed and separated from the rest of the building by walls, partitions, floor and ceiling of not less than one-hour fire resistive construction. Not more than two Central Heating Boilers and/or Furnaces shall be permitted in any one tenancy in any building unless all are enclosed and separated by walls, partitions, floors and ceilings of one-hour fire resistive construction.

(b) A Central Heating Boiler and/or Furnace of any input capacity installed in a building having a capacity for its particular use of more than seventy-five persons, or Group 'D' Institutional Occupancy, or Group 'E' Theater or Assembly Occupancy, or Group 'H' Special Hazardous Occupancy, shall be separated from the rest of such building by walls or partitions, floor and ceiling construction having a fire resistive rating of not less than two hours, except as required below.

(c) Every steam boiler carrying more than fifteen (15) pounds per square inch pressrre with a rating in excess of ten (10) boiler horsepower, installed in a building other than one of Group G, Industrial Occupancy, shall be located in a separated room, or compartment, and separated from the rest of the building by walls or partitions having at least two-hour fire-resistance and by floor or ceiling construction having not less than two and one-half hour fire-resistance, provided, however, that when in the opinion of the Building Official it is desirable to provide for the venting of a possible explosion upward this rating may be reduced in accordance with the hazard existing.

SECTION 813 — STEAM AND HOT WATER

813.1 — STEAM AND HOT WATER PIPES

(a) Steam pipes and hot water heating pipes shall be installed with a clearance of at least one (1) inch to all combustible construction or material, except that at the points where pipes carrying steam or hot water at not over fifteen (15) pounds gauge pressure emerge from a floor, wall or ceiling the clearance at the opening through the finish floor boards or wall or ceiling boards may be less than one (1) inch but not less than one-fourth (¼) inch. Each such opening shall be covered with a plate of non-combustible material.

(b) Such pipes passing through stock shelving shall be covered with not less than one (1) inch of approved insulation.

(c) Wooden boxes or casings enclosing steam or hot water pipes, or wooden covers to recesses in walls in which such pipes are placed, shall be lined with metal or asbestos millboard.

813.2 — PIPE COVERINGS

Covering or insulation used on steam or hot water pipes shall be of non-combustible material.

813.3 — SEALING OPENING AROUND PIPES

Where heating pipes pass through floors or partitions of fire-rated construction, the openings around them shall be filled with non-combustible material to prevent the passage of fire.

SECTION 814 — FIRE PROTECTION OF MISCELLANEOUS HEATING DEVICES

814.1 — COOKING EQUIPMENT

(a) Cooking equipment used in commercial, industrial, institutional and similar cooking applications shall be equipped with an exhaust system installed, protected and maintained in accordance with the standards of "Removal of Smoke and Grease-Laden Vapors from Commercial Cooking Equipment, NFPA 96-1971".

(b) When required as part of the assembly, ventilation ducts from residential range hoods shall exhaust to the outside air through an independent system.

814.2 — BURNING SOLID OR LIQUID FUELS

(a) Every cooking stove, range, water heater or any other heating appliance burning solid or liquid fuel without an open space beneath its base, shall be securely and firmly set on not less than two-hour fire-resistive construction as set forth in Chapter X. Such construction shall extend not less than twelve (12) inches beyond the appliance on all sides and not less than eighteen (18) inches on side from which ashes are removed.

(b) Where appliances approved for installation on floors of combustible construction are set on supports providing not less than four (4) inches open space under the base, are mounted on wood floors, such floors shall be protected with not less than No. 24 U. S. Gauge sheet metal extending at least six (6) inches beyond the appliance on all sides and eighteen (18) inches on side where ashes are removed.

(c) Every heating appliance burning solid or liquid fuel shall be connected to a chimney. (See Section 802.1 (a).

814.3 — CLEARANCES

There shall be a minimum clearance between combustible material and heating appliances burning solid or liquid fuel in accordance with Table 811-A.

SECTION 815 — AUXILIARY EQUIPMENT

Auxiliary equipment used in connection with combustion equipment shall be installed in accordance with good engineering practice and in accordance with the manufacturers recommendations.

SECTION 816 — COMBUSTION EQUIPMENT — GENERAL

816.1 — CAPACITY OF COMBUSTION EQUIPMENT

Combustion equipment shall be of adequate capacity to serve the load without violating good engineering practice.

816.2 — AIR FOR COMBUSTION (SOLID AND LIQUID FUELS)

(a) Combustion equipment shall be installed in a location in which the facilities for ventilation permits satisfactory combustion of the fuel and proper elimination of the products of combustion under normal conditions of use.

(b) The standards set forth in this section define the minimum requirements of air for combustion. These standards shall not be decreased, however, manufacturers specifications shall be followed where their requirements exceed these standards. The minimum dimension of rectangular air ducts shall be not less than three (3) inches.

(c) Operations of exhaust fans, kitchen ventilating systems or fireplaces may create conditions requiring special attention to avoid unsatisfactory combustion equipment operation.

(d) For combustion equipment with an in-put of 400,000 BTU/hr or less located in an equipment room, there shall be provided two openings for supplying combustion air, one to be located near the floor and one to be located near the ceiling of the equipment room, each having an area of not less than one square inch per 1,000 BTU/hr of in-put rating. Such opening shall be directly connected to the outside except as otherwise may be specifically approved by the Building Official.

(e) The size of combustion air openings specified in 809.2 (a) shall not necessarily govern any installation having an in-put in excess of 400,000 BTU per hour. Sizes of openings for such installations shall be in accordance with good engineering practice.

(f) These standards establish requirements for combustion air only, where necessary to prevent space temperatures exceeding 120 degrees Fahrenheit additional ventilation air shall be supplied to the equipment room. Under no circumstances shall equipment rooms be ventilated by any means which would reduce pressures in the space sufficiently to draw gases of combustion from the furnace into the equipment room.

816.3 — ACCESSIBILITY

Combustion equipment shall be located so that it will be readily accessible for operation and servicing.

816.4 — GAS APPLIANCES

Unless otherwise specifically noted, gas appliances shall be installed in accordance with the Southern Standard Gas Code.

816.5 — PERMISSIBLE TEMPERATURES ON COMBUSTIBLE MATERIAL

(a) All combustion equipment and their chimney connectors shall be installed so that operation will not create a hazard to persons or property.

(b) Minimum clearances between combustible walls and the back and sides of various conventional types of combustion equipment and their chimney connectors shall be not less than as specified in Table 811-A and Table 811-B.

816.6 — SEALED COMBUSTION SYSTEM APPLIANCE

Sealed combustion system appliances shall be listed and shall be installed in accordance with their listings and the manufacturer's instructions.

SECTION 817 – WATER HEATERS AND HOT WATER STORAGE TANKS – BURNING SOLID OR LIQUID FUEL

817.1 — PROHIBITED INSTALLATIONS

Water heaters, with the exception of those having sealed combustion systems, shall not be installed in bathrooms or bedrooms. However, water heaters of the automatic storage type may be installed as a replacement in a bathroom when specifically authorized by the administrative authority, provided they are properly vented and are supplied with adequate combustion air.

817.2 — LOCATION

Water heaters shall be located as close as practicable to the chimney or vent. Water heaters shall be located in an approved ventilated and lighted space so as to be accessible for cleaning and repairing.

817.3 — CLEARANCE

Water heaters shall be positioned in relation to combustible construction with a clearance in accordance with Table 811-A. In no case shall the clearances be such as to interfere with the requirements for combustion air and accessibility.

817.4 — CONNECTIONS

Water heaters shall be connected in a manner to permit observation, maintenance and servicing.

817.5 — WATER PRESSURE RELIEF VALVES AND TEMPERATURE RELIEF VALVES REQUIRED

See Southern Standard Plumbing Code. (Chapter XII)

817.6 — CONNECTION TO BOILER OR TANK

The method of connecting a circulating water heater to the tank shall assure proper circulation of water through the heater, and permit a safe and useful temperature of water to be drawn from the tank.

817.7 — SEDIMENT DRAINS

A suitable water valve or cock, through which sediment may be drawn off or the tank emptied, shall be installed at the bottom of the tank and heater.

817.8 — ANTI-SYPHONING DEVICES

Means acceptable to the Building Official shall be provided to prevent syphoning in any boiler or tank to which any water heater is attached. A cold water tube with a hole near the top is commonly accepted for this purpose.

SECTION 818 — ROOM OR SPACE HEATERS SOLID OR LIQUID FUEL

818.1 — CLEARANCE

(a) A room or space heater shall be placed so as not to cause a hazard to walls, floors, curtains, furniture, doors when open, etc., and to the free movements of persons within the room. Appliances designed and marked "For use in non-combustible fire-resistive fireplace only," shall not be installed elsewhere. Room or space heaters shall be installed with clearances not less than specified in Table 811-A, except that appliances approved for installations at lesser clearances may be installed in accordance with their approval. In no case shall the clearances be such as to interfere with the requirements of combustion air and accessibility.

(b) There shall be a minimum clearance between combustible construction for room or space heaters in accordance with Table 811-A.

818.2 — WALL TYPE ROOM HEATERS (SOLID OR LIQUID FUEL)

Wall type room or space heaters shall not be installed in walls of combustible construction unless specifically approved for such installation.

SECTION 819 — CENTRAL HEATING BOILERS AND FURNACES SOLID OR LIQUID FUEL

819.1 — CLEARANCES

(a) Central heating boilers and furnaces shall be installed with clearances not less than specified in Table 811-A except that appliances approved for installation at lesser clearances may be installed in accordance with their approval. In no case shall the clearance be such as to interfere with the requirements for combustion air and accessibility.

(b) There shall be a minimum clearance between combustible construction for central heating boilers and furnaces in accordance with Table 811-A.

819.2 — ERECTING AND MOUNTING

A central heating boiler or furnace shall be erected in accordance with the manufacturer's instructions and shall be installed on a firm level, fire-resistive floor unless approved for installation on a combustible floor, or the floor is protected in a manner acceptable to the Building Official.

819.3 — ACCESSIBILITY

The installation of central heating boilers and furnaces shall be such as to make them accessible for cleaning of heating surfaces, removal of burners and/or stokers, replacement of sections, motors, controls, filters, and other working parts and for adjustment and lubrication of parts requiring such attention.

819.4 — CONNECTION OF FLOW AND RETURN PIPES TO BOILERS

The method of connecting the flow and return pipes on steam and hot water boilers shall facilitate a rapid circulation of steam or water. For common piping systems reference may be made to the American Society of Heating, Refrigeration and Air Conditioning Engineer's Guide and Data Book, and to the Institute of Boiler and Radiator Manufacturer's (IBR) Guides.

819.5 — FEED WATER AND DRAIN CONNECTION TO BOILERS

A steam or hot water boiler shall be provided with a direct connection to a water supply through an individual control valve. A drain valve, by means of which the boiler may be flushed or drained, shall also be provided.

819.6 — TEMPERATURE AND PRESSURE CONTROL DEVICES AND PRESSURE RELIEF DEVICES ON BOILERS

(a) Steam and hot water boilers shall be provided with automatic devices, approved by the Building Official, to shut down the burners and/or stokers in the event of undue pressure or low water or in the event of overheating in a hot water boiler. Each automatically fired steam or hot water boiler shall be provided with a high limit shutoff and a separate operating shutoff in conjunction with other controls for the firing or operating controls.

(b) Steam and hot water boilers shall be fully safeguarded by means of ASME approved pressure relief devices complying with the ASME Boiler Code.

819.7 — PLENUM CHAMBERS AND AIR DUCTS

A plenum chamber when not a part of a furnace shall be constructed in accordance with the manufacturer's instructions. Reference may be made to the Standard for installation of "Air Conditioning and Ventilating Systems—NFPA 90A-1971" and standards for the installation of "Residence Type Warm Air Heating and Air Conditioning Systems, NFPA 90B-1971," and to the Design and Installation

Manuals of the National Warm Air Heating and Air Conditioning Association.

819.8 — RETURN AIR CONNECTIONS

Return air connections in the furnace room or directly below the furnace room shall be made so as to preclude any possibility of air being drawn from the furnace room into the return air duct by action of the furnace fan.

819.9 — FURNACES USED WITH REFRIGERATION SYSTEMS AND BLOWERS

A furnace when used in conjunction with a refrigeration system shall be installed on a discharge side of the blower. The furnace shall be installed parallel with or on the air inlet side of the refrigeration coil. If the refrigerating coil is installed in the air discharge duct from the furnace, this coil shall be installed in or provided with a by-pass duct unless the furnace is listed for operation with at least 0.5" static pressure at the outlet air connection.

819.10 — DUCT CONNECTIONS

All flexible duct connections shall be made of fire-resistant materials.

SECTION 820 – FLOOR FURNACES (SOLID OR LIQUID FUEL)

820.1 — INSTALLATION

Floor furnaces shall not be installed in floors of combustible construction unless approved specifically for such installation by a recognized laboratory and are installed in accordance with the conditions of such approval.

820.2 — COMBUSTION AIR

Fixed ventilation by means of a duct or grill arranged to supply air from a permanently ventilated attic or underfloor space, shall be provided in accordance with Section 816.2.

820.3 — PLACEMENT

The following are requirements that will serve in properly placing the furnace or furnaces to serve one story:

(a) Floor furnaces shall not be installed in the floor or any aisle or passageway of any auditorium, public hall, or place of assembly, or in any exitway from any such room or space.

(b) Walls and Corner — With the exception of wall-register models, a floor furnace shall not be placed closer than six (6) inches to a wall or corner.

(c) The furnace shall be so placed that a drapery or similar object cannot be nearer than twelve inches to any portion of the furnace. Floor furnace shall not be installed in concrete floor construction built on ground. Doors that can be closed shall not be permitted between the floor furnace installation and the controlling thermostat, unless a high limit control thermostat is installed in the same room as the floor furnace; nor shall any floor furnace be installed in such a manner that will permit any door to swing over it.

820.4 — PROHIBITED LOCATION

Floor furnaces shall not be installed above the ground floor of any building.

820.5 — BRACING

The floor around the furnace shall be braced and headed with an adequate frame work of material not smaller in dimension than the joists.

820.6 — SUPPORT

Means shall be provided to support the furnace when the floor grille is removed.

820.7 — CLEARANCE

The lowest portion of the floor furnace shall have at least a twelve (12) inch clearance from the general ground level. When this clearance is not present, the ground below and to the sides shall be excavated to form a "basin-like" pit under the furnace so that the required clearance is provided beneath the lowest portion of the furnace. Said pit shall have masonry or concrete floors and walls, with walls being not less than four (4) inches above ground level and a self-operating floor drain installed. Not less than an eighteen (18) inch clearance shall be provided on all sides except the control side, which shall have a minimum twenty-four (24) inch clearance.

820.8 — ACCESS

Provision shall be made for access to the floor furnace by means of an opening in the foundation wall of at least twenty-four (24) by thirty (30) inches or through a trap door of at least twenty-four (24) by thirty (30) inches, located at some convenient point, and a clear and unobstructed passageway to the floor furnace of at least twenty-four (24) inches high by thirty (30) inches wide.

820.9 — WIND PROTECTION

Floor furnaces shall be protected, where necessary, against severe wind conditions.

820.10 — CHIMNEY OR VENT CONNECTORS

Except as hereinafter otherwise provided, chimney or vent connectors shall be installed with clearances to woodwork or other combustible material whether plastered or unplastered, of not less than eighteen (18) inches.

SECTION 821 — ATTIC FURNACES (SOLID OR LIQUID FUEL)

Heating furnaces shall not be installed in attics unless specific written approval has been given by the Building Official, and such installation shall be made in accordance with terms of such approval.

SECTION 822 — UNIT HEATERS, CEILING-TYPE, DIRECT-FIRED (SOLID OR LIQUID FUEL)

822.1 — SUPPORT

Unit heaters shall be safely and adequately supported with due consideration given to their weight and vibration characteristics.

822.2 — CLEARANCES

Except as hereinafter otherwise provided, ceiling-type direct-fired unit heaters shall be installed to provide a clearance in any direction to woodwork or other combustible material whether plastered or unplastered, of not less than eighteen (18) inches.

822.3 — DUCTWORK

A unit heater shall not be attached to a warm air duct system unless approved for such installation.

822.4 — GARAGE INSTALLATION

Unit heaters installed in Commercial garages, (Maintenance, Repair, Storage) or in airplane hangars shall be installed at least eight (8) feet above the floor.

SECTION 823 — UNDERFLOOR HORIZONTAL FURNACES (SOLID OR LIQUID FUEL)

823.1 — INSTALLATION

Underfloor horizontal furnaces shall not be installed unless of a type approved for such installation with clearances provided as specified in Section 823.4.

823.2 — COMBUSTION AIR

Fixed ventilation by means of a duct or grille arranged to supply air from a permanently ventilated underfloor space, shall be provided in accordance with Section 816.2.

823.3 — SUPPORT

The underfloor horizontal furnace shall be braced and headed with a frame-work of material not lighter than the joists or they may be suspended from overhead structural members of adequate strength by means of hanger rods as recommended by the manufacturer.

Underfloor horizontal furnaces shall be supported independently of the floor grills.

823.4 — CLEARANCE

The lowest portion of underfloor horizontal floor furnaces shall have at least six (6) inch clearance from the ground except that where the lower six (6) inch portion of the floor furnace is sealed by the manufacturer to prevent the entrance of water, the clearance may be reduced to not less than two (2) inches. Where the ground must be excavated to provide this clearance, the excavation shall extend at least twelve (12) inches beyond the furnace on all sides and not less than eighteen (18) inches on the control side. Where such excavation exceeds twelve (12) inches or ground contour and ground moisture conditions are such that water seepage is likely, a water-tight pan constructed of copper, galvanized iron or other suitable corrosion-resistant material, and properly anchored in place, or a waterproof concrete pit, shall be provided under the furnace. The side of the pan or pit shall extend four (4) inches above ground level. When equipment is sealed by the manufacturer to meet this condition the pan or pit may be omitted if not required for maintaining a dry condition for service access.

823.5 — ACCESS

Provision shall be made for access to the underfloor horizontal furnace by means of an air opening in the foundation wall of at least 18" x 24" or through a trap door at least 18" x 24" located at some convenient point, and a clear unobstructed passageway to the furnace of at least 24" high by 24" wide. In all cases, access opening and passageway shall be of sufficient height and width to permit removal of the furnace.

823.6 — WIND PROTECTION

Underfloor horizontal furnaces shall be protected, where necessary, against severe wind conditions.

SECTION 824 – HEATING, AIR CONDITIONING AND VENTILATING DUCTS

This section applies to air conditioning, warm air heating, air cooling and ventilating systems employing either gravity or mechanical means for the movement of air through ducts. For ventilating systems for the removal of dust, smoke, fumes, gases, vapors, odors or other hazardous, obnoxious, or injurious impurities see Section 2001.

824.1 — MATERIALS

(a) All air ducts shall be constructed entirely of iron, steel or other approved non-combustible material. Only approved non-combustible material or approved fire-resistive linings shall be used inside of ducts.

(b) Work involving torches shall not be undertaken on ducts until the system has been shut down, the duct cleaned and all combustible lining and covering material has been removed from the portion being repaired.

(c) Ducts shall be so constructed as to provide ample strength to meet the conditions of service for which they are used. In no case shall their structural strength and durability be less than the equivalent of galvanized sheet iron or steel of the thickness indicated in Table 824. Wire glass may be used for inspection windows in ducts.

TABLE 824 — THICKNESS OF METAL FOR AIR DUCTS

(a) Residential

	Ducts Not Enclosed In Partitions Round Ducts	
Diameter Inches	Minimum Thickness Galv. Sheet Gauge	Minimum Thickness Aluminum B & S Gauge
14 or less	30	26 (.020)
Over 14	28	24 (.025)

	Rectangular Ducts	
Width Inches	Minimum Thickness Galv. Sheet Gauge	Minimum Thickness Aluminum B & S Gauge
14 or less	28	24 (.025)
Over 14	26	23 (.023)

(Table continued on next page)

Table 824—Continued

	Rectangular Ducts Enclosed in Partitions	
Width Inches	Minimum Thickness Galv. Sheet Gauge	Minimum Thickness Aluminum B & S Gauge
14 or less	30	26 (.020)
Over 14	28	24 (.025)

(b) Other than Residential

Duct Dimension Longest Side Inches	*LOW PRESSURE	
	Gauge Sheet Galv.	Aluminum B & S Gauge
Thru 12"	26	24 (.020)
13" thru 18"	24	22 (.025)
19" thru 30"	24	22 (.025)
31" thru 42"	22	20 (.032)
43" thru 54"	22	20 (.032)
55" thru 60"	20	18 (.040)
61" thru 84"	20	18 (.040)
85" thru 96"	18	16 (.051)
Over 96"	18	16 (.051)

Duct Dimension Longest Side Inches	Med. Pressure	High Pressure
	Galv. Sheet Gauge	Galv. Sheet Gauge
Thru 12"	24	22
13" thru 18"	24	22
19" thru 24"	22	22
25" thru 36"	22	22
37" thru 48"	22	22
49" thru 60"	20	20
61" thru 72"	20	20
73" thru 84"	18	18
85" thru 96"	18	18
97" and over	18	16

(Table continued on next page)

Table 824—Continued

HIGH VELOCITY ROUND DUCT—MEDIUM AND HIGH PRESSURE

Duct Diameter	Galvanized Steel Sheet Gauge		
	Spiral Lock Seam Duct	Longitudinal Seam Duct	Round Duct Fittings
Up thru 8"	26	24	22
9" thru 22"	24	22	20
23" thru 36"	22	20	20
37" thru 50"	20	20	18
51" thru 60"	—	18	18
61" thru 84"	—	16	16

*The Static pressures in the duct systems for the classifications are: Low pressure, not over 2" of water; Medium pressure, through 6" of water; High pressure, up to 10" of water. It shall be noted that these values are the pressures in the ducts, not the total pressure at the fan.

824.2 — FIRE PROTECTION

(a) If ducts are formed by parts of the building structure, their construction shall consist of not less than three-quarter (¾) inch cement or gypsum plaster on metal lath or equivalent protection, applied to those parts, if such parts are combustible.

(b) Ducts shall be tight throughout with no openings except those essential to the required functions of the system. Ducts shall be substantially supported by metal hangers or brackets.

(c) Where ducts pass through walls, floors or partitions, the space around the duct shall be sealed with rope asbestos, mineral wool or other non-combustible material to prevent the passage of flame and smoke.

(d) Where ducts pass through concealed ceiling spaces of combustible construction or are located inside combustible partitions or walls, either the ducts or the interior surface of such enclosing space shall be protected with one-quarter (¼) inch asbestos or other equivalent approved insulating material, or a minimum clearance of one-half (½) inch shall be provided. However, warm air ducts which run approximately horizontal near combustible construction which is either plastered or unplastered, shall have a clearance of not less than six (6) inches except that the minimum clearance shall be one (1) inch where ducts are of metal covered with one-half (½) inch or more of approved non-combustible insulating material.

(e) Where ducts pass through floors, (except in the case of dwellings) fire protection shall be provided as specified in Section

701—Protection of Vertical Openings. Such construction, however, shall not be required for branches which are cut off from the main duct by approved fire dampers.

824.3 — RETURN AND SUPPLY DUCTS

(a) Return ducts, and supply ducts, unless air passes through either water spray or filters, other than vertical, shall be so constructed that their interior is accessible for cleaning except that the Building Official may waive this requirement if the occupancy is not productive of lint, greasy vapors, or other combustibles. (For occupancies such as banks, office buildings, churches, institutions and hotels, the requirements may be waived except for kitchens, service rooms, and rooms used for manufacturing purposes.)

(b) Ducts shall not pass through fire walls unless unavoidable. When ducts or the outlets from or inlets to them pass through fire walls, they shall be provided with automatic fire doors approved for the protection of openings in fire walls (Class A openings) on both sides of the walls through which they pass. (See "Fire Doors and Windows," NFPA 80-1970 and (Standard for Installation of "Air Conditioning and Ventilating Systems, NFPA 90A-1971.")

(c) Fresh air intakes shall be protected against exterior fire and smoke exposure by approved doors, dampers or other protectives.

(d) A room, attic, void, hollow or concealed space shall not be used as an integral part of a duct system, unless the component parts of such space is constructed entirely of non-combustible material, properly protected by means of fire-damper or controls so as to restrict the spread of fire, and arranged so as to protect the fire-resistiveness of the assembly.

(e) Public corridors shall not be used as a return plenum in hotels, multi-family residences and institutional occupancies.

(f) Structural members of a building may be used as a duct to carry air provided such members are designed to fully satisfy the structural and fire requirements of this code, and are of non-combustible material and are properly designed, engineered and installed to provide a continuous duct of proper size.

(g) Duct systems shall be constructed and installed to meet the Standard for Installation of of "Air Conditioning and Ventilating Systems, NFPA 90-A-1971" and standards for the installation of "Residence Type Warm Air Heating and Air Conditioning Systems, NFPA 90-B-1971," and the "Duct Manual and Sheet Metal Construction for Ventilating and Air Conditioning Systems" by the Sheet Metal and Air Conditioning Contractors National Association, Inc. (SMACNA), except as otherwise provided herein.

(h) In systems other than residence type, discharge and exhaust air openings and re-circulating air intakes shall be located at least 3 inches above the floor except as allowed hereinafter. When located less than 7 feet above the floor, inlet and outlet openings shall be protected by a substantial grille or screen, through the openings in which a half inch sphere will not pass.

Exceptions:

1. Protected floor inlets may be permitted, under seats, in theatres.

2. Supply registers or grilles may be located in the floor discharging air upward provided the following conditions are met:

   (aa) The connecting ductwork is constructed of metal as previously specified.

   (bb) The duct lines, if used, adhesives, and accessories shall have a flame spread rating of not over 25 and with a smoke developed rating not over 50.

   (cc) The registers or grilles shall be protected so as not to pass a half-inch sphere through the openings.

   (dd) There shall be no direct-fired heat exchanger of any nature in the supply air system to the floor registers or grilles, and that the system shall not operate with an air temperature exceeding 150° F. entering the supply duct to the registers and grilles.

   NOTE: Direct-fired heat exchangers shall include all gas fired units such as furnaces, unit heaters, duct heaters, and shall also be considered to include "open-wire" element, "bare" element, "sheathed" element or "encased" element in electric duct heaters. Low-pressure steam coils supplied by low-pressure steam boilers having a gauge pressure of not over 15 pounds and hot and cold water circulating coils are permitted in the supply air system to the floor registers and grilles.

   (ee) The space in which the registers are located does not present a hazardous condition when so served by floor registers or grilles.

(i) Where ducts penetrate ceiling membrane fire protection which is a required part of the floor/ceiling or roof/ceiling assembly, openings shall be protected with approved fire dampers or duct opening protection and fire dampers shall be in compliance with the designs listed by an approved laboratory.

824.4 — UNDER FLOOR SLAB AIR DUCTS — GROUND LEVEL

(a) Where air ducts are used under concrete floor slabs at ground level, careful preparation of site shall be exercised to insure a well-drained and continuously dry condition. All organic material shall be removed.

(b) The slab shall be laid over a coarse fill of not less than six (6) inches in thickness. In no case shall sand, cinders or unscreened rock or gravel be used as a filler.

(c) Ducts shall be of non-combustible material, corrosive-resistant and shall have a compressive strength of not less than three hundred (300) pounds per lineal foot.

821.5 — HEATING AND AIR CONDITIONING DUCT COVERING AND LININGS

(a) All heating supply duct work exposed in unheated attic or crawl spaces shall be insulated.

(b) Covering, insulation, liners, coating, sealers and adhesives used on heating and air conditioning ducts shall be of non-combustible material.

SECTION 825 — PLENUM FLOOR SYSTEMS

825.1 — GENERAL

The use of a crawl space as a plenum shall be restricted as provided herein.

(a) Such spaces shall have not more than 24 inches nor less than 3 inches clearance between the bottom of the floor joists and sand or concrete cover. Such spaces shall be cleaned of all excess combustible material.

(b) The enclosing material of the under-floor space, including the side wall insulation, shall have a flame spread classification of not more than 200.

(c) Access, if provided to such spaces, shall be through an opening in the floor and shall not be greater than 24 inches by 24 inches.

825.2 — CONSTRUCTION PRACTICES

(a) Framing. Framing shall comply with the requirements of this Code.

(b) Decay and Termite Protection. Where required, preservatives shall be of approved water borne type.

(c) Soil Poisoning. Chemical soil treatment shall be applied to both sides of the foundation wall from the footing to the grade level. Water emulsions of approved chemicals shall be used.

(d) All excavations for plumbing and other services shall be completed at the time of the chemical soil treatment, or retreatment shall be necessary.

(e) Vapor Barrier. After the soil has been treated, a vapor barrier shall be provided within the foundation perimeter, from wall to wall, with joints lapped two inches but not sealed. The vapor barrier membrane shall be carefully fitted around pipes and drains and turned up at the foundation wall.

(f) The vapor barrier membrane shall be equal or greater than polyethylene film in 4 mil thickness.

(g) The vapor barrier membrane shall be covered with a 2 inch layer of clean sand or other non-combustible material.

(h) Insulation and Sill Sealer. The foundation wall shall be insulated along its inner face from the sill vertically to the under-floor plenum grade level and horizontally over the vapor barrier, a distance of two feet. The insulation shall provide a maximum heat loss of 35 btu per linear foot of perimeter wall above the under-floor space, assuming a plenum air temperature of 70 degrees for return and 110 degrees Fahrenheit for supply.

(i) A positive seal shall be provided at the junction of the house wall and the foundation.

(j) Outlets. Outlets from the plenum shall be provided by one of the following methods:

(1) An air slot at the base of the exterior wall or interior partitions in accordance with the recommendations of the National Forest Products Association.

(2) Floor registers shall be designed for easy removal in order to give access for cleaning.

(3) Wall registers shall be connected to the plenum space with a duct or boot complying with the requirements of this Chapter.

825.3 — APPLIANCES

(a) The furnace supplying warm air to the plenum space shall be equipped with an automatic control that will start the air circulating fan when the air in the furnace bonnet reaches a temperature not higher than 150°F. Such control shall be one that cannot be set higher than 150° F.

(b) The furnace supplying warm air to such plenum space shall be equipped with an approved temperature limit control that will limit outlet air temperature to 200° F.

(c) Furnaces, boilers or other heat producing appliances shall not be installed in such a plenum system.

CHAPTER IX

SPRINKLERS AND STANDPIPES

SECTION 901 – SPRINKLERS

901.1 — APPROVED EQUIPMENT AND LAYOUT

Only approved sprinklers and devices shall be used in automatic sprinkler systems and the complete layout of the system shall be submitted to the Building Official for approval before installation.

901.2 — REQUIREMENTS

Every automatic sprinkler system required by this Code shall conform with the requirements of the standards for the installation of "Sprinkler Systems NFPA 13-1972," except that a single water supply of adequate pressure, capacity and reliability, equal to the primary supply required by those standards, may be permitted by the Building Official.

901.3 — OCCUPANCY CLASSIFICATIONS

For the purpose and application of, and comparison with, the standards for installing "Sprinkler Systems NFPA 13-1972," occupancies shall be classified as follows:

LIGHT HAZARD
GROUP A, B-1, C and D

ORDINARY HAZARD
GROUP B-2, E, F and G

EXTRA HAZARD
GROUP H

901.4 — MATERIAL

Piping shall be specified in the standards for installation of "Sprinkler Systems NFPA 13-1972."

901.5 — HOSE THREADS

All hose threads in connections shall be uniform with that used by the Fire Department of the City or authority having jurisdiction.

901.6 — GENERAL

Approved automatic sprinkler equipment meeting the requirements of this Section shall be installed in buildings as follows:

(1) Basements or cellars with ceiling less than 4' 6" above grade having floor areas exceeding 2,500 square feet when used as workshops or for the manufacture, repair, sale or storage of combustible materials.

(2) In buildings which do not have suitable access as set forth in Section 703.1 (b) to each story above the basement on at least one accessible side of the building.

901.7 — COMMERCIAL GARAGES

Approved automatic sprinkler systems shall be provided in the following garages:

(1) Enclosed parking garages over 65 feet in height and exceeding 10,000 square feet per floor.

(2) Repair garages two (2) stories or more in height, and exceeding 10,000 square feet per floor, or located below another accupancy.

(3) One story repair garages exceeding 15,000 square feet.

(4) Basement or sub-basement garages below other occupancies having a capacity of more than 3 motor vehicles or exceeding 5,000 square feet in area.

(5) Garages used for the storage of commercial trucks and having an area exceeding 5,000 square feet.

(6) Bus garages three (3) or more stories in height.

901.8 — OTHER OCCUPANCY SPRINKLER REQUIREMENTS

(a)—GROUP "B-2"—BUSINESS—MERCANTILE

An approved automatic sprinkler system shall be provided in stores and similar occupancies where stocks of combustible materials are on display for public sale and where the story floor area exceeds 20,000 square feet.

(b)—GROUP "D-2"—INSTITUTIONAL

Approved automatic sprinkler systems shall be provided in all convalescent or nursing homes two or more stories in height and having more than ten patients.

(c)—GROUP "E-1"—LARGE ASSEMBLY OCCUPANCY

An approved automatic sprinkler system shall be provided in Group E-1 Large Assembly Occupancies over areas which could be used for the display, sale or storage of combustible materials when such display, sale or storage floor area exceeds 20,000 square feet.

901.9 — SUPERVISORY FACILITIES

(a) The automatic sprinkler system shall wherever possible be provided with approved facilities to assure that it is in proper operative condition, such as by electrical connections to a continuously manned central station or fire department headquarters to give automatic notice of any closed water supply valve or other condition that might interfere with the operation of the system; also notice of any flow of water in the system due to fire or other cause. Such facilities shall include provision for immediate alarm to the fire department in case of fire or suspected fire, and appropriate immediate action to restore the sprinkler system to operative condition in case of any impairment.

(b) Subject to the approval of the authorities concerned, sprinkler supervision may also be provided by direct connection to fire departments, or in the case of very large establishments, to a private headquarters providing similar functions.

See "Central Station Protective Signaling Systems, NFPA 71-1972."

"Proprietary Signaling Systems, NFPA 72D-1972."

"Municipal Fire Alarm Systems, NFPA 73-1967."

SECTION 902 – STANDPIPES

902.1 — REQUIREMENTS

Unless otherwise provided herein, standpipes, standpipe systems, hose, water supply, pumps, connections, etc., shall be constructed and installed to meet the requirements of the standard for the installation of "Standpipe and Hose Systems, NFPA 14-1971", except that the single source of water supply, if reliable and capable of automatically supplying the required service, may be approved by the Building Official.

902.2 — APPROVAL

The complete layout of the standpipe and hose system shall be submitted to the Building Official before installation.

902.3 — STANDPIPES REQUIRED

Buildings shall be equipped with standpipes as follows:

(1) Buildings two (2) stories or more and 50 feet in height.

(2) In Group E-1, Large Assembly occupancies, a standpipe outlet with hose attached shall be provided on each side of the rear of each balcony and gallery on each side of the stage, on each tier of dressings rooms, and within 50 feet of all property rooms, store rooms and workrooms. See Section 512.15.

(3) In Group "D"—Institutional occupancies three stories or more in height.

(4) Hotels and motels three stories or more in height.

902.4 — DRY STANDPIPES

In buildings requiring wet standpipes in accordance with Section 902.1 where in the opinion of the Building Official and the Chief of the Fire Department such constant and automatic water supply is not necessary because of the occupancy and type of construction, with their approval dry standpipes may be substituted for one or more of the required wet standpipes.

902.5 — STANDPIPES DURING CONSTRUCTION

See Section 2101.9 (c).

SECTION 903 — SPRINKLER AND STANDPIPE REFERENCES

903.1 — OTHER CODE REFERENCES

Hereunder are listed the Section numbers and subjects of references in other portions of this Code pertaining to Sprinklers and Standpipes.

402.6 — Height Increase for Sprinklers.
403.2 — Area Increase Not Permitted (With Exception).
403.6 — Area Increase for Sprinklers.
403.7 — Unlimited Areas.
406.5 — Group "C" School Occupancy.
408.6 — Group "E" Assembly Occupancy.
409.4 — Group "F" Storage Occupancy.
410.4 — Group "G" Industrial Occupancy.
501.1(b)— Group "H" Hazardous Occupancies.
501.2(d)— Dry Cleaning or Similar Occupancy.
501.3 — Handling or Storage of Combustible film.
501.4 — Grain Elevators.
502 — Airplane Hangars.
508 — Public Garages.
512.14 — Group E-1, Large Assembly—Sprinklers.
512.15 — Group E, Assembly Occupancy—Standpipes.
516 — Covered Malls, Walkways and Tunnels.
604.5 — Table 604.5 (See Note 4).
605.5 — Table 605.5 (In Note: *).
606.5 — Table 606.5 (In Note: *).
702.2 — Partition Requirements.
703.1(b)— Exterior Walls Without Openings.
704.3 — Ceilings and Interior Wall Finish.
1103.1 — Distance of travel increase for sprinklers.
1126 — Fire Alarm.

CHAPTER X

FIRE RESISTANCE STANDARDS FOR MATERIALS AND CONSTRUCTION

SECTION 1001 — GENERAL

(a) Fire protection requirements of this Code are based on fire resistance ratings. Materials, thicknesses, and assemblies which have successfully performed under tests made by a recognized laboratory in accordance with the requirements of the "Standard Methods of Fire Tests of Building Construction and Materials, ASTM E119-71," shall be accepted by the Building Official for specific ratings.

(b) Thicknesses as established by said tests shall be construed as establishing minimum requirements for fire resistance only, and shall not preclude the application of other requirements of this code where consideration of strength, durability or stability require greater thicknesses.

(c) No combustible materials shall enter into the construction of assemblies except as provided in the foregoing prescribed tests.

(d) Fire doors, curtains, shutters, windows, or other protection required for openings in fire resistive walls, shall be in accordance with the requirements of Section 703.

(e) The penetration of fire resistant walls or partitions, floors, or floor ceiling assemblies for electrical, telephone, plumbing, air conditioning, intercommunication systems or similar facilities shall not be permitted unless such openings are installed in such a manner that the required fire resistance is not decreased.

1001.1 — MATERIALS FOR FIRE PROTECTION

Materials prescribed herein for fire-resistance and fire protection shall conform with the requirements of this Chapter.

1001.2 — CONCRETE

(a) Concrete used for fire-protection shall consist of one part (by volume) portland cement and not more than two parts of sand and four parts of approved aggregate, not over ¾" diameter, reinforced with wire or metal fabric.

(b) Grade A Concrete is concrete in which the coarse aggregate consists of blast-furnace slag, limestone, calcareous gravel, trap rock, burnt clay or shale, cinders containing not more than 25% of combustible material and not more than 5% of volatile material, and other materials meeting the requirements of this Code and containing not more than 30% quartz, chert, flint, and similar materials.

(c) Grade B Concrete is concrete in which the coarse aggregate consists of granite, quartzite, siliceous gravel, sandstone, gneiss, cinders containing more than 25% but not more than 40% of combustible material and not more than 5% of volatile material, and other materials meeting the requirements of this Code and containing more than 30% quartz, chert, flint, and similar materials.

1001.3 — BRICK

Brick shall be laid in Type M, S, N or O mortar. Solid clay and shale brick shall conform to the "Standard Specifications for Facing Brick", ASTM C216-71 or "Standard Specifications for Building Brick," ASTM C62-69. Hollow clay and shale brick shall conform to the "Standard Specification for Hollow Brick", ASTM C652-70. Concrete brick shall conform to the "Standard Specification for Concrete Building Brick", ASTM C55-71. Sand-lime brick shall conform to the "Standard Specification for Calcium Silicate Face Brick (Sand-Lime Brick)", ASTM C73-67.

1001.4 — CLAY OR SHALE TILE

Hollow clay or shale tile shall be laid in Type M, S, N, O or gypsum mortar. Clay or shale tile used in non-bearing partitions, and for fire-proofing shall meet the requirements of the "Standard Specifications for Structural Clay Non-Load Bearing Tile, ASTM C56-71." Clay or shale tile used in exterior walls and in all load-bearing walls or load-bearing partitions, shall comply with the requirements of the "Specifications for Structural Clay Load-Bearing Wall Tile, ASTM C34-62."

1001.5 — GYPSUM

(a) Gypsum partition tile blocks shall contain not more than 12½ per cent by weight of binding material, shall be laid in gypsum mortar, and shall meet the requirements of the "Specifications for Gypsum Partition Tile or Block, ASTM C52-65."

(b) Poured gypsum used for fire-proofing and floor and roof construction shall contain not more than 12½ per cent of wood chips, shavings or fiber, measured in a dry condition, as a percentage, by weight, of the dry mix. Gypsum mortar shall be composed of one part gypsum and not more than three parts clean, sharp, well-graded sand, by weight.

(c) Fibered plaster may be used where unsanded or neat gypsum plaster is prescribed.

(d) All plaster mixes for sanded gypsum plasters shall be measured by dry weight.

(e) When gypsum plaster is used with an aggregate, the proportions shall be as required in Section 1803(b). These proportions take precedence over the proportions for plastering shown in the Tables of Appendix B.

1001.6 — GYPSUM LATH, GYPSUM WALLBOARD, AND GYPSUM SHEATHING BOARD

(a) Gypsum lath shall comply with the provisions of the "Specifications for Gypsum Lath, ASTM C37-69." Perforated gypsum lath shall have perforations not less than ¾" in diameter, with one perforation for not more than 16 square inches of lath surface.

(b) Gypsum lath shall be nailed to wood studs or joists in all constructions required to be fire-resistive, with No. 13 gage, 1⅛", 19/64" flat-head blued nails at intervals not exceeding 4" on centers, (Five nails per lath for support of 16 inch lath), or equivalent attachment.

(c) Gypsum wallboard shall comply with the provisions of the "Specifications for Gypsum Wallboard, ASTM C36-68."

(d) Gypsum wallboard finishes shall be in accordance with applicable provisions of "Specifications for the Application and Finishing of Gypsum Wallboard, ANSI A97.1-65."

(e) Gypsum sheathing board shall comply with the provisions of the "Specifications for Gypsum Sheathing Board, ASTM C79-67."

1001.7 — METAL OR WIRE LATH

(a) Wherever metal lath or wire lath and plaster are used as required protection against the spread of fire, the weight of lath shall be not less than 2.5 lbs. per square yard when used in vertical position, and not less than 2.75 lbs. per square yard when used in horizontal position. Wire lath shall be not lighter than 2½ meshes per inch, or equivalent.

(b) Weight tags shall be left on all metal lath or wire lath until inspected and approved by the Building Official.

(c) Metal lath for ceilings below wood joists in construction which is required to be fire-resistant shall be attached with 1½", 11 gage, 7/16" head barbed roofing nails spaced at intervals not to exceed 6" on centers, or equivalent attachment.

(d) Whenever the word wire lath is used in fire-resistive plastering it shall not preclude the use of paper back wire lath. Whenever paper back wire lath is used it shall be in accordance with the Requirements for Paper-Backed Wire Fabric as prescribed in Chapter XVIII of this Code.

1001.8 — CONCRETE BLOCK

Hollow concrete masonry units used in exterior walls and in all walls or partitions shall comply with the requirements of the "Specifications for Hollow Load-Bearing Concrete Masonry Units, ASTM C90-66T" and "Nonload-Bearing, C-129-64T."

Solid load-bearing concrete masonry units shall comply with the requirements of "Specifications for Solid Load-Bearing Concrete Masonry Units, ASTM C-145-66T."

1001.9 — VERMICULITE

Vermiculite, when used as an aggregate with plaster, shall conform in particle size to the "Specifications for Inorganic Aggregates for use in Gypsum Plaster, ASTM C35-70." The weight of vermiculite shall be not less than 6 nor more than 10 lbs. per cu. ft., as determined by measurement in a cubic-foot box, using the shoveling procedure as outlined in the "Tentative Method of Test for Unit Weight of Aggregate, ASTM C29-71."

1001.10 — PERLITE

Perlite, when used as an aggregate with plaster, shall conform in particle size to the "Specifications for Inorganic Aggregates for use in Gypsum Plaster, ASTM C35-70." The weight of perlite shall be not less than 7½ nor more than 15 lbs. per cu. ft., as determined by measurement in a cubic-foot box, using the shoveling procedure as outlined in the "Tentative Method of Test for Unit Weight of Aggregate, ASTM C29-71."

SECTION 1002 – FIRE RESISTANCE RATINGS FOR MATERIALS AND CONSTRUCTION

1002.1 — REFERENCE TABLES

(a) Fire resistance ratings for materials, construction and assemblies of construction materials are shown in the Tables of Appendix B of this Code.

(b) The fire resistance ratings as stipulated in Appendix B are not intended to be all inclusive but rather those which are acceptable as the result of having successfully performed under accepted tests as prescribed in Section 1001. Other fire resistance ratings may be accepted by the Building Official as fire protection on compliance with Section 1001 of this Code.

CHAPTER XI

MEANS OF EGRESS REQUIREMENTS
(Exits and Exit Access)

SECTION 1101 – GENERAL PROVISIONS

(a) In every building hereafter erected means of egress shall comply with the minimum requirements of this Chapter.

(b) Means of egress shall consist of continuous and unobstructed paths of travel to the exterior of a building at all times. Means of egress through any room or space used as a kitchen or preparation of food shall not be permitted.

(c) Where unusually hazardous conditions exist, additional means of egress facilities shall be provided as required by the Building Official, when necessary to assure the safety of the occupants.

(d) No building shall hereafter be altered so as to reduce the capacity of the means of egress to less than required by this Chapter nor shall any change of occupancy be made in any building unless such building conforms with the requirements of this Chapter.

(e) Stairways, ramps and passageways used for required exits shall be of non-combustible construction except where otherwise specifically permitted by Sections 1115—Stairways; 1112—Exit Outlets; and 1118—Ramps.

SECTION 1102 – DEFINITION

(a) A MEANS OF EGRESS is a continuous path of travel from any point in a building or structure to the open air outside at ground level and consists of two separate and distinct parts: (1) the exit access, and (2) the exit. A means of egress comprises the vertical and horizontal means of travel and may include the room space, doorway, corridor, hallway, passageway, stairs, ramp, lobby, fire escape, escalator, and other paths of travel.

(1) EXIT is that portion of a means of egress which is separated from the area of the building from which escape is to be made, by walls, floors, doors or other means which provide the protected path necessary for the occupants to proceed with safety to the exterior of the building.

(2) EXIT ACCESS is that portion of a means of egress which leads to an entrance to an exit.

NOTE: An interior aisle, corridor, hallway, or other means of travel used to reach an exit stair or doorway is not an exit, except where the maximum allowable distance of travel to an exit is exceeded—at which point it shall be treated as part of the exit or is so located, arranged, and enclosed as to constitute an integral part of an exit facility.

SECTION 1103 – ARRANGEMENTS

1103.1 — ARRANGEMENTS

(a) Exits shall be so located that the distance from the most remote point in the floor area, room or space served by them to the nearest exit shall be not more than 150 feet (in office buildings, hotels and apartments where floor areas are subdivided into rooms, the distance of travel to an exit shall be measured from the corridor entrance to such rooms, however the travel distance to the corridor entrance within any room shall not exceed the maximum distance of travel to an exit measured along the line of travel). In buildings provided with a complete automatic fire sprinkler system the distance may be increased to 200 feet. In Group H, Hazardous occupancies the distance shall not exceed 75 feet. In Group G, Industrial occupancies and in public parking decks having at least 50% of its perimeter open to the air at each story, the distance may be increased to 200 feet in unsprinklered buildings and 300 feet in sprinklered buildings.

(b) Where more than one exit is required, they shall be located as remote from each other as is practicable.

1103.2 — MINIMUM NUMBER OF EXITS

(a) Every room or floor space of a building, occupied by seventy-five (75) persons or more, or occupied by a Group H, Hazardous occupancy, shall have not less than two (2) independent exits.

(b) In Group E-2, Small Assembly Places, there shall be at least two exit ways, and in Group E-1, Large Assembly Places, there shall be not less than three (3) exitways, except that where more than 1,000 persons are accommodated there shall be at least four (4) exitways.

(c) There shall be not less than two (2) exits serving every floor area, except that in the following cases there may be access to one (1) exit provided such exit is enclosed by construction as specified in Section 1106, but affording in no case less than one-hour fire-resistance:

Where one exit is permitted:

1. In Group A, Residential Buildings having no floor over three thousand five hundred (3,500) sq. ft. in area, of Type I, or of Type II Construction, or of other types of construction provided not over two stories in height. Maximum distance of travel to reach an exit from the entrance door to any living unit shall not exceed 30 feet.

2. In Group B-1, Office Buildings having no floor over three thousand five hundred (3,500) sq. ft. in area and not over two stories in height provided the occupant content shall not exceed 40 persons above the street floor. Maximum distance of travel to an exit shall not exceed 75 feet.

3. In Group B-2 Occupancies at street floor level having a floor area less than 2250 sq. ft. and a distance of travel to an exit not exceeding fifty (50) feet.

4. In Group F storage occupancies, one story only, and having a floor area less than 2,500 sq. ft., with a distance of travel 50 ft. or less.

(d) Sufficient exit facilities shall be provided so that the aggregate capacity of all such exits, determined in accordance with this Chapter, shall not be less than the occupant content as determined from Section 1105.1.

(e) It shall be unlawful to occupy any part of a building by a greater number of persons than that for which exit capacity, as prescribed in this Chapter, has been provided.

SECTION 1104 — SPECIAL EXIT REQUIREMENTS

1104.1 — INSTITUTIONAL OCCUPANCIES

(a) Group D-2, Institutional occupancies, all doorways to areas housing bedridden patients, and doorways between patient occupied spaces and the required exit, and all exit doorways leading to the exterior shall be not less than 44 inches in clear width except that exit doors so located as not to be subject to use by patients, may be not less than 36 inches in clear width. Required corridors, ramps, or passageways shall be not less than 8 feet in clear width in all areas occupied by patients or serving as part of the means of egress from patient areas.

(b) Corridors in all Group D, Institutional occupancies shall be subdivided by smoke-tight partitions at intervals not to exceed 150 feet. Doors in such partitions shall be tight fitting. Such doors may have wire glass panels of not more than 720 square inches. Smoke stop doors shall open in the direction of exit travel only and shall be provided with approved door holding devices of the fail safe type which will release the door(s) causing it to close upon the detection of products of combustion other than heat by a device complying with the requirements of "Smoke Detectors for Fire Protective Signaling Systems, U.L. Standard 168." (See Section 1117.1)

1104.2 — BOILER ROOMS, ETC.

In rooms in which are located steam boilers, oil fired incinerators, or apparatus using or producing gas or vapor, the maximum distance of travel to an exit shall not exceed 50 feet.

1104.3 — DEAD END POCKETS OR HALLWAYS

Exits and exit access shall be so arranged that no dead end pockets or hallways shall occur in excess of 20 feet in depth.

1104.4 — COVERED MALLS

(a) One half of the required unit of exit width for buildings connected by a covered mall shall lead to the outside by means other than through the mall. The covered mall connecting buildings shall have not less than two (2) independent exits located as remotely as practical from each other and shall have a total number of units of exit width equal to that required for the exits from the buildings which are within a 100 foot travel distance to the exits from the mall plus that required for 1 person per foot of mall width. The maximum distance of travel to an exit measured within the mall shall not exceed 200 feet. In order to provide free and unencumbered travel in the mall to the outside, each side of the mall floor area shall be provided with an unobstructed space, not less than 10 feet in width, parallel to the building lines and extending to the exit from the mall.

(b) Enclosed and tunneled walkways shall not be accepted as a required means of egress unless they comply with the provisions of this chapter. When the length of enclosed or tunneled walkways not meeting the provisions of this chapter for required exits, is more than one and one-half (1½) times the maximum allowable distance of travel of the most restrictive occupancy being connected, one or more exits from the enclosed or tunneled walkway shall be provided. Such exits shall be located as remotely from the points of connection between the enclosed or tunneled walkway and the buildings as is practicable.

1104.5 — EXIT ACCESS CORRIDORS

(a) It shall be prohibited to use public corridors, separated from building use areas by fire rated partitions and providing access to exit, for return and/or exhaust from adjoining air conditioned spaces through louvers or other devices mounted in corridor doors or partitions.

(b) Except in institutional or residential occupancies, Paragraph 1104.5 (a) may be waived by the authority having jurisdiction providing corridors are equipped with approved fire detectors sensing products of combustion other than heat arranged to automatically stop supply, return and exhaust fans and close louvers or other devices mounted within the corridor doors or partitions.

SECTION 1105 — MEANS OF EGRESS CAPACITY REQUIREMENTS

1105.1 — OCCUPANT CONTENT

For determining the exits required, the minimum number of persons or the occupant content of any floor area shall in no case be taken less than specified below:

Occupancy	Minimum Occupant Content Floor Area per Person*
Group A—Residential	125 Sq. Ft.
Group B—Stores—street floor and sales basements	30 Sq. Ft.
—upper sales floors	60 Sq. Ft.
Office Buildings and other Group B occupancies	100 Sq. Ft.
Group C—Schools—classrooms and recreation	20 Sq. Ft.
—laboratories, museums, libraries, and similar rooms	30 Sq. Ft.
—shops, vocational, administrative rooms	100 Sq. Ft.
—gymnasiums	15 Sq. Ft.
Group D—Institutional—sleeping departments	125 Sq. Ft.
—treatment departments	250 Sq. Ft.
Group E—Assembly—with fixed seats	6 Sq. Ft.
—Assembly—without fixed seats	15 Sq. Ft.
Group F—Storage	300 Sq. Ft.
Group G—Industrial	100 Sq. Ft.
Group H—Hazardous	100 Sq. Ft.

*The occupant content of floor areas of the building shall be computed on the basis of the specific occupancy classification of the building. Where mixed occupancies occur, the occupant content of each occupancy area shall be computed on the basis of that specific occupancy.

NOTE: The above Minimum Occupant Content Floor Area per person applies to NET area of the listed rooms, or similar rooms of prime occupant content for Group C—Schools and NET area for the assembly rooms or areas of Group E—Assembly occupancies. GROSS floor area shall apply to all other above listed occupancies. In computing net areas, it is not intended to include areas such as corridors, stairs, toilet rooms and other similar rooms or areas. Where fixed seats are installed or are to be installed, the occupant content may be established as determined by the actual number of fixed seats installed or to be installed to a seating plan accepted and approved by the Building Official.

1105.1.1 — SEATING CAPACITY POSTED

Signs stating the maximum seating capacity shall be conspicuously posted by the owner of the building in each assembly room, auditorium or room used for a similar purpose where fixed seats are not installed. It shall be unlawful to remove or deface such notice or to permit more than this legal number of persons within such space.

1105.2 — MEASUREMENT OF MEANS OF EGRESS WIDTH

(a) The width of the means of egress shall be measured in units of 22 inches. Fractions of a unit shall not be counted except that 12 inches added to one or more full units shall be counted as one-half a unit.

(b) The width shall be measured in the clear at its narrowest point. Handrails may project 3½ inches and door jambs 1 inch on each side of the measured width.

1105.3 — CAPACITY OF MEANS OF EGRESS

(a) The capacity or number of persons per unit (22 inches) of means of egress through doors, corridors, stairs and other paths of exit travel shall be in accordance with the following Table:

OCCUPANCY	Person Per Unit (22 inches) of Exit Width	
	Level Travel (Corridors, doors, ramps, etc.)	Stairs
Group A—Residential	60	45
B—Business	100	60
C—Schools	100	60
D—Institutional	30	22
E—Assembly	100	75
F—Storage	60	45
G—Industrial	100	60
H—Hazardous	60	45

(b) The minimum aggregate width of main entrance doorways for Group E Assembly occupancies shall be sufficient to accommodate 50 percent of the occupant content but in no case less than 36 inches. Main entrance doorways shall be considered as part of the requirements for the means of egress.

(c) The capacity of exit stairways constructed in accordance with Section 1115 shall not exceed the limits specified herein and may be used as a required exit from all floors which they serve. If, for example, three (3) stairways are required to serve the third floor of a building and a like number are required for the second floor, the total number of stairways required shall be three, not six, and the capacity of the stairway shall be determined by the floor having the highest occupant content and not the total occupant content of the building.

(d) The aggregate width of passageways, aisles or corridors serving as access to exits shall be at least equal to the required width of the exit. Where all travel to any exit is along the same access to the exit, the width of the access shall be at least equal to the exit; where there are several accesses to an exit each shall have a width suitable for the travel which it may be called on to accommodate.

(e) The minimum width of any means of egress shall be 30 inches in the clear.

(f) Where exits serve more than one floor, only the occupant content of each floor, considered individually, need be used in computing the required capacity of the exits at that floor; provided that such capacity shall not be decreased at any point along the exit facility in the direction of exit travel. When exits from floors above and below converge at an intermediate floor, the capacity of the exit from such intermediate floor shall not be less than the sum of the widths of the exits converging on such intermediate floor. There shall be no reduction in the capacity of the exits along the means of egress from the building.

1105.4 — CAPACITY OF ESCALATORS OR MOVING STAIRS

The width and exit capacity of escalators complying with the requirements of Section 1122 shall be as specified for stairways except that the maximum width of escalators shall not exceed 48 inches.

SECTION 1106 — EXIT ENCLOSURES

(a) In all buildings, four (4) stories or more in height, except one and two family dwellings, and except in those occupied by forty (40) people or less above or below the story at street level, all interior stairways including platforms, landings and hallways connecting them to the doorway leading to the outside, shall be completely enclosed with partitions of not less than 2-hour fire-resistance. Structural members supporting all such enclosing walls and partitions, and floors or roofs that form a part of the enclosure shall have at least 2-hour fire resistance also.

(b) In all buildings not over three (3) stories in height, except in one and two family dwellings, all required interior stairways shall be enclosed in partitions of at least one-hour fire resistance.

1. Private interior stairways located within a dwelling unit need not be enclosed.

(c) In Group E, Assembly occupancies, all exit enclosures shall be not less than 2-hour fire-resistance.

(d) Stairways in buildings of Group G, Industrial occupancies that are not required for exits and that serve only one floor above the first floor may not be required to be enclosed, provided the occupancy of the building is of low fire hazard and provided the omission of such stair enclosure is approved by the Building Official.

(e) Basement or cellar stairs: Except in one and two family dwellings, basement or cellar stairways located under stairways from upper stories shall be completely enclosed by construction providing fire resistance not less than required for the stair enclosure above the basement but in no case less than 1-hour fire resistance.

(f) In stair enclosure walls or partitions protecting the stair from the interior of the building, no openings except the necessary doorways shall be permitted. (This shall not, however, prohibit the use of fire windows of approved type, in stair enclosures provided they open to the exterior of the building and are located at least ten (10) feet from any other wall opening.) Such doorways shall be equipped with approved self-closing fire doors, except that when enclosing partitions are not required to provide over 1-hour fire-resistance, approved self-closing metal or metal-covered doors or solid core wooden doors of the flush type of nominal thickness of at least one and three quarters (1¾) inch in all parts, may be used.

SECTION 1107 – MONUMENTAL STAIRS

No enclosure shall be required for a flight of "monumental" stairs (as used in public buildings, stores, hotels, office buildings, etc.) from the main street entrance floor to the floor next above or floor next below or for stairs leading to a mezzanine or balcony from the main floor when:

(a) Such stairs are not a required part of the building exit facilities, and

(b) Such stairs are not connected with corridors providing access to exits.

SECTION 1108 – EXTERIOR BALCONIES

(a) Any exterior balcony, porch, or gallery may serve as a means of egress if it complies with all the requirements as to width, arrangement and materials of construction that are specified in this Chapter for means of egress and provided they comply with the requirements of the following paragraphs of this section.

(b) All balconies and other open spaces whether serving as a means of egress or not, shall have solid floors substantially level, and when located more than 3 feet above grade, shall be provided with guard rails supported for not less than 36 inches, nor more than 42 inches above the floor. The space between guard rails and floors shall form an enclosure of solid, slatted, grill or screen construction, in which the openings, if any, have a dimension, in not less than one direction, of not more than 8 inches, except that a bottom rail or curb shall be employed no more than 2 inches above the floor. Construction of guard rails shall be adequate in strength, durability, and attachment for its purpose as prescribed in Section 1204.2.

(c) Balconies, porches or galleries serving as means of egress in climates subject to snow or ice shall have a roof to protect against the accumulation of snow and ice.

(d) Balconies or other open spaces serving as a means of egress shall be maintained as a required path of travel without obstruction so as to maintain the required minimum width of exit travel.

(e) Balconies, porches or galleries having structural concrete floors shall have all supporting framing members of non-combustible materials.

SECTION 1109 – MEANS OF EGRESS FOR INTERIOR BALCONY AND GALLERY

(a) For balconies or galleries of Group E, Assembly occupancies having a seating capacity of over 50, at least two means of egress shall be provided, one from each side of every balcony or gallery, leading directly to a street or exit court.

(b) All interior stairways and other vertical openings shall be enclosed and protected as provided in this Chapter, except that stairs may be open between balcony and main assembly floor in occupancies such as theaters, churches and auditoriums. The means of egress capacity required for balconies or galleries shall be determined on the same basis as those required for the occupancy use.

(c) The maximum distance of travel for balcony or gallery from any seat to an exit shall be determined on the same basis as the building occupancy.

SECTION 1110 – MEANS OF EGRESS FOR STAGE AND DRESSING ROOM AREAS OF GROUP E-1 LARGE ASSEMBLY

Not less than one exit to a street, exit court, or passageway to a street, 3 feet or more in width, shall be provided from each side of the stage of every Group E-1, Large Assembly Place, and from each side of the sub-stage or basement or cellar under the stage, and an exit not less than 30 inches wide shall be provided from each fly-gallery and from the gridiron. An iron ladder shall be provided leading from the gridiron to a scuttle in the stage roof; such scuttle shall be not less than 2 feet x 3 feet in size and shall be provided with a metal-covered or non-combustible trap door. Each tier of dressing rooms shall be provided with at least two means of egress, each not less than 2 feet-6 inches wide, one of which shall lead directly into an exit court or street. All exit stairs shall be constructed of non-combustible material as prescribed in Section 1115, Stair Construction. Stair exits from stage and dressing rooms need not be enclosed.

SECTION 1111 — AISLES AND SEATING

For the requirements in Group E, Assembly occupancies, see Section 512, and for Church occupancies see Section 514.

SECTION 1112 — EXIT OUTLETS

(a) Every required exit shall provide continuous and protected egress discharging finally into a street, an open space leading to a street, or into an exit court or passageway leading to a street or into an approved open space having access to a street.

(b) Such exit courts or passageways shall be enclosed with construction providing not less than 2-hours fire resistance.

(c) The width of such courts or passageways shall be not less than the width of the exits tributary thereto. There shall be no reduction of width in the direction of exit travel. Such courts or passageways shall be not less than 8 feet in height.

(d) Slope of floors in exits shall not exceed one foot in ten feet.

SECTION 1113 — FOYER — REQUIRED

(a) In every Group E-1, Large Assembly Places, a foyer consisting of a space at the main entrance of the auditorium or place of assembly shall be provided. Such foyer, if not directly connected to a public street by all the main entrances or exits, shall have a straight and unobstructed corridor or passage to every such main entrance and exit.

(b) The width of foyer at any point shall not be less than the combined width of aisles, stairways, and passageways tributary thereto. The foyer shall be at the same level as the back of the auditorium, and exits leading therefrom shall not have a steeper gradient than one foot in ten feet.

SECTION 1114 — WAITING SPACES — REQUIRED

In theaters and similar Group E, Assembly occupancies, where persons are admitted to the building at times when seats are not available and are allowed to wait in a lobby or similar space, such use of lobby or similar space shall not encroach upon the required clear width of exits. Such waiting areas shall be separated from the required exitways by substantial permanent partitions or by fixed rigid railings not less than 42 inches high.

SECTION 1115 — STAIRWAY CONSTRUCTION

1115.1 — GENERAL

(a) Exterior and interior exit stairways shall be constructed of non-combustible materials throughout in the following buildings:

(1) All buildings of Type I and of Type II Construction.

(2) All buildings of Group C—Schools three (3) stories or more; of Group D—Institutions; Group E-1 Assembly Occupancy, and Group E-2 three (3) stories or more.

(3) All other buildings three (3) stories or more in height or occupied by more than forty (40) persons above or below the first story at street or grade level, except 1 and 2 family dwellings and buildings of Type VI Construction.

(b) Except when located within a dwelling unit, all interior stairways shall have solid risers. Exterior stairs may have open risers.

(c) Interior stairs constructed of wood, except those with open risers, shall be firestopped as specified in Section 705.

(d) Except in 1- and 2-family dwellings which are less than three (3) stories in height, no closet shall be located beneath stairs that are in whole or part of combustible construction; such space shall be left entirely open and free from encumbrance.

(e) Except in 1- and 2-family dwellings the underside of interior stairways, if of combustible construction, shall be protected to provide not less than 1-hour fire-resistance.

(f) In buildings two or three stories in height with balconies, porches or galleries where each room opens directly into such areas, exit stairways may extend from floor to floor on the outside provided such stairways are protected by a one-hour fire-resistive separation from the building. In buildings four (4) or more stories in height, two (2) hour protection shall be provided.

1115.2 — BASEMENT STAIRS

(a) In Group E, Theaters and Assembly occupancies, no exit stair from a lower story shall lead to an exit doorway serving an exit stair from an upper story.

(b) In no case shall a stair from a lower story lead to an exit doorway serving an exit stair from an upper story, unless such stair from below is separated at its upper end from the stair above by partitions equal to the fire rating of the stair enclosure.

1115.3 — TREADS AND RISERS

(a) Treads and risers of required stairs shall be so proportioned that the sum of two (2) risers and a tread, exclusive of projection of nosing, is not less than twenty-four (24) inches nor more than twenty-five (25) inches. The height of riser shall not exceed seven and three-quarter (7¾) inches, and treads, exclusive of nosing, shall be not less than nine (9) inches wide. Every tread less than ten (10) inches wide shall have a nosing, or effective projection, of approximately one (1) inch over the level immediately below that tread.

(b) Treads shall be of uniform width and risers of uniform height in any one flight of stairs.

(c) The use of winders and/or spiral stairways, is prohibited in stairways serving as required exits.

1115.4 — LANDINGS

(a) No flight of stairs shall have a vertical rise of more than twelve (12) feet between floors or landings; provided that in stairways serving as exits in buildings of Group E Theater and Assembly occupancies, such vertical rise shall not exceed eight (8) feet between landings.

(b) The length and width of landings shall be not less than the width of stairways in which they occur.

(c) In buildings of Group E Assembly occupancies, flights of less than three risers shall not be used in stairways, interior or exterior, passageways, at entrance or elsewhere in connection with required exits. To overcome lesser differences in level, gradients not exceeding one foot in ten feet may be used.

1115.5 — HANDRAILS

(a) All stairs shall have walls with handrails, or well secured handrails or guards on both sides of stairs of not less than thirty (30) nor more than thirty-four (34) inches high.

Stairs of less than forty-four (44) inches in width may have handrails on one side only. Horizontal runs of rails around open wells shall be not less than 36 inches high.

(b) When the required width of a flight of stairs exceeds eighty-eight (88) inches, one or more intermediate handrails, continuous between landings, substantially supported and terminating at the upper end in newels or standards shall be provided and there shall be not more than sixty-six (66) inches between such adjacent handrails.

1115.6 — WIDTH

(a) Stairs serving as required means of egress shall be clear of all obstructions except that handrails attached to walls may project not more than 3½ inches at each side within the required width.

(b) Width of stairs shall not decrease in the direction of exit travel.

(c) The minimum width of any stair serving as a means of egress shall not be less than 36 inches, except that stairs in Group D, Institutional Occupancies, shall be a minimum of 44 inches.

1115.7 — HEADROOM

(a) Stairs serving as required means of egress shall have a minimum headroom clearance of six feet eight inches (6′ 8″), measured vertically from the nearest nosing to the nearest soffit. This minimum shall be maintained for the full required width of stairs and landings.

SECTION 1116 — FIRE ESCAPES

(a) Fire escapes, or outside stairs not meeting the requirements of this chapter, shall not be permitted except as approved by the Building Official for existing buildings not over four (4) stories in height, where additional exits are necessary and conditions do not permit the use of more adequate exit facilities.

(b) Exterior fire escapes constructed with the approval of the Building Official on buildings heretofore erected, shall conform so far as possible with the requirements of this chapter. Fire escapes shall be constructed of non-combustible material and shall be arranged and located so that they can readily be reached by occupants of the building and so that safe egress is provided at the foot of the fire escape.

(c) All openings located within ten (10) feet of exterior stairways or fire escapes shall be protected with approved self-closing fire doors or approved fire windows.

(d) Exterior stairways, unless otherwise enclosed by non-combustible materials, shall be provided throughout with metal mesh or other rigid guards at least three (3) feet high on each unenclosed side of such stairway. All glass used in the construction of such enclosures shall be wired glass.

SECTION 1117 — DOORWAYS

1117.1 — DOORWAYS, GENERAL

(a) Every exit doorway shall open into an enclosed stairway, a horizontal exit, a fire protected corridor or passageway, meeting the requirements of this chapter and providing continuous protected egress to a street, or to an exterior open space leading to a street.

The clear height of exit doorways shall be not less than six feet and six inches (6′ 6″).

(b) No exit doorway shall be less than thirty-six (36) inches in width except that in Group D, Institutional occupancies, doorways serving as exits for areas housing bedridden patients shall be not less than forty-four (44) inches in width.

(c) Exit doorways shall swing in the direction of exit and shall not obstruct the travel along any required exit, except that doors swung flat against the walls may project not more than six (6) inches. No door shall at any point in its swing reduce the required width of an exit stairway or landing to less than thirty (30) inches nor interfere with full use of the stairs. Doors from individual rooms occupied by less than 50 people may open into such room or space.

(d) No exit doorway shall open immediately upon a flight of stairs. A landing of at least the width of door shall be provided.

(e) All doors designed to be kept normally closed in connection with exits, such as doors on stair enclosures and smoke stop doors, shall be provided with reliable self-closing mechanism and shall not at any time be secured in the open position, except stair doors in schools and smoke stop doors in hospitals may be kept normally open for operating convenience provided that qualified personnel is continually available to assure prompt closing of doors in case of fire or other emergency.

(f) Smoke barriers, horizontal exits, stairway enclosures and other fire doors opening on exitways shall be self-closing and so maintained or shall be provided with approved door holding devices of the fail safe type which will release the door(s) causing it to close upon the detection of products of combustion other than heat by a device complying with the requirements of "Smoke Detectors for Fire Protective Signaling Systems, U.L. Standard 168."

(g) Required exit doors shall be openable from the inside without the use of a key, tool, special knowledge or effort. When exit doors are in pairs, manually operated edge or surface mounted flush bolts and surface bolts are prohibited. If approved automatic flush bolts are used, that door leaf shall have no door knob or surface mounted hardware. The unlatching of any leaf shall not require more than one (1) operation.

(h) For required width of doorways, serving exit stairways and the exit capacity of doorways, see Sections 1105.2 and 1105.3.

(i) Locks, if provided, shall not require any key to operate from the inside, except as may be required for mental and penal institutions, except this requirement shall not apply to exterior exit doors in a Group B, E-2, F or G Occupancy if there is a readily visible durable sign on or adjacent to the door stating "THIS EXIT TO REMAIN UNLOCKED DURING BUSINESS HOURS." The sign shall be in letters no less than 1" high on a contrasting background. The locking device must be of a type that will be readily distinguishable as locked. The use of this exception may be revoked by the Building Official for due cause.

1117.2 — PANIC HARDWARE

(a) The exit doors of schools (except doors of individual school rooms), motion picture theaters, and theaters of any capacity shall be equipped with latches (fire exit bolts) which release when pressure of not to exceed 15 pounds is applied to the releasing devices in the direction of the exit travel. Such releasing devices may be bars or panels extending not less than two-thirds of the width of the door and placed at heights suitable for the service required, but not less than 30 nor more than 44 inches above the floor.

(b) The exit doors of all other places of public assembly having capacity in excess of 600 persons shall be equipped with latches (fire exit bolts) as provided in the preceeding paragraph.

1117.3 — POWER OPERATED DOORS

Where required doors are operated by power which is activated by a photo-electric device, floor mat, wall switches or other approved device or as well as doors with power assisted manual operation, the design, installation and maintenance shall be such that, in the event of power failure, the door may be manually opened to permit exit travel. These doors shall be openable as is required for other non-power operable doors.

1117.4 — REVOLVING DOORS

(a) Approved revolving doors may be used between street floor and street as required exits except as noted in paragraph (e) below, but not within five (5) feet of the swing of the wings at foot of stairs from upper floors nor within 3 feet of the swing of the wings at head of basement stairs. Where used there shall be at least one swing door exit within 20 feet of each revolving door, with there being no fewer swing doors than revolving doors as individual exits, except as provided in paragraph (c) below.

(b) Each revolving door shall receive egress credit equal to the dimension of the clear opening between the extreme ends of the enclosure walls, less that space occupied by all of the wings when collapsed in a "book-fold" manner and moved to the extreme egress position.

(c) Revolving doors may serve as exits, without swinging doors, for street floor elevator lobbies if no stairways or doors from other parts of the building discharge through the lobby, and the lobby has no occupancy other than as a means of travel between elevator and street.

(d) All approved revolving doors shall be:

1. Equipped with means to prevent their rotation at too rapid a rate to permit orderly egress. (A rate of 12 revolutions per minute is recommended), and

2. Equipped with emergency collapsing devices such that each of the wings will collapse in either direction when a force of not more than 180 pounds is applied on the outer stile of the wings at push bar level, and all of the wings must collapse together into a "book-fold" position.

(e) Revolving doors may be used in the following occupancy classifications in accordance with this section:

Group A—Residential

Group B—Business

Group F—Storage

Group G—Industrial

Group C—Schools—Only at main entrances of administrative buildings where not subject to emergency use.

Group D—Institutional—Only at main entrances of administrative buildings where not subject to emergency use.

Group E—Not permitted as required exits.

1117.5 — SPECIAL DOORWAY REQUIREMENTS

(a) No door, when opening or when fully open shall project beyond the building line. (See Chapter XXII, Use of Public Property).

(b) Every door used as a means of egress or ingress in cafes, restaurants, or in any building of Group E, Assembly Occupancy, shall be considered as an exit doorway and shall meet all the requirements as set forth in this chapter.

SECTION 1118 — RAMPS

(a) The width and enclosure of exit ramps shall be as required in Section 1112—Exit Outlets.

(b) The slope of ramps shall not exceed one foot in ten feet.

(c) Surface of ramps shall be of non-slip material.

(d) Exit ramps shall be of non-combustible construction except as otherwise permitted for stairs.

(e) Ramps shall comply with all requirements for stairways so far as those requirements are applicable.

(f) In all public buildings such as, but not limited to, public libraries, city halls, court houses and public schools that have their main floor level above or below ground level shall have at least one ramp at a public entrance to the building accessible and available for use by the Physically Handicapped using wheel chairs, braces or crutches. Ramps for Physically Handicapped shall not have a slope greater than one foot rise in twelve feet, or 8.33 percent, or 4 degrees 50 minutes. Ramps shall have handrails on at least one side, 32 inches in height, measured from the surface of the ramp, that are smooth and extend one foot beyond the top and bottom of the ramp. Ramps shall have a level platform at the top at least three (3) feet wide and shall extend at least one foot beyond each side of the doorway. It shall have at least six (6) feet of straight clearance at the bottom.

SECTION 1119 — HORIZONTAL EXITS

(a) A horizontal exit is a horizontal passageway through a fire wall or through a two-hour fire-resistive partition into another building or into another section of the same building, provided however, that horizontal exits may comprise not more than fifty percent (50%) of the required exits from any building or floor area except for exits in mechanical or equipment rooms.

(b) The width of horizontal exits shall not be less than required for exit doorways. The exit capacity of horizontal exits shall be as specified in Section 1105.3.

(c) Horizontal exits shall be equipped with at least one (1) approved fire door of a self-closing type. No automatic sliding fire door shall be used on horizontal exits. Door openings shall be protected as specified in Section 703.

(d) Floor area on either side of a horizontal exit shall be sufficient to hold the occupants of both floor areas served, allowing not less than three (3) square feet net clear area per person.

(e) The area into which a horizontal exit leads shall be provided with exits adequate to meet the requirements of this Chapter, but not including the added capacity imposed by persons entering it through horizontal exits from another area. At least one of its exits shall lead directly to the exterior.

(f) Where there is a difference of level between connected areas, ramps, not steps, shall be used, meeting the requirements of Section 1118—Ramps.

(g) Doors in horizontal exits shall be kept unlocked and unobstructed.

SECTION 1120 – EXIT TO ROOF

(a) In all buildings four (4) stories or more in height, of other than Type I or Type II Construction, one enclosed exit stairway shall be continued from street grade level to or through the roof level except where roofs have a slope greater than one in four. Such stairway shall be marked at street and floor levels with a sign indicating that it continues to the roof.

(b) Where roofs are used for roof gardens or for other purposes, stairways shall be provided as required for such use or occupancy.

(c) Where no stairway extends to the roof, scuttles shall be provided of size not less than two (2) feet by three (3) feet, giving access to the roof.

SECTION 1121 – ELEVATORS

(a) Elevators shall conform with the requirements of the "Safety Code for Elevators, Dumbwaiters and Escalators, ANSI A17.1-1971," as applicable unless otherwise specified.

(b) Elevator shafts shall be enclosed and protected from the rest of the building as specified by Section 701.3.

(c) Elevators shall not be located in a common enclosing shaft with an exit stairway. See Section 701.3.

SECTION 1122 – ESCALATORS

(a) Unless otherwise specified, escalators shall comply with the requirements for escalators in the "Safety Code for Elevators, Dumbwaiters and Escalators, ANSI A17.1-1971."

SECTION 1123 – EXIT SIGNS

(a) Exits shall be indicated by approved signs or lights at all times when the building is occupied.

(b) All exits shall be marked with illuminating signs bearing the word "Exit" in letters at least 6 inches high. No battery operated electric light or any type of portable lantern shall be used for primary exit illumination. No luminescent or fluorescent, or reflective materials may be used as a substitute for any of the required illumination for exit signs. All required illumination shall be so arranged that the failure of any single light unit, such as the burning out of an electric bulb, will not leave any area in darkness. All exit signs shall be illuminated at all times when the building is occupied by a reliable light source of not less than 25 watts or equivalent photometric rating that will be readable easily at a distance of 100 feet. Where a main entrance serves as an exit and is visible to the occupants, no exit sign is required over the main entrance door.

(c) Where exit lights or signs or the exits themselves are not visible from the exit approach, directional signs indicating the way of egress shall be provided. The level at which there is direct exit to the exterior shall also be clearly indicated.

(d) Lighting of exit signs shall be from a source independent of the general building lighting in buildings of Group E, Assembly occupancy or department stores having over five thousand (5000) sq. ft. of area on any one sales floor, in hotels with sleeping accommodations for more than one hundred (100) persons and in Group D, Institutional occupancies.

SECTION 1124 – ILLUMINATION OF EXITS

(a) Exits shall be illuminated at all times when the building is occupied with light of not less than one foot candle intensity at the floor level.

(b) Artificial lighting when necessary to meet the requirements of this Section, shall be from a source independent of the general building lighting in buildings of Group E, Assembly occupancy, of Group D, Institutional occupancy, in hotels with sleeping accommodations for more than one hundred (100) persons and in department stores of over five thousand (5,000) sq. ft. of area on any one sales floor.

SECTION 1125 — EXIT OBSTRUCTIONS

(a) No obstruction shall be placed in any aisle, exit, foyer, passageway or corridor.

(b) Where the floor space of a Group E, Assembly occupancy, is occupied by tables, chairs or other movable furniture, aisles at least 36 inches in clear width shall be maintained to provide ready access to exit doorways.

SECTION 1126 — FIRE ALARM

(a) A manual fire alarm system in accordance with provisions of "Local Protective Signaling Systems, NFPA 72-A, 1972," shall be installed in all the following buildings unless the building is equipped with an automatic fire alarm system or an automatic sprinkler system:

Group A: Hotels having accommodations for more than 15 guests, apartment houses four stories or more in height, dormitories, lodging or rooming houses having more than 20 sleeping accommodations.

Group B: Business buildings having a total occupancy of 500 or more persons or more than 100 persons above or below the street floor.

Group C: Schools.

Group D2: Hospitals, sanitariums, nursing homes, homes for the aged and orphanages.

Group E: Assembly having a capacity of 1,000 persons.

Group G: Industrial occupancies of buildings over two stories or more in height, and having a total capacity of 500 or more above or below the street floor level.

Group H: Hazardous occupancies.

CHAPTER XII

MINIMUM DESIGN LOADS

SECTION 1201 – GENERAL

Every building and structure shall be of sufficient strength to support the imposed live, dead and wind loads and impact loads, if any, without exceeding, in any of its structural elements, the stresses prescribed elsewhere in this code.

SECTION 1202 – DEAD LOADS

The dead load of a building or other structure is the weight of all permanent construction, such as floors, roofs, permanent partitions, stairways, and walls. (See Appendix for weights of construction materials.)

SECTION 1203 – LIVE LOADS

1203.1 (a) — FLOOR LOADS

Uniformly Distributed Loads—The live loads assumed for purposes of design shall be the greatest loads that probably will be produced by the intended uses and occupancies; provided that the minimum live loads to be considered as uniformly distributed shall be as given in the following table.

OCCUPANCY	Minimum Live Loads Lbs. Per Sq. Ft.
Apartments	40
Armories	150
Assembly Places—Fixed Seats	50
Assembly Places—Movable Seats	100
Balconies and Galleries—Fixed Seats	60
Balconies and Galleries—Movable Seats	60
Cornices	60
Corridors, Public	100
Dance Halls	120
Drill Rooms	150
Driveways and Yards	250**
Dwellings—Sleeping Rooms	30
Attics with Storage	30
All other Rooms	40
Exterior Balconies—Public	60
Exterior Balconies—Private	40***
Fire Escapes	100
Garages and Trucking Spaces	120*
Gymnasiums, Main Floor and Balcony	100
Hospitals—Wards and Rooms	40
Hotels—Guest Rooms and Private Corridors	40

Libraries—Reading Rooms	60
Libraries—Stack Rooms	125
Manufacturing—Light	100
Manufacturing—Heavy	150
Offices	50
Open Parking Decks—Passenger Cars Only	50*
Printing Plants—Press Room	150
Printing Plants—Composing and Linotype Rooms	100
Public Rooms	100
Rest Rooms	50
Reviewing Stands and Bleachers	100
Roof Loads (See Sections 1203.2 and 1205.3)	
Schools—Class Rooms, including Sunday Schools	40
Sidewalks	200**
Skating Rinks	100
Stairways, Public	100
Storage—Light	125
Storage—Heavy (Load to be determined from proposed use or occupancy, but never less than)	250
Stores—Retail (Light Merchandise)	75
Stores—Wholesale (Light Merchandise)	100
Theaters	
Corridors, Lobbies, and Standing Space	100
Orchestra Floor and Aisles	50
Balconies	50
Stage Floor	150

*Design shall provide for maximum wheel loads. See Section 1203.1 (d) Concentrated Loads.

**See also Section 1203.3 Sidewalks.

***Exterior balconies, private, are those balconies for use by one or two families.

1203.1 (b) — ITEMS NOT SPECIFICALLY COVERED

For occupancies not listed above, the live load shall be approved by the Building Official.

1203.1 (c) — PROVISION FOR PARTITIONS

The actual weight of all permanent partitions shall be included in the dead load. Where partitions are likely to be used, although not definitely located, or where they are likely to be shifted, twenty (20) pounds per square foot shall be added to the dead load in the areas supporting them, except in the case of light partitioning.

1203.1 (d) — CONCENTRATED LOADS

In the design of floors, probable concentrated loads shall be considered. Where such loads may occur, the supporting beams, girders and slabs shall be designed to carry either the concentrated loads or the live load described in Section 1203.1 (a), whichever produces the greater stresses.

1203.1 (e) — REDUCTION OF LIVE LOAD

(1) No reduction shall be applied to the roof live load.

(2) No reduction of the live load shall be allowed in the design of any slabs, or joists.

(3) In designing a column, girder, truss, wall, pier or foundation carrying more than one floor, the live loads of the floors which are supported by such members may be reduced, except in buildings used for storage or warehouse purposes and in open parking decks. The reduced load shall be calculated over the entire tributary floor area, and shall be not less than the following percentages of the live load for which such floors were designed: . . .

100 per cent for members carrying one (1) floor,
90 per cent for members carrying two (2) floors,
80 per cent for members carrying three (3) floors,

and at corresponding decreasing percentages for each successive floor. In no case, however, shall the load be less than seventy per cent of the live load for any floor in industrial buildings, stores or garages, or fifty per cent for any floor in buildings of occupancies other than those for which specific provision is made herein.

(4) Except as above provided, beams, girders and trusses shall be designed to support the full dead and live loads; provided that in buildings other than those used for storage or warehouse purposes and open parking decks, beams, girders or trusses carrying three hundred square feet or more of tributary floor area may be designed to carry eighty-five (85) per cent of the live load and the full dead load.

This load reduction shall not be used in addition to reduction set forth in subsection (3) above.

1203.1 (f) — RESTRICTIONS ON LOADING

It shall be unlawful to place, or cause or permit to be placed, on any floor or roof of a building or other structure a load greater than is permitted by these requirements.

1203.2 — ROOF LIVE LOADS

(a) Where the rise is less than thirty degrees (30°), roofs shall be designed for a vertical live load only of not less than twenty (20) pounds per square foot of horizontal projection applied to any and all slopes; when the rise is greater than thirty degrees (30°) the roof shall be designed for wind load only.

(b) Design requirements for wind pressures shall be as specified in Section 1205.3.

(c) Roofs intended for use as floors, shall be designed for the floor live load, determined by the intended occupancy, Section 1203.1(a).

(d) The foregoing provisions of this section shall not apply to glazed greenhouse roofs.

1203.3 — OTHER LIVE LOADS

SIDEWALKS—Sidewalks shall be designed to carry either a uniformly distributed load of two hundred (200) pounds per square foot or a concentrated load of eight thousand (8,000) pounds on a space two and one-half (2½) feet square and placed in any position, whichever will produce the greater stresses. (This does not apply to sidewalks on grade).

1203.4 — IMPACT AND CRANE RUNWAYS

(a) All moving loads shall be increased by not less than twenty-five (25) per cent to provide for impact except that supports for elevators shall provide for one hundred (100) per cent increase of live load.

(b) The lateral force on crane runways to provide for the effect of crane trolleys shall be not less than twenty-five (25) per cent of the lifted load plus weight of the crane trolley but exclusive of other parts of the crane, applied at the top of rail, one-half (½) on each side of runway, and considered as acting in either direction normal to the runway rail. The longitudinal force on crane runways shall be taken as not less than twelve and one-half (12½) per cent of the wheel loads of the crane, applied at top of rail.

1203.5 — LOAD TESTS

The Building Official may require a load test of any construction whenever there is reason to question its safety for the intended use. Such tests are to be made at the expense of the owner or his agent. The construction shall sustain a superimposed load equal to twice the design live load and shall recover at least seventy-five (75) per cent of its maximum deflection within twenty-four (24) hours after the load is removed.

SECTION 1204 — SPECIAL LOADS

1204.1 — SOIL PRESSURES ON BASEMENT WALLS AND FLOORS

(a) In the design of basement walls and similar approximately vertical structures below grade, provisions shall be made for lateral pressure of adjacent soil. Due allowances shall be made for possible surcharge from fixed or moving loads.

(b) In the design of basement floors and similar approximately horizontal constructions below grade, the upward pressure of water, if any, shall be taken as the full hydrostatic pressure applied over the entire area. The hydrostatic head shall be measured from the underside of the construction.

1204.2 — RAILINGS

(a) Stairway railings, both exterior and interior shall be designed to resist a horizontal thrust of twenty (20) pounds per linear foot applied at the top of the railing.

(b) Balcony railings, both exterior and interior, shall be designed to resist a horizontal thrust of fifty (50) pounds per linear foot applied at the top of the railing. (See Section 1108).

1204.3 — SUPPORTS FOR WALKWAYS

Where walkways are to be installed above ceilings, supports shall be designed to carry a load of two hundred (200) pounds occupying a space two and one-half (2½) square feet, so placed as to produce maximum stresses in the affected members.

SECTION 1205 — WIND LOADS

1205.1 — MINIMUM DESIGN LOADS

(a) Buildings or other structures shall be capable of withstanding the horizontal loads shown in the following table and, applied in each zone, allowing for wind from any direction. The first height zone shall be measured above the average level of the ground adjacent to the building and the subsequent height zones shall be added progressively upward to the overall height of the building.

DESIGN WIND PRESSURE FOR VARIOUS HEIGHT ZONES OF BUILDINGS OR OTHER STRUCTURES

Height Zone Ft.	Lb./Sq. Ft. Horizontal Loads For Southern Inland Regions	Lb./Sq. Ft. Horizontal Loads For Southern Coastal Region*
Less than 30	10	25
31 to 50	20	35
51 to 99	24	45
100 to 199	28	50
200 to 299	30	50
300 to 399	32	50
Over 400	40	50

*Coastal region is defined as that area lying within 125 miles of the coast and subject to hurricanes, tropical disturbances and occasional winds attaining exceptionally high wind velocities.
(See Appendix "D" for Hurricane Requirements.)

(b) Allowance to be made for shape factors are as follows:

(1) Round or eliptical structure sixty percent (60%) of basic allowable pressure.

(2) Hexagonal or octagonal structures eighty percent (80%) of basic allowable pressure.

(3) Rectangular or square structure 1.00 times basic allowable pressure.

(c) Correction factors for trussed towers with flat, angular or cylindrical members shall be determined by qualified structural designers but in no case shall be less than those recommended by the United States of America Standards Institute.

1205.2 — EXTERIOR WALLS

Every exterior wall shall be capable of withstanding the loads specified in the above table, acting either inward or outward.

1205.3 — ROOFS — WIND LOADS

(a) The roofs of all buildings or other structures shall be designed to withstand loads acting outward normal to the surface equal to one and one-quarter (1¼) times the horizontal loads specified for the corresponding height zone in which the roof is located. The height is to be taken as the mean height of the roof structure above the average level of the ground adjacent to the building or other structure. The load is to be applied over the entire roof.

(b) Roofs with slopes greater than thirty degrees (30°) shall withstand resulting loads acting inward normal to the surface equal to those specified for the height zone in which the roof is located, the load to be applied to the windward slope only.

(c) Overhanging eaves and cornices shall be capable of withstanding upward loads equal to twice those specified.

(d) Adequate anchorage of the roof to the walls and columns and of walls and colmuns to the foundations shall be required in all cases.

1205.4 — CHIMNEYS

Chimneys of circular cross section shall be capable of withstanding sixty (60) per cent of the loads specified. Chimneys of square or rectangular cross section shall be capable of withstanding the full loads specified.

1205.5 — SIGNS

(a) Ordinary solid signs erected on ground level shall be capable of withstanding a load of fifteen (15) pounds per square foot. Solid signs on the roofs of buildings or other structures shall be capable of withstanding the loads specified in Section 1205.1 corresponding to the height of the center of the sign. Signs in which the actual projected area on a plane normal to the direction of the wind, is at least seventy per cent (70%) of the gross projected area shall be considered solid signs and the gross area shall be used in computing the load.

(b) An open sign is one for which the total normal projected area of letters, figures, strips and structural framing members is less than seventy per cent (70%) of the gross area and whose effective area shall be the sum of the projected areas of its exposed parts, including the structural framing on a plane perpendicular to the direction of the wind.

(c) Open signs erected at ground level shall be capable of withstanding a pressure of thirty (30) pounds per square foot, applied to the projected area of the members. Open signs on the roofs of buildings or other structures shall be capable of withstanding twice the loads specified in Section 1205.1 corresponding to the height of the center of the sign, applied to the projected area of the members.

1205.6 — OTHER STRUCTURES

The Building Official may require evidence to support the values of the wind load used in the design of structures not specifically covered by this section.

1205.7 — SHIELDING AND UNUSUAL EXPOSURES

(a) No allowance shall be made for the shielding effect of other buildings or structures.

(b) If the building or other structure is on an ocean promontory, or in any other location considered by the Building Official to be unusually exposed, higher wind loads may be specified by the Building Official.

1205.8 — OVERTURNING MOMENT

The overturning moment due to the wind load shall not exceed sixty-six and two-thirds (66⅔) per cent of the moment of stability of the building or other structures due to the dead load only unless the building or other structure is anchored to its foundations so as to resist the excess overturning moment without exceeding the allowable working stresses for the materials used.

1205.9 — STRESSES DURING ERECTION

Provision shall be made for wind stresses during erection of a building or other structure.

SECTION 1206 – EARTHQUAKE LOADS

1206.1 — SEISMIC DESIGN REQUIREMENTS

Where seismic design is required by local authorities, all buildings and structures shall be designed to withstand seismic forces in accordance with the requirements of ANSI A58.1-72—Building Code Requirements for Minimum Design Loads in Buildings and Other Structures. (See Figure 1)

SEISMIC RISK MAP

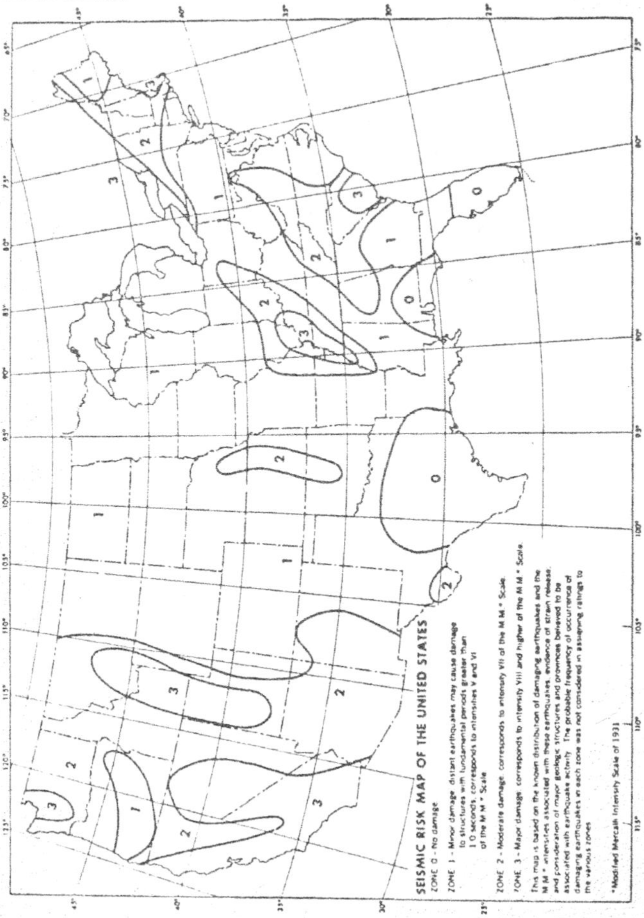

FIGURE NO. 1 — SEISMIC ZONE MAP OF THE UNITED STATES

SECTION 1207 — OCCUPANCY PERMITS FOR CHANGED LOADING

Plans for other than residential buildings filed with the Building Official with applications for permits shall show on each drawing the live loads per square foot of area covered, for which the building is designed, and occupancy permits for buildings hereafter erected shall not be issued until the floor load signs, required by Section 110, have been installed. No changes in the occupancy of a building now existing or hereafter erected shall be made until a revised occupancy permit has been issued by the Building Official certifying that the floors are suitable for the loads characteristic of the proposed occupancy. (See Section 109).

CHAPTER XIII

FOUNDATIONS

EXCAVATIONS, FOOTINGS AND FOUNDATIONS

SECTION 1301 — EXCAVATIONS

1301.1 — GENERAL

When excavating for buildings or excavations accessory thereto, such excavations shall be properly assured against any danger to life and property. Permanent excavations shall have retaining walls of sufficient strength made of steel, masonry, or reinforced concrete to retain embankments, together with any surcharged loads. Excavations for any purpose shall not extend within one (1) foot of the angle of repose or natural slope of the soil under any footing or foundation, unless such footing or foundation is first properly underpinned or protected against settlement.

1301.2 — SUPPORT OF ADJOINING BUILDINGS AND STRUCTURES

(a) When an excavation extends not more than 10 feet below the established curb grade nearest the point of excavation under consideration, the owner of any adjoining building or structure, the footings or foundations of which are to be underpinned or protected under the requirements of this section, shall be notified in writing by the one causing the excavation to be made. The owner of the adjoining structure or building shall be afforded the necessary license to enter the premises where the excavation is to be made, and at his own expense, shall provide the necessary underpinning or protection.

(b) Notice to the owner of adjoining buildings or structures shall be served at least 10 days before an excavation is commenced, and it shall state the depth and location of the proposed excavation.

(c) When an excavation extends more than 10 feet below the established curb grade nearest the point of excavation under consideration, the one causing the excavation to be made, if given the necessary license to enter the adjoining premises, shall provide at his own expense the underpinning and protection required by that part of the excavation which extends to a depth greater than 10 feet below the established curb grade nearest the point of excavation under consideration, whether or not the existing footings or foundations extend to the depth of 10 feet or more below curb grade; or he may shore and brace the sides of his excavation so as to prevent effectively any soil movement into his excavation. If permanent lateral support is provided, the method used must satisfy requirements of the Building Official. If the necessary license is not afforded the person causing the excavation to be made, it shall be the duty of the owner failing to afford such license to provide the required underpinning or protection, for which purpose he shall be afforded the necessary license to enter the premises where such excavation is to be made.

(d) If there is no established curb grade, the depth of excavation shall be referred to the level of the ground at the point under consideration. If an existing building or structure, the footings or foundations of which are required to be underpinned or protected, is so located that the curb grade or level to which it is properly referred is at a higher level than the level to which the excavation is properly referred, then such part of the required underpinning or protection that is necessary due to the difference in these levels shall be made and maintained at the joint expense of the owner of the building or structure and the person causing the excavation to be made. For the purpose of determining such part of the underpinning, or protection that is necessary due to such difference in levels, the level to which a building more than five feet back of the street line is properly referred shall be considered to be the level of the natural ground surface adjoining the building or structure.

(e) A party wall which is in good condition and otherwise suitable for continued use, shall be underpinned or protected as required at the expense of the person causing the excavation to be made.

(f) Where the necessary license has been given to the person making an excavation to enter any adjoining structure for the purpose of underpinning or protecting it, the person receiving such license shall provide for such adjoining structure adequate protection against injury due to the elements resulting from such entry.

(g) Only approved granular materials shall be used for backfill. It shall be properly compacted in order to prevent lateral displacements of the soil of the adjoining property after the removal of the shores or braces.

SECTION 1302 – FOOTINGS AND FOUNDATIONS

1302.1 — GENERAL

Except in the case of temporary structures or secondary buildings not over 1 story in height and not exceeding 400 square feet in area, footings and foundations, unless specifically provided, shall be constructed of grillages of steel, of masonry, reinforced concrete or of other approved materials, (one and two family dwellings may not be required to have reinforced concrete footings or grillage of steel) in no case less than 12 inches below grade. Masonry units used in foundation walls and footings shall be laid up in Type M, S, or N mortar. The base areas of all footings and foundations shall be proportioned as specified in Section 1302.3.

1302.2 — BEARING CAPACITY OF SOIL

(a) Footings shall be so designed that the allowable bearing capacity of the soil in pounds per square foot as given below shall not be exceeded unless the particular soil on which the building is to be placed shows a greater bearing capacity than that specified in this Section, under tests as provided herein.

BEARING CAPACITIES OF VARIOUS SOILS

Foundation-bed	Pounds per sq. ft.
Soft Clay	2,000
Firm Clay	3,000
Wet Sand	4,000
Sand and clay, mixed or in layers	4,000
Fine and dry sand	4,000
Coarse Sand	4,000

(b) Where the bearing capacity of the soil is not definitely known or is in question, the Building Official may require load tests or other adequate proof as to the permissible safe bearing capacity at that particular location. To determine the safe bearing capacity of soil, it shall be tested at such locations and levels as conditions warrant, by loading an area not less than 4 square feet to not less than twice the maximum bearing capacity desired for use. Such double load shall be sustained by the soil for a period of not less than 48 hours with no additional settlement taking place, in order that such desired bearing capacity may be used. Examination of sub-soil conditions shall be made at the expense of the owner, when deemed necessary by the Building Official.

(c) Foundations shall be built upon natural solid ground. Where solid natural ground does not occur at the foundation depth, such foundations shall be extended down to natural solid ground or piles shall be used. Foundations may be built upon mechanically compacted earth or fill material subject to approval by the Building Official upon submittal of evidence that proposed load will be adequately supported.

(d) Where footings are supported by soils of widely different bearing capacity, the allowable bearing values of the more yielding soil shall be reduced or special provisions shall be made in the design to prevent serious differential settlements.

(e) When it is definitely known the top or sub-soils are of a shifting or moving character, all footings shall be carried to a sufficient depth to insure stability. The excavation around piers shall be back filled with soils or materials which are not subject to such expansion or contraction.

1302.3 — FOOTING DESIGN

(a) The base area of the footings of all buildings shall be designed in the following manner: The area of the footing which has the largest percentage of live load to total load shall be determined by dividing the total load by the allowable soil load. From the area thus obtained the dead load soil pressure of such footing is determined and the areas of all other footings of the building shall be determined on the basis of their respective dead loads only and such dead load soil pressure. In no case shall the load per square foot under any portion of any footing, due to the combined dead, live, wind and/or any other loads, exceed the safe sustaining power of the soil upon which the footing rests. The total reduced live load occurring in the column immediately above the footing shall be the live load used in the above computation.

(b) Footings shall be proportioned to sustain the applied loads and induced reactions without exceeding the allowable stresses specified in this code.

(c) Concrete in footings shall have an ultimate compressive strength of not less than 2500 pounds per square inch at 28 days.

1302.4 — DOWELS

(a) Dowels of the same number as the vertical bars in the column, but not less than four, shall extend into the column a distance of not less than 20, 24 and 30 bar diameters for specified yield strengths of 50,000 and under, 60,000, and 75,000 psi, respectively, nor not less than 12 inches with concrete having a strength of 3,000 psi or more. When the specified concrete strengths are less than 3,000 psi the amount of lap shall be one-third greater than the values given above. For plain bars, the minimum amount of lap shall be twice that specified for deformed bars. Dowels shall extend down into the supporting pedestal or footing the distance required to transfer the full working value of the dowel to the concrete without exceeding the bond stress and to support all vertical loads without exceeding the unit stresses permitted in Chapter XVI, Section 1601. Hooks shall not be considered effective in adding to the compressive resistance of bars.

(b) The thickness of concrete above the reinforcement shall be not less than 8 inches for footings bearing on soil, nor less than 12 inches for footings on piles. The thickness of concrete protecting the steel reinforcement shall in no case be less than 3 inches.

(c) Design of footings shall be in accordance with Chapter XVI.

1302.5 — FOUNDATION WALLS

(a) Foundation walls shall be designed and constucted in accordance with accepted engineering practice, provided however the provisions of paragraphs (b), (c) and (d) may be used without additional engineering design.

(b) Foundation walls shall be not less in thickness than the walls immediately above them and not less than 12 inches for unit masonry walls, or 8 inches for cast-in-place concrete walls; except that solid masonry walls extending not more than 5 feet, and hollow walls of masonry or walls of hollow units extending not more than 4 feet below the adjacent finished ground level may be 8 inches in thickness. These depths may be increased to a maximum of 7 feet with the approval of the Building Official when he is satisfied that soil conditions warrant such an increase. The total height of 8 inch foundation walls and walls supported shall not exceed that permitted by this Code for 8 inch walls. In all cases, however, foundation walls shall have sufficient strength and thickness to resist all lateral pressures permitted by this code.

(c) Foundation walls of 8 inch thickness [except as provided for in Section 1302.5(d)] and conforming to the provisions of Section 1302.5 (a) may be used as foundations for dwellings with walls of brick veneer on frame walls or with 10 inch cavity walls, provided that the dwelling is not more than 1½ stories in height and the total height of the wall, including the gable, is not more than 20 feet. Foundation walls of 8 inch thickness supporting brick veneer or cavity walls, shall be corbeled with solid units to provide a bearing the full thickness of the wall above. The total projection shall not exceed 2 inches with individual corbels projecting not more than ⅓ the height of the unit. The top corbel course shall not be higher than the bottom of floor joists and shall be a full header course.

(d) Foundation walls of cast-in-place concrete when supporting one story basementless structures may be 6 inches thick if the total height of the foundation wall and the wall supported is within the allowable height permitted by this Code for 6 inch walls.

(e) Crawl spaces under buildings without basements shall be ventilated by approved mechanical means or by openings in foundation walls. Openings shall be arranged to provide cross ventilation and shall be covered with corrosion resistant wire mesh of not less than ¼" nor more than ½" in any dimension. Openings in foundation walls shall not be less than the following:

(1) Where wood floor systems are used such openings shall have a net area of not less than 2 sq. ft. for each 100 linear feet of exterior walls, plus 1/3 sq. ft. for each 100 sq. ft. of crawl space.

(2) Where other than wood floor systems are used, such openings shall be not less than 1½ sq. ft. of net opening for each 15 lineal feet or major fraction thereof of exterior wall.

(3) Where an approved vapor barrier is installed over the ground surface, the required net area of openings may be reduced by 50 per cent.

(4) Where combustion equipment is installed within a crawl space, air for combustion shall be provided in accordance with Section 816.2.

(f) Foundation walls of hollow masonry supporting Type VI construction shall be capped with 4 inches of solid masonry or concrete or shall have cavities of top course filled with concrete or grout unless a sill plate of 2" nominal thickness bears on both face shells.

1302.6 — TIMBER FOOTINGS

Footings of wood may be used if they are entirely below permanent water level, or if they are pressure impregnated with an approved preservative as established in Appendix "C".

1302.7 — WOOD FOUNDATION SYSTEMS

The foundation system may be of wood when the engineering design is based upon the bearing capacity of the soil (see Section 1302.2) and the design and construction complies with the provisions of "All-Weather Wood Foundation System—Design and Construction Methods"—1971 with the March 30, 1973 Addendum as published by the National Forest Products Association.

SECTION 1303 — PILES

1303.1 — GENERAL REQUIREMENTS

(a) **Spacing** — The minimum center-to-center spacing of piles not driven to rock shall be not less than twice the average diameter of a round pile, nor less than 1.75 times the diagonal dimension of a rectangular or rolled structural steel pile, nor less than 2 ft.-6 in. Piles in groups or in abutting groups that receive their principal support in sand or clay or a mixture of those soils shall have their spacing increased by 10% for each interior pile up to a maximum increase of spacing of 40%.

(1) The minimum center-to-center spacing of piles driven to rock shall be not less than twice the average diameter of a round pile, nor less than 1.7 times the diagonal dimension of a rectangular or rolled structural steel pile, nor less than 2 ft.-0 in.

(2) A column or pier supported by piles, unless connected to permanent construction which provides adequate lateral support shall rest on not less than three piles. When the supporting capacity of a single row of piles is adequate for the wall of a structure, effective measures shall be taken to provide for eccentricity and lateral forces, or the piles shall be driven alternately in lines spaced at least one foot apart and located symmetrically under the center of gravity of the loads carried. A single row of piles without lateral bracing may be used for private dwellings not exceeding two stories in height, provided the centers of the piles are located within the width of the foundation wall.

(3) In no case shall the spacing of piles be such that the average load on the supporting strata will exceed the safe bearing value of those strata, as determined by test borings or other approved methods.

(b) **Allowable Loads** — All piles used to support any building or part thereof shall be driven in such a manner as not to impair their strength. The allowable load on piles may be determined by the applicable formulas of this section or by load tests as prescribed in Section 1303.6.

(1) Subject to the limitations prescribed in this section for the various types of piles, the allowable load up to a maximum of forty (40) tons per pile shall be determined by the value

of R obtained by one of the following formulas or by tests as provided in Section 1303.6. When the allowable load is determined by one of the following formulas, piles with an average diameter or side of 8 inches or less shall be driven by a hammer which delivers a blow of at least 7,000 foot-pounds; piles with an average diameter or side greater than 8 inches and not more than 18 inches shall be driven by a hammer which delivers a blow of at least 15,000 foot-pounds; and piles with an average diameter or side of more than 18 inches shall be driven by a hammer which delivers a blow of at least 22,000 foot-pounds. Double-acting hammers shall be operated at full rated speed, pressure, and stroke as shown in the manufacturer's catalog. The minimum hammer blow for piles intended to carry 25 tons or more shall be 15,000 foot-pounds.

For drop hammers: $$R=\frac{2\,W\,H}{s+1}$$

For single-acting steam hammer: $$R=\frac{2\,W\,H}{s+0.1}$$

For double-acting steam hammers: $$R=\frac{2\,E}{s+0.1}$$

In which:

R=Allowable pile load in pounds

W=Weight of striking part of hammer in pounds

H=Effective height of fall in feet

E=Actual energy delivered by the hammer per blow in foot pounds

s=Penetration of pile per blow, inches, determined under conditions required by accepted good practice.

(2) Piles shall be designed as short columns except that where piles extend above permanent ground or where piles below ground level receive negligible lateral support from the surrounding soil, they shall be designed as long columns throughout their unsupported length. Any soil other than water or fluid soil shall be deemed to afford sufficient lateral support to permit the design of any pile in accordance with normally accepted engineering practice and the applicable provisions of this code.

(c) Protection of Piles—To such depths or at such horizons as boring records or site conditions indicate possible deleterious action on pile materials because of soil constituents or water levels, such (pile) materials shall be adequately protected by approved preservatives or impervious encasements which will not be rendered ineffective by driving and which will prevent such deleterious action.

(d) Splices — Splices shall be such that the resultant vertical and lateral loads at the splices are adequately transmitted. Splices shall be so constructed as to provide and maintain true alignment and position of the component parts of the pile during installation and subsequent thereto. Except for piles which can be visually inspected after driving, splices shall develop not less than fifty per cent of the value of the pile in bending. Proper consideration shall be given to the design of splices at sections of piles which may be subject to tension or to bending.

(e) Substantiation of Higher Allowable Pile Loads—Allowable loads greater than those specified in Sections 1303.2, 1303.3, and 1303.4 shall be permitted when substantiating data justifying such higher loads is submitted to the Building Official by a foundation designer knowledgeable in the field of soil mechanics and pile foundations and familiar with the locale of the proposed project. Such substantiating data shall include adequate test borings, soil profiles, pile load tests and such other information as may be required by the Building Official. However the maximum pile loads contemplated shall not exceed the provisions of Section 1303.1 (b) (2).

1303.2 — STEEL PILES

(a) Definition — Steel piles may consist of rolled shapes, pipe, or built-up structural shapes.

(b) Steel Pipe Piles—Steel pipe piles shall consist of steel pipe conforming to the "Specifications for Welded and Seamless Steel Pipe Piles, ASTM A252-69." They may be driven either open-ended or with ends closed. Steel pipe piles driven open-ended shall have a nominal outside diameter of not less than 10 inches and a minimum wall thickness of not less than 0.25 inch for diameters less than 14 inches and a minimum wall thickness of not less than .375 inch for diameters 14 inches and over. Pipe of less wall thickness may be driven open-ended if a suitable cutting shoe is provided. If steel pipe piles are to be driven with closed ends, a forged, cast steel, flat plate or other end closure of approved design shall be used. Steel pipe piles driven with ends closed may be of smaller sizes and wall thickness than specified above but no such pile of uniform section shall have a nominal outside diameter of less than 8 inches.

(1) The concrete in concrete-filled steel pipe piles shall conform to the requirements for concrete as given elsewhere in this Code. Concrete shall have a minimum compressive strength at twenty-eight days of 2,500 pounds per square inch. No concrete shall be placed through water except with the written approval of the building official, who may specify the proportions of the concrete to be so placed and the method of placing. The pipe shall be cleaned of all dirt or other foreign matter before the concrete is placed.

(2) If reinforcement is to be used in concrete-filled steel pipe piles it shall be assembled and tied together so that it may

be placed in the pile as a unit. No reinforcement shall be placed within 1 inch of the steel shell.

(3) The maximum allowable load on concrete-filled steel pipe piles not driven open-ended to bearing on rock shall be as specified in Section 1303.1(b) or Section 1303.6. The maximum allowable unit stress in the steel pipe shall be 35% of minimum specified yield strength. The minimum specified yield strength shall not be assumed greater than thirty-six thousand pounds per square inch for computation purposes. No stress shall be allowed on tubular sections less than 0.10 inch thick. The maximum allowable compressive stress in the concrete shall be 33% of its 28 day compressive strength. The maximum allowable unit stress in the reinforcement if used shall be as specified in Chapter XVI.

(4) The maximum allowable load on concrete-filled steel pipe piles driven open-ended to bearing on rock, such that the net penetration for the last five blows total ¼ inch or less under the hammers specified in Section 1303.1(b), shall be 80 percent of the load limited by the allowable stresses specified in paragraph (3) above unless a greater load is permitted by the provisions of Section 1303.6.

(c) Rolled Structural Steel Piles—Steel used in rolled structural steel piles shall conform to one of the following standard specifications:

"Specification for Structural Steel, ASTM A36-69"

"Specification for High-Strength Low Alloy Columbium Vanadium Steels of Structural Quality, ASTM A572-68."

Sections of such piles shall be of H form, with flange projection not exceeding fourteen times the minimum thickness of metal in either web or flange and with total flange width at least 85% of the depth of the section. No section shall have a nominal thickness of metal less than ⅜ inch, nor a nominal depth in the direction of the web less than 8 inches. Other structural sections or combinations of sections having flange widths and depths of not less than 10 inches and thickness of metal not less than ⅜ inch may also be used.

(1) The allowable load on rolled structural steel piles not driven to bearing on rock shall be as specified in Section 1303.1(b) or Section 1303.6 but in no case shall the unit stress exceed 35% of the minimum specified yield strength. The minimum specified yield strength shall not be assumed greater than thirty-six thousand pounds per square inch for computation purposes. The allowable load on rolled structural steel piles driven to bearing on rock such that the net penetration for the last five blows totals ¼-inch or less under the hammers specified in Section 1303.1(b) is 70 tons for piles of 12 inch nominal depth and 90 tons for piles of 14 inch and over nominal depth unless a greater load is permitted by the provisions

of Section 1303.6 provided the unit stress does not exceed the allowable stress specified herein.

1303.3 — CONCRETE PILES

(a) Concrete piles shall be of material complying with the requirements for Portland cement, fine aggregates, coarse aggregates and reinforcement as specified in Chapter XVI and steel as specified in Chapter XV. The compressive strength of the concrete shall be determined by tests as specified in Chapter XVI. The maximum allowable stresses shall be as follows:

(1) Uncased Cast-in-Place Concrete Piles: The allowable compressive stress in the concrete shall not exceed 33 percent of the 28 day compressive strength of the concrete. The allowable stress in the reinforcement shall not exceed that specified in Chapter XVI.

(2) Steel-Cased Concrete Piles: Allowable stresses shall not exceed the values specified in Section 1303.3 (a) (1), except that the allowable concrete stress may be increased to a maximum value of 40 percent of the 28 day compressive strength of the concrete for that portion of the pile meeting the following conditions:

1. The thickness of the steel casing is not less than 14 gauge.
2. The casing is seamless or is provided with seams of equal strength and is of a configuration which will provide confinement to the cast-in-place concrete.
3. The ratio of steel yield strength to concrete 28 day strength shall not be less than 6.
4. The nominal diameter of the pile is not greater than 16 inches.

(3) Precast Concrete Piles: The allowable stresses shall not exceed the values specified in Section 1303.3 (a) (1).

(4) Precast-Prestressed Concrete Piles (Pretensioned): The allowable compressive stress in the concrete due to externally applied load shall not exceed

$$f_C = 0.33\ f_{IC} - 0.27\ f_{PE}$$

In which:

f_C = Allowable compressive stress in the concrete due to externally applied load.

f_{IC} = 28 day compressive strength of the concrete.

f_{PE} = Effective prestress stress on the gross section.

The allowable stresses in the prestressing steel shall not exceed the values specified in Chapter XVI.

(b) The maximum allowable load on concrete piles shall be as specified in Section 1303.1(b), provided that in no case shall the unit stresses exceed the maximum allowable stresses specified in paragraph (a) above, and provided that except where the load capacity is determined by tests in accordance with Section 1303.6, the allowable load shall not exceed 40 tons for precast concrete piles and steel-cased concrete piles with steel shells driven in intimate contact with the soil and left permanently in place. The maximum allowable load on uncased cast-in-place concrete piles shall be 25 tons.

(c) Concrete piles cast-in-place shall be made in such a manner as to insure the exclusion of any foreign matter and to secure a full-sized shaft. The diameter of cast-in-place piles with steel shells left in place shall be not less than 8" at the point and not less than 12" at the butt.

(d) No precast concrete pile shall be driven before the concrete has attained a compressive strength at least 3,000 lbs. per sq. in., but in all cases concrete shall be sufficiently cured to attain the ultimate strength upon which its use is based, before piles are driven. Such piles shall be reinforced to withstand conditions of handling, driving, and loading, and shall be so handled and driven as not to cause injury or overstressing which will affect their durability or strength. Precast concrete piles shall have a diameter of not less than 10 inches.

(e) Pile reinforcement when required shall have a protective covering of not less than 1½" of concrete except that where a pile has a metal casing, reinforcement shall be kept not less than 1 inch clear of such exterior casing. Reinforcing for cast-in-place concrete piles shall be considered necessary only when uplift, unbalanced lateral forces, or unsupported lengths (see Sec. 1303.1 (b) are to be considered.

1303.4 — WOOD PILES

(a) Wood piles used to support permanent structures shall be treated in accordance with this section unless it is established that the cut-off of untreated wood piles will be below lowest ground water level assumed to exist during the life of the structure.

When used in fresh water or land the treatment, preservative and minimum final retention shall be in accordance with American Wood Preservers Association Standard C-3.

When used in salt water the treatment, preservative and minimum final retention shall conform to American Wood Preservers Bureau Standard MP 1, MP 2 or MP 3 depending upon the degree or marine borer attack anticipated.

The treated pile cut-off shall have at least two (2) successive coats of the preservative liberally applied and 1) be encased in masonry footings so that no part of the pile will be exposed to the air or 2) the cut-off shall be exposed and accessible for inspection. The cut-off on all wood piles shall be along a horizontal plane.

(b) Wood piles shall conform to the "Specification for Round Timber Piles"—ASTM D25-70.

(c) The maximum allowable load on a timber pile shall meet the requirements of Section 1303.1(b) provided that the allowable working stresses of the timber is not exceeded.

ALLOWABLE WORKING STRESSES FOR PRESSURE TREATED TIMBER PILING; VALUES AT TIP OF PILE

(Allowable working stresses are determined in accordance with "Establishing Design Stresses for Round Wood Piles" ASTM D 2899)

Species	Compression** Parallel to Grain, psi	Bending** psi	Shear** Horiz. psi	Compression** Perp. to Grain, psi	Modulus of Elasticity
Pacific Coast D. Fir*	1200	2150	110	260	1,600,000
Southern Pine*	1200	2150	130	260	1,600,000
Red Oak	1100	2000	150	400	1,500,000
Red Pine	850	1550	100	180	1,200,000

*Working stresses of Douglas Fir and Southern Pine shall be increased by 0.2 percent for each foot of distance from the tip of the pile to the critical section. For compression parallel to grain, an increase of 2.5 psi per foot is recommended.

**The allowable stresses tabulated above have been reduced in accord with heat processes used to condition piling before pressure treatment. Where pressure-treated piles have been air dried, or where untreated piling are used, working stresses shall be increased by dividing tabulated values by the following factors:

Pacific Coast D. Fir, Red Oak, Red Pine	0.90
Southern Pine	0.85

1303.5 — SPECIAL TYPES OF PILES

The use of types of piles not specifically mentioned herein, and the use of piles under conditions not specifically covered herein, may be permitted, subject to the approval of the Building Official, upon the submission of acceptable test data, calculations and other information relating to the structural properties and/or load-carrying capacity of such piles. Prior to giving such approval, the Building Official may require any information or demonstrations which he deems necessary for the determination of the adequacy of the design or of the suitability of the method of installation. In no case, however, shall the allowable load exceed the limitations specified in the various subsections of Section 1303.

1303.6 — TEST OF PILES

(a) When greater loads per pile than permitted by Section 1303.1 (b) are desired, or when the design load for any pile foundation is in doubt, control-test piles shall be tested in each area by maintaining constant load under increasing settlements. The resulting allowable load shall be not more than one-half of that test load which produces a permanent net settlement per ton of test load of not more than 0.01", but in no case more than one-half inch. In subsequent driving of the balance of foundation piles, all piles shall be deemed to have a supporting capacity equal to the control-pile, when the rate of penetration of such piles is equal to or less than that of the control pile through a comparable driving distance. At least one test pile shall be driven and test loaded in each area of uniform foundation materials and additional piles shall be driven and test loaded if deemed necessary to establish safe pile loading.

(b) When any doubt exists as to the safe load-carrying capacity of any pile the Building Official may order a loading test to be made on the pile. Subject to the limitations prescribed in the various sub-sections of Section 1303, the allowable pile load shall be determined as prescribed in the foregoing paragraph.

SECTION 1304 — CAISSONS

The footings of any structure may be carried down to a firm foundation by isolated piers of reinforced concrete or by open or pneumatic caissons either with or without enlarged base or bell at the bottom. The safe carrying capacity of such shafts or caissons shall not exceed the allowable unit bearing capacity of the soil multiplied by the area of the base or bell at bottom, provided such bell shall have at least twelve (12) inch thickness of concrete at its edge and the sides shall slope at an angle of not less than sixty (60°) degrees with the horizontal unless of approved design properly reinforced. In no case shall such piers be of less than two (2) feet minimum horizontal dimension.

CHAPTER XIV

MASONRY CONSTRUCTION

SECTION 1401 — GENERAL

(a) All masonry construction shall conform to the provisions of this chapter and other applicable sections of this Code.

(b) In all cases masonry shall be of adequate thickness, strength and proportions to support all superimposed loads within the allowable working stresses prescribed.

(c) All masonry materials are required to meet the specifications as outlined in Section 1402. If the Building Official has reason to doubt the materials meeting the applicable specification he may require tests on the materials.

(d) Masonry units may be re-used when clean, whole and conforming to the other requirements of this chapter, except that the allowable working stresses shall be 50 per cent of those permitted for new masonry units.

Masonry units to be reused as structural units in areas subject to the action of the weather or soil shall not be permitted unless representative samples are tested for compliance with the applicable requirements of Section 1402.

(e) The wall thickness and other specified dimensions are nominal dimensions. The actual masonry or wall dimensions may vary from the nominal dimensions by not more than one-half (½) inch.

SECTION 1402 — MATERIALS

1402.1 — BRICK

Brick of clay, shale, concrete or calcium silicate (sandlime) shall be of a quality at least equal to that required by ASTM Standards: solid clay and shale brick shall conform to ASTM C216-71 for facing brick or ASTM C62-69 for building brick, hollow clay and shale brick shall conform to ASTM C652-70, concrete brick shall conform to ASTM C55-71, calcium silicate brick shall conform to ASTM C73-67.

When in contact with the ground, such as foundation work, or when used in retaining walls, the brick shall be of at least Grade SW for clay or shale or sand-lime brick, or Grade P for concrete brick. When exposed to the weather above ground the brick shall be of at least Grade MW for clay or shale or sand-lime brick, or Grade P for concrete brick.

1402.2 — STRUCTURAL CLAY TILE

Structural clay tile shall be of a quality at least equal to that required by "Specifications for Structural Clay Load-Bearing Wall Tile, ASTM C34-62," Grade LB when used for bearing walls or piers or Grade LBX when exposed to the weather or soil, or equal to the "Specifications for Structural Clay Non-Load-Bearing Tile, ASTM C56-71," when used for interior non-load-bearing purposes, or equal to "Specifications for Structural Clay Floor Tile, ASTM C57-57," when used as a structural unit in combination reinforced floor or roof construction. (Tile meeting specifications of "Structural Clay Load-Bearing Wall Tile, ASTM C34-62," Grade LB may be used in on-grade floors.)

1402.3 — CERAMIC AND SALT GLAZED STRUCTURAL CLAY FACING TILE OR BRICK

All glazed masonry building units shall conform to the applicable requirements for solid or hollow clay masonry units of Sections 1402.1 and 1402.2.

1402.4 — HOLLOW AND SOLID CONCRETE MASONRY UNITS

(a) Hollow concrete masonry units shall be of a quality at least equal to that required by "Specifications for Hollow Load-Bearing Concrete Masonry Units, ASTM C90-70," or "Specifications for Solid Load-Bearing Concrete Masonry Units, ASTM C145-66T," when used for bearing walls or piers or when in contact with the ground or exposed to the weather, or equal to "Specifications for Hollow Non-Load-Bearing Concrete Masonry Units, ASTM 129-64T," when used for non-load-bearing purposes and not exposed to the weather.

(b) Structural concrete filler-block or floor tile when included in strength calculations in ribbed floor construction shall have webs and shells not less than one inch thick, unless otherwise designed, and shall develop an average compressive strength on the net area not less than that of the rib concrete.

1402.5 — CAST STONE

Cast stone shall be of a quality at least equal to that required by "Specifications for Cast Stone, ACI 704-44."

1402.6 — PLAIN CONCRETE

Cast-in-place concrete construction reinforced only for shrinkage or temperature changes shall be classed as plain concrete. Plain concrete, other than fill, shall conform to the requirements for reinforced concrete (Chapter XVI) and have a minimum ultimate compressive strength at 28 days of 2,000 psi.

1402.7 — NATURAL STONE

Natural stone for masonry shall be sound and free from loose or friable inclusions; and shall meet the strength, fire-resistance, durability, and resistance to impact for the proposed use. The Building Official may require satisfactory written evidence to this effect.

1402.8 — ARCHITECTURAL TERRA COTTA AND CERAMIC VENEER

All architectural terra cotta and ceramic veneer shall have a strong homogenous body and shall conform to the applicable requirements of Section 1402.2. All units of the anchor type shall have the necessary anchor holes and shall be so formed as to engage properly with the supporting structure. All units of the adhesion type shall have keyed or scored back surfaces.

1402.9 — GYPSUM MASONRY UNITS

Gypsum partition tile or block shall be of a quality at least equal to that required by "Specifications for Gypsum Partition Tile or Block, ASTM C52-65."

1402.10 — STRUCTURAL GLASS BLOCK UNITS

Structural glass block shall comply with the provisions of Section 1413.

1402.11 — MORTAR AND GROUT MATERIALS, PROPORTIONS AND WORKABILITY

(a) Mortar and its ingredients as delivered to the mason shall be of a quality at least equal to that required by "Standard Specifications for Mortar for Unit Masonry", ASTM C270-71, or "Standard Specifications for Mortar and Grout for Reinforced Masonry", ASTM C476-71.

(b) Masonry cement shall be of a quality at least equal to that required by "Masonry Cement, ASTM C91-68."

(c) The type of mortar delivered to the mason for any specific job shall be as required in 1402.12.

(d) Grout for non-reinforced and reinforced masonry shall conform to "Standard Specifications for Mortar and Grout for Reinforced Masonry", ASTM C476-71.

(e) Where mortar type is determined in accordance with Table 1, the volume of aggregate in mortar shall be not less than two and one-fourth times but not more than three and one-half times the volume of cementitious material. When mortar type is determined by proportions, the aggregate ratio shall comply with Table 2.

(f) Gypsum mortar shall be composed of one part unfibered calcined gypsum to not more than three parts sand by weight. Only gypsum mortar shall be used with gypsum tile and block units.

1402.12 — TYPES OF MORTAR REQUIRED

Masonry shall be laid in mortar of the types specified in Tables 1, 2 and 3.

TABLE 1 — TYPES OF MORTAR

Type	Average Compressive Strength 2 inch cubes at 28 days, psi
M	2500
S	1800
N	750
O	350

TABLE 2 — MORTAR PROPORTIONS BY VOLUME*

Mortar Type	Portland Cement Cu. Ft.	Masonry Cement Cu. Ft.	Hydrated Lime or Lime Putty Cu. Ft.	Aggregate Measured in Damp Loose Condition Cu. Ft.
M	1	None	¼	Not less than 2¼ and not more than 3 times the sum of the volumes of cement and lime used.
	1	1	None	
S	1	None	Over ¼ to ½	
	½	1	None	
N	1	None	Over ½ to 1	
	None	1	None	
O	1	None	Over 1 to 2	

* For the purpose of these specifications, the weight of one cubic foot of the respective materials used shall be considered to be as follows:

Portland Cement .... 94 pounds
Masonry Cement .... weight printed on bag
Hydrated Lime .... 40 pounds
Lime Putty (Quicklime) .... 80 pounds
Sand, damp and loose .... 80 pounds of dry sand

TABLE 3 — TYPES OF MORTAR REQUIRED

Type of Masonry	Types of Mortar Permitted
Foundations: (below grade masonry)	
Footings	M or S
Walls of Solid Units	M, S or N
Walls of Hollow Units	M or S
Hollow Walls	M or S
Masonry Other Than Foundation Masonry	
Piers of Solid Masonry	M, S or N
Piers of Hollow Units	M or S
Walls of Solid Masonry	M, S, N or O
Walls of Hollow Masonry	M, S or N
Hollow Walls and Cavity Walls	
(a) Design Wind Pressure Exceeds 20 psf.	M or S
(b) Design Wind Pressure 20 psf or less.	M, S or N
Glass Block Masonry	M, S or N
Non-Bearing Partition and Fireproofing	M, S, N, O or Gypsum
Gypsum Partition Tile or Block	Gypsum
Fire Brick	Refractory Air Setting Mortar
Masonry Other Than Above	M, S or N

1402.13 — CERAMIC TILE

(a) Ceramic tile units shall be as defined in "Specifications for Ceramic Tile, ANSI A137.1-1967," and shall be of quality at least equal to "Standard Grade," graded and marked in conformance with that Specification.

(b) Ceramic tile set in cement mortar shall be installed in accordance with one of the following Standard Specifications with mortar mixes as specified therein for particular uses:

"Glazed Ceramic Wall Tile Installed With Portland Cement Mortar, ANSI A108.1-1967"

"Ceramic Mosaic Tile Installed With Portland Cement Mortar, ANSI A108.2-1967"

"Quarry Tile and Paver Tile Installed With Portland Cement Mortar, ANSI A108.3-1967"

(c) Organic adhesives to be used in installing ceramic tile shall conform to requirements of "Organic Adhesives for Installation of Ceramic Tile, ANSI A136.1-1967." Installation of ceramic tile with such adhesives shall conform to "Specifications for Ceramic Tile Installed With Water-Resistant Organic Adhesives, ANSI A108.4-1968."

(d) Ceramic tile set in dry-set mortar shall be installed in accordance with "Ceramic Tile Installed With Dry-Set Portland Cement Mortar, ANSI A108.5-1967," and the dry-set mortar shall conform with requirements of "Dry-Set Portland Cement Mortar, ANSI A118.1-1967."

(e) Ceramic tile set in epoxy shall be installed in accordance with "Ceramic Tile Installed With Chemical Resistant, Water Cleanable Tile-Setting and -Grouting Epoxy, ANSI A108.6-1969," and the epoxy shall conform with "Chemical Resistant, Water Cleanable Tile-Setting and -Grouting Epoxy, ANSI A118.3-1969."

SECTION 1403 — WORKING STRESSES

1403.1 — GENERAL REQUIREMENTS

(a) In determining the stresses in masonry, the effects of all loads and conditions of loading and the influence of all forces affecting the design and strength of the several parts shall be taken into account.

(b) The thickness of masonry walls shall be sufficient at all points to withstand all vertical and horizontal loads as specified in Chapter XII, Minimum Design Loads.

(c) Stresses shall be calculated on actual rather than nominal dimensions.

(d) The maximum allowable stresses in masonry shall not exceed those set out in this chapter, unless it can be determined by accepted engineering analysis that the design meets all safety requirements; see Sections 1403.3, 1403.7 and 1411.1.

1403.2 — WORKING STRESSES IN UNREINFORCED MASONRY

Except as permitted in other sections of this Code, the compressive stresses in unreinforced masonry shall not exceed the values given in Table 4.

TABLE 4 — ALLOWABLE COMPRESSIVE STRESSES IN UNIT MASONRY

Construction; grade of unit	Allowable compressive stresses gross cross-sectional area (except as noted)			
	Type M Mortar	Type S Mortar	Type N Mortar	Type O Mortar
Solid masonry of brick and other solid units of clay or shale; sand-lime or concrete brick:	psi	psi	psi	psi
8,000 plus, psi	400	350	300	200
4,500 to 8,000, psi	250	225	200	150
2,500 to 4,500, psi	175	160	140	110
1,500 to 2,500, psi	125	115	100	75
Grouted[1] solid masonry of brick and other solid units of clay or shale; sand-lime or concrete brick:				
4,500 plus, psi	350	275	200	......
2,500 to 4,500, psi	275	215	155	......
1,500 to 2,500, psi	225	175	125	......
Solid masonry of solid concrete masonry units:				
1,800 plus, psi	175	160	140	100
1,200 to 1,800, psi	125	115	100	75
Masonry of hollow units	85	75	70	......
Piers of hollow units, cellular spaces filled, as in Section 1405.6	105	95	90	......
Hollow walls (cavity or masonry bonded)[2].				
Solid units:				
2,500, plus psi	140	130	110	......
1,500 to 2,500, psi	100	90	80	......
Hollow units	70	60	55	......
Stone ashlar masonry:				
Granite	800	720	640	500
Limestone or Marble	500	450	400	325
Sandstone or cast stone	400	360	320	250
Rubble stone, coursed, rough or random	140	120	100	80

[1]See Section 1410.

[2]On gross cross-sectional area of wall minus area of cavity between

wythes (leaves). The allowable compressive stresses for cavity walls are based upon the assumption that the floor loads bear upon but 1 of the 2 wythes. When hollow walls are loaded concentrically, the allowable stresses may be increased by 25 per cent.

1403.3 — HIGHER WORKING STRESSES

Higher stresses than herein specified may be used, but only if it is clearly established to the satisfaction of the Building Official, by test, or other approved evidence, that material of a higher grade or a superior workmanship than is generally provided in accepted practice will be employed under approved inspection. The use of higher stresses, however, shall not be allowed until a statement, giving the reasons for such permission together with the facts and circumstances on which it is based, has been placed on file and made a part of the official record of the permit.

1403.4 — ALLOWABLE STRESSES IN COMPOSITE WALLS

In composite walls or other structural members composed of different kinds or grades of masonry units or mortars, the maximum stress shall not exceed the allowable stress for the weakest of the units and mortars of which the wall or member is composed.

1403.5 — ALLOWABLE STRESSES IN UNREINFORCED CONCRETE

The allowable stresses shall not exceed 25 per cent for compression and 3 per cent for tension in extreme fiber in flexure of the compressive strength of the concrete. When the ratio of height to thickness of structural members of plain concrete exceeds 10, the percentage for compressive stress shall be reduced proportionately to 18 per cent for a ratio of height to thickness of 20.

1403.6 — CONCENTRATED LOADS

(a) A fifty percent increase in the allowable working-stresses shown in Table 4 shall be permitted for concentrated loads meeting the bearing requirements of Section 1409.3 where such loads are supported upon not less than a 4-inch height of solid masonry units or hollow masonry units with the cells filled solidly with mortar or grout. The clear distance between the application of two adjacent concentrated loads on a wall shall be not less than the sum of the widths of the bearings or bearing plates.

(b) For piers and columns the bearing plate shall not exceed sixty percent of the cross-sectional area of the pier or column and the resultant reaction of all vertical and horizontal loads shall fall within the middle third of the member.

(c) Concentrated loads shall not be considered as distributed in masonry laid in stacked bond unless one continuous ¼-inch round reinforcing bar or its equivalent is placed in the horizontal bed joints for each 4 inches of wall thickness and spaced not less than 16 inches on centers vertically.

1403.7 — STRUCTURAL ANALYSIS OF UNREINFORCED MASONRY

(a) Where wall thickness and lateral support requirements are determined by a structural analysis based upon accepted engineering practice, the requirements contained in Sections 1404 and 1405 may be waived. The Building Official may require a copy of the structural analysis by the architect or engineer and he may require the designer or other qualified person to inspect the construction.

(b) For brick masonry of solid masonry units made from clay or shale, the design and construction may conform to the provisions of "Building Code Requirements for Engineered Brick Masonry, BIA August 1969".

(c) For load-bearing concrete masonry units, the design and construction may conform to the provisions of "Specification for the Design and Construction of Load-Bearing Concrete Masonry, NCMA-1970."

SECTION 1404 — WALL THICKNESS

1404.1 — GENERAL

The minimum thickness of all masonry bearing or non-bearing walls shall be sufficient to resist or withstand all vertical or horizontal loads required by this Code and the fire-resistance requirements set out in Chapter X. (See Section 1403.7.)

1404.2 — THICKNESS OF BEARING WALLS

The minimum thickness of masonry bearing walls shall be at least 12 inches in thickness for the uppermost 35 feet of their height and shall be increased 4 inches in thickness for each successive 35 feet or fraction thereof measured downward from the top of the wall.

Exceptions:

(a) **Stiffened Walls:** Where solid masonry bearing walls are stiffened at distances not greater than 12 feet apart by masonry cross-walls or by reinforced concrete floors, they may be of 12-inch thickness for the uppermost 70 feet, measured downward from the top of the wall, and shall be increased 4 inches in thickness for each successive 70 feet or fraction thereof.

(b) **Top Story Walls:** The top-story bearing wall of a building not exceeding 35 feet in height may be of 8-inch thickness, provided the roof construction imparts no lateral thrust to the walls, and providing the walls meet lateral support requirements of Section 1405.1 or Appendix "D", whichever is applicable.

(c) **One-Story Walls:** The walls of a one-story building may be not less than 6 inches in thickness, provided the masonry units meet the minimum compressive strength requirement of 2500 psi for the gross area and that the masonry be laid in Type M, S or N mortar.

Exception: The walls of one-story single-family dwellings and private garages may be not less than 6 inches in thickness provided the masonry units meet the minimum requirements of Section 1402 for masonry units and mortar.

(d) **Walls of Residence Buildings:** In residence buildings not more than three stories in height, walls other than coursed or rough or random rubble stone walls, may be of 8-inch thickness when not over 35 feet in height, provided the roof is designed to impart no horizontal thrust. Such walls in one-story buildings or private garages may conform to exception (c) above, and the provisions of Section 1405.1.

(e) **Penthouses and Roof Structures:** Masonry walls above roof level, 12 feet or less in height, enclosing stairways, machinery rooms, shafts, or penthouses, may be of 8-inch thickness and may be considered as neither increasing the height nor requiring any increase in the thickness of the wall below.

(f) **Walls of Plain Concrete:** Plain concrete walls may be 2 inches less in thickness than required otherwise in this section but not less than 8 inches except that they may be 6 inches in thickness when meeting the provisions of exception (c) above.

(g) **Cavity Walls:** Cavity walls shall not exceed 35 feet in height. The cavity between wythes shall be not less than 2 inches (actual) nor more than 4 inches in width, and the minimum wythe thickness shall be not less than 4 inches, except where 3-inch thick wythes are specifically permitted. The backing wythe shall be at least as thick as the facing wythe. (See Section 1401.)

(1) Where both the facing and backing wythes have a thickness of 4 inches, the height of such cavity walls shall not exceed 25 feet.

(2) Where both the facing and backing wythes are composed of solid masonry units, the wythes may be 3-inches thick but the height of such cavity walls shall not exceed 20 feet.

(h) **Masonry Bonded Hollow Walls:** Masonry bonded hollow walls shall not exceed 35 feet in height. The cavity between wythes shall not be less than 2 inches (actual) nor more than 4 inches in width, and the minimum wythe thickness shall not be less than 3 inches. The backing wythe shall be at least as thick as the facing wythe. (See Section 1401.)

(i) **Composite or Faced Walls:** Neither the height of faced (composite) walls nor the distance between lateral supports shall exceed that prescribed for the masonry of either of the types forming the facing or the backing.

(j) **Stone Walls:** Rough or random or coursed rubble stone walls shall be 4 inches thicker than required for solid masonry walls of the same height, but in no case less than 16 inches in thickness.

1404.3 — THICKNESS OF NON-BEARING WALLS

(a) **Exterior Non-Bearing Walls.** Non-bearing exterior masonry walls may be 4 inches less in thickness than required for bearing walls but the thickness shall be not less than 8 inches except where 6-inch walls are specifically permitted.

(b) **Exterior Panel, Apron or Spandrel Walls.** Panel, apron or spandrel walls that do not exceed 13 feet in height above their support shall not be limited in thickness, provided they meet the fire-resistance requirements of Chapter X and are so anchored to the structural frame as to insure adequate lateral support and resistance to wind or other lateral forces (See Section 608.3).

1404.4 — FOUNDATION WALLS

See Section 1302.5.

SECTION 1405 — LATERAL SUPPORT

1405.1 — EXTERIOR WALLS

Exterior masonry walls, whether they be bearing or non-bearing shall be supported either horizontally or vertically (whichever distance is the lesser) at right angles to the face of the wall at intervals not exceeding those shown in Table 5 except that an additional 6 feet will be permitted for gables in residential structures and private garages that do not exceed one story in height. (See Section 1403.7.)

TABLE 5 — LATERAL SUPPORT-EXTERIOR WALLS

Type of Masonry	Maximum Ratio of Unsupported Height or Length to Thickness			
Mortar Type	M	S	N	O
Grouted Brick Masonry	22	22	22	—
Plain Solid Masonry	20	20	20	16
Hollow Unit Masonry	18	18	18	12
Masonry Bonded Hollow Walls	18	18	18	—
Cavity Walls*	18	18	18	—

*In computing the ratio for cavity walls the value for thickness shall be the sum of the nominal thickness of the inner and outer wythes.

1405.2 — BEARING PARTITIONS

Masonry bearing partitions shall be supported either vertically or horizontally (whichever distance is the lesser) at right angles to the face of the wall at intervals not exceeding 24 times the wall thickness for solid masonry units, and 20 times the wall thickness for hollow masonry units when laid in Type M, S or N mortar. Gypsum partition tile or block shall not be used in bearing walls. (See Section 1403.7.)

1405.3 — NON-BEARING PARTITIONS

(a) Non-bearing partitions shall be supported either vertically or horizontally (whichever distance is the lesser) at right angles to the face of the wall at intervals not exceeding 45 times the nominal wall thickness exclusive of plaster. (See Section 1403.7.)

(b) Gypsum partition tile shall not be used for partitions to receive portland cement plaster, ceramic tile, marble or structural glass, unless self-furring metal lath is placed over the gypsum tile. Gypsum partition tile or block shall not be used where they will be subjected to continuous dampness.

(c) Only gypsum cement mortar shall be used in the erection of gypsum partition tile or block.

1405.4 — METHOD OF SUPPORT

(a) Lateral support shall be provided by intersecting walls, pilasters, columns, or other vertical members of sufficient strength to provide the required support when the distance is measured horizontally; or by floors, roofs, or other horizontal structural elements which are of sufficient strength to provide the required support when the distance is measured vertically.

(b) Sufficient bonding or anchorage shall be provided between the walls and its supports to resist the assumed wind or other horizontal forces acting either inward or outward. All structural elements relied upon for lateral support shall have sufficient strength and stability to transfer the horizontal force acting in either direction to adjacent structural members or to the ground. When floors or roofs are depended upon for receiving horizontal forces, provisions shall be made in the buildings to transfer the lateral forces to the ground.

(c) When horizontal structural elements of a building (such as floors, roof, spandrel beams) are depended upon for lateral support, vertical bracing of bearing or non-bearing walls shall also be provided at intervals of not more than 75 times the wall thickness. Such vertical bracing may be provided by cross-walls, pilasters, buttresses or other equivalent structural members.

1405.5 — PILASTERS

When relied upon to provide the required lateral support, the width of pilasters shall be not less than one-tenth (1/10) the spaces between such pilasters. All pilasters shall be not less than four (4) inches thicker than the wall supported. In no case shall the distance between such pilasters exceed the lateral support provisions of Table 5.

1405.6 — PIERS

The unsupported height of masonry piers shall not exceed 10 times their least dimension. When structural clay tile or hollow concrete masonry units are used for isolated piers to support beams and girders, the cellular spaces shall be filled solidly with concrete or Type M or S mortar, except that unfilled hollow piers may be used if their unsupported height is not more than 4 times their least dimension. When hollow masonry units are solidly filled with concrete or Type M, S or N mortar, the allowable compressive stress may be increased as provided for in Table 4.

Hollow piers shall be capped with 4 inches of solid masonry or concrete or shall have cavities of top course filled with concrete or grout or other methods approved by the Building Official.

SECTION 1406 — PARAPET WALLS

1406.1 — GENERAL

(a) Parapet walls may be of plain solid masonry, hollow masonry units, cavity wall design or reinforced masonry as provided in this section.

(b) All parapet walls shall be properly coped with non-combustible, weatherproof material of a width not less than the thickness of the parapet wall.

(c) Proper flashings shall be installed in such a manner as to prevent moisture entering the wall through the joints in the coping.

(d) There shall be placed in all parapet walls scuppers or relief openings as close as practical to each downspout of not less than 4 to 6 inches in size and spaced not more than 6 inches above the roofline unless the roof is especially designed for water cooling, in which case the scuppers may be raised to provide for retaining the water.

1406.2 — PLAIN MASONRY PARAPET WALLS

Parapet walls of plain solid masonry construction shall be not less than 8 inches in thickness and their height shall not exceed 4 times the nominal wall thickness.

1406.3 — CAVITY WALL PARAPETS

Cavity wall parapets may be used when the facing and backing are constructed of solid masonry units and when they conform to all of the other requirements for cavity walls, except their height shall not exceed 4 times the combined nominal thickness of the facing and backing masonry units.

1406.4 — REINFORCED MASONRY PARAPET WALLS

Unless reinforced to withstand safely wind loads to which they may be subjected, reinforced masonry parapet walls may be considered adequate if they conform to one of the following:

(a) When solid masonry parapet walls are reinforced both horizontally and vertically with not less than ¼-inch rods spaced not more than 2 feet on centers, the height shall be not more than 6 times the nominal wall thickness.

(b) When solid masonry parapet walls are reinforced both horizontally and vertically with a minimum of ¼-inch round rods horizontally on 16-inch centers and ⅜-inch round rods vertically on 24-inch centers, the height shall not exceed 8 times the nominal wall thickness.

SECTION 1407 — BONDING

1407.1 — GENERAL

The facing and backing of masonry walls and partitions shall be bonded in such a manner to provide for common action of the wythes of the material used. Bonding may be accomplished as outlined in Sections 1407.2, 1407.3, 1407.4, 1407.5, or 1407.6.

1407.2 — BONDING WITH HEADERS

When the facing and backing are bonded by means of masonry headers, no less than 4 per cent of the wall surface of each face shall be composed of headers extending not less than 4 inches into the backing. The distance between adjacent full length headers shall not exceed 24 inches either vertically or horizontally. In walls in which a single header course does not extend through the wall, bonders from the opposite sides shall lap at least 4 inches, or bonders from opposite sides shall be covered with another bonder course overlapping the bonder below at least 4 inches.

1407.3 — BONDING WITH METAL TIES

(a) The facing and backing of cavity walls shall be bonded with corrosion-resistant 3/16-inch diameter steel rods or metal ties of equivalent stiffness embedded in the horizontal joints. There shall be one metal tie for not more than each 3 square feet of wall area. Ties in alternate courses shall be staggered. Rods bent to rectangular shape shall be used with hollow masonry units laid with the cells vertical; in other walls, the ends of ties shall be bent to 90° angles to provide hooks not less than 2 inches long. Additional bonding ties shall be provided at all openings, spaced not more than 3 feet apart around the perimeter and within 12 inches of the opening.

(b) The facing and backing of solid masonry or hollow masonry walls may be bonded with corrosion-resistant metal ties conforming to the requirements for cavity walls. Where the space between metal tied wythes is solidly filled with mortar, the allowable stresses and other provisions for masonry bonded walls shall apply. Where the space between metal tied wythes is not filled with mortar, metal tied walls shall conform to the allowable stress and lateral support requirements for cavity walls.

1407.4 — BONDING WITH HOLLOW MASONRY UNITS

When two or more hollow masonry units are used to make up a thickness of a wall, the inner and outer courses may be bonded at vertical intervals not exceeding 34 inches by lapping at least 4 inches over the units below, or by lapping with units at least 50 per cent greater in width than the unit below at vertical intervals not exceeding 17 inches.

1407.5 — ASHLAR, NATURAL OR CAST STONE

(a) **Construction:** In ashlar masonry, bond stones uniformly distributed shall be provided to the extent of not less than 10 per cent of the area of exposed faces.

Rubble stone masonry 24 inches or less in thickness shall have bond stones with a maximum spacing of 3 feet vertically and 3 feet horizontally and, if the masonry is of greater thickness than 24 inches, shall have one bond stone for each 6 square feet of wall surface on both sides.

(b) **Minimum Thickness:** Stone masonry walls shall in no case have a minimum thickness of less than 16 inches.

1407.6 — MASONRY BONDED HOLLOW WALLS

(a) In masonry bonded hollow walls, the facing and backing shall be bonded so that not less than 4 per cent of the wall surface of each face is composed of masonry bonding units extending not less than 3 inches into the backing. The distance between adjacent bonders shall not exceed 24 inches either vertically or horizontally.

(b) Where the bonding units have a compressive strength of 4500 psi gross area, the facing and backing may be bonded so that not less than 2 per cent of the wall area is composed of bonders.

SECTION 1408 — ANCHORAGE

1408.1 — GENERAL

All structural elements depending upon one another for continuity or support shall be securely anchored in such a manner as to resist all forces which might tend to separate the structural elements. It is particularly important that adequate anchorage be provided between the roof structure and load-bearing masonry walls to prevent the uplifting and subsequent separation of the roof from the masonry wall as the result of wind pressures.

1408.2 — ROOF ANCHORAGE

Roof structures shall be securely anchored to load-bearing masonry walls. Anchorage shall be considered adequate if provided in one of the following manners or its equivalent:

(a) Anchorage may be provided by ½-inch bolts extending a minimum of 15 inches into the masonry and spaced not more than 6 feet on centers. A steel plate having a minimum surface area of 6 square inches shall be securely attached to the head of the bolt and completely embedded in the masonry.

(b) Where a continuous bond beam at least 8 inches deep and having a minimum continuous reinforcing of 0.2 square inches is provided at the top of the wall, anchorage may be provided by one of the following methods:

(1) Welding ½-inch anchor bolts to longitudinal reinforcing.

(2) Hooking tightly around the longitudinal reinforcing through 180°.

With this type of anchorage, bolts need only extend into the wall a minimum of 6 inches.

1408.3 — FLOOR ANCHORAGE

(a) Wood floor joists or beams resting on masonry walls shall be anchored at intervals of not more than six (6) feet by approved metal fasteners attached in a manner to be self-releasing. When joists are parallel to masonry, such walls shall be secured to three (3) or more joists by approved metal anchors at intervals of not more than six (6) feet.

(b) Steel floor joists not supporting a concrete slab floor shall be anchored in a manner providing anchorage equivalent to that required for wood floor joists. Concrete slabs bearing on masonry walls shall be considered to provide adequate anchorage without additional anchorage.

1408.4 — ANCHORING INTERSECTING WALLS AND PARTITIONS

Masonry walls that meet or intersect shall be adequately bonded or anchored as follows:

(a) Intersecting bearing walls may be bonded either by laying a true bond of at least 50 per cent of the units at the intersection or by using corrosion-resistant metal ties embedded in the bed joints. When metal ties are used for such bonding, they shall be corrosion-resistant 3/16-inch diameter steel rods bent to a rectangular shape and spaced at intervals not exceeding 16 inches vertically. They shall be placed in such a manner as to extend at least 3 inches into each intersecting wall.

(b) Non-bearing partitions, when intersecting walls or partitions, shall be anchored with metal ties or clips at least ⅞ inch wide and not less than 16 gauge galvanized iron at intervals of not more than 32 inches vertically.

(c) Brick or tile facing against concrete shall be anchored to the concrete by the use of dovetailed anchors inserted in slots built into the concrete. Anchors shall be at least ⅞ inch wide and not less than 16 gage galvanized iron. They shall be spaced not more than 18 inches vertically and 24 inches horizontally.

(d) Two-inch split furring and 2-inch open back (split) tile (soaps) shall be anchored to the backing with hardware cloth ties consisting of ½ inch mesh, No. 20 gage galvanized iron fabric, at least 4 inches wide and extending at least 3 inches into the masonry and to within ½ inch of the face of the furring, or by other approved ties. Ties shall be spaced not farther apart than 24 inches vertically and 36 inches horizontally.

SECTION 1409 – MISCELLANEOUS DETAILS

1409.1 — CHANGE IN WALL THICKNESS

Except for permissable chases and recesses, walls shall not vary in thickness between their lateral supports. Where cavity walls or walls of hollow masonry units are decreased in thickness, a course of solid masonry not less than 4 inches in thickness shall be interposed between the wall below and the thinner wall above, or the hollow units in the top course of the thicker wall shall be filled solidly with concrete or Type M, S or N mortar.

1409.2 — CHASES

(a) Chases in masonry walls shall not be deeper than ⅓ the wall thickness, nor longer than 4 feet horizontally, except that chases below windows may equal the width of the opening above.

(b) No chase shall be cut or built in an 8-inch wall or within the required area of a pier, except that in buildings of residential occupancy not more than 2 stories in height, chases not more than 4 inches deep may be built in 8 inch walls.

(c) Chases shall not be cut in cavity walls, hollow walls or walls of hollow units but, when permitted, may be built in.

1409.3 — SUPPORTED STRUCTURAL MEMBERS

(a) When combustible structural members frame into walls of thicknesses not greater than 12 inches, they shall project not more than 4 inches into the wall and shall be so spaced that the distance between embedded ends is not less than 4 inches. The space above, below, and between such members shall be filled solidly with burnt-clay materials, mortar, concrete, or equivalent fire-resistive material to a depth of not less than 4 inches on all sides of the members.

(b) Beams, joists, girders or other concentrated loads supported by a wall or pier shall have bearing at least 4 inches in length upon solid masonry or upon a bearing plate of adequate design and dimensions to distribute safely the loads on the wall or pier and in no case shall the stresses be greater than allowed in Section 1403.

1409.4 — SUPPORT ON WOOD

No masonry shall be supported on combustible construction, except that prefabricated partitions, weighing not more than 30 pounds per square foot, properly strapped or reinforced and provided with proper nailing devices for attachment may be supported on combustible construction, provided the supporting construction has been designed to carry such loads. Concrete decks for roofs or floors may be supported on wood construction provided such decks and their supporting members have been designed in accordance with accepted engineering practices. When exposed to the weather the wood supporting members shall be of approved wood of natural decay resistance or pressure treated wood or shall be separated from the concrete by the use of membrane covering.

1409.5 — CORBELLING

The maximum horizontal projection of corbelling from the face of the wall from which it projects shall not exceed ½ of the nominal wall thickness. Individual corbells or the maximum projection of one unit shall not exceed ½ the height of the unit nor ⅓ its bed depth. For corbeling of chimneys see Section 803.1(b).

1409.6 — ARCHES AND LINTELS

The masonry above openings shall be supported by well buttressed arches or lintels of metal or masonry, plain or reinforced, which shall bear on the wall at each end for not less than 4 inches. In addition, the bearing area shall be sufficient to prevent a concentration of compressive stresses greater than those allowed in Table 4, Section 1403.

1409.7 — COLD WEATHER CONSTRUCTION

Masonry may be laid when the temperature of the outside air is below 40 degrees F when protection requirements are in compliance with "Recommended Practices for Cold Weather Masonry Construction", as published by the International Masonry Industry All-Weather Council, August, 1970.

1409.8 — WETTING OF MASONRY UNITS

(a) Brick of clay or shale shall be wetted when laid unless their gain in weight is less than 0.025 ounce per square inch of surface when immersed flatwise in ⅛ inch of water for one minute.

(b) Structural clay tile having absorptions (1-hour boil) of 12 per cent or more shall be wetted before laying.

1409.9 — CONSTRUCTION PRECAUTIONS

Except when carried independently by girders at each floor, no wall shall be built up more than 25 feet in height in advance of other walls of the building. Walls shall be adequately braced during erection. Masonry walls in locations where they may be exposed to high winds during erection shall not be built higher than 10 times their thickness unless adequately braced or until provision is made for the prompt installation of permanent bracing at the floor or roof level immediately above the story under construction. Back fill shall not be placed against foundation walls until they have been braced to withstand the horizontal pressure.

1409.10 — USE OF EXISTING WALLS

(a) An existing masonry wall may be used in the alteration or extension of a building, provided that under the new conditions it meets the requirements of this Code and is structurally sound or can be made so by reasonable repairs.

(b) No existing wall shall be used for the alteration or extension of a building, or increased in height without specific written permission from the Building Official.

SECTION 1410 — PLAIN GROUTED MASONRY

1410.1 — GENERAL

(a) Grouted masonry shall conform to all requirements of Sections 1401 to 1409 inclusive, except as modified by this Section.

(b) Plain grouted masonry is that form of construction made with clay, shale or concrete masonry units in which the interior joints are filled by pouring grout therein as the work progresses.

(c) The masonry units in either the facing or backing, but not necessarily both, at the time of laying, shall absorb in 24 hours of cold immersion an amount of water weighing at least 5 per cent of the dry weight of the unit.

1410.2 — MORTAR AND GROUT

(a) Only Types M, S or N mortar shall be used. Grout shall be Types M, S or N mortars with the addition of sufficient water to give the required pouring consistency without segregation of the constituents of the mortar.

(b) In grout spaces of 2 inches or more in both horizontal dimensions, the grout may contain an addition of pea gravel equal to not more than 2 parts by volume of the cement. Such pea gravel shall be graded with not more than 5 per cent passing the No. 8 sieve and with not less than 95 per cent passing the ⅜-inch sieve.

(c) Brick pieces or chips may be embedded into grout in such spaces, provided each piece or chip is surrounded by not less than ½ inch of grout.

(d) Where the minimum continuous clear openings of a grout space exceeds 6 inches, it may be filled and treated as unreinforced monolithic concrete.

1410.3 — BOND

Where all interior joints are filled with grout, masonry headers shall not be used, but metal wall ties may be used to prevent spreading of the wythes and to maintain the vertical alignment of the wall.

1410.4 — CONSTRUCTION REQUIREMENTS

All masonry units in the 2 outer tiers or wythes shall be laid plumb with full bed and head joints. All interior joints shall be filled with grout. One of the outer tiers may be carried up not more than 8 inches before grouting, but the other face tier shall be carried up not more than 5½ inches above the grout. Each pour of grout shall be stopped at least 1½ inches below the top and properly puddled. The longitudinal vertical joints (collar joints) shall be not less than ¾ inch in width.

SECTION 1411 — REINFORCED MASONRY

1411.1 — GENERAL

All reinforced masonry, except reinforced brick masonry, shall conform to the provisions of "Building Code Requirements for Reinforced Masonry, ANSI A41.2-1960," (Handbook 74). Reinforced brick masonry shall conform to the provisions of "Recommended Building Code Requirements for Engineered Brick Masonry, BIA—August 1969." Reinforced load-bearing concrete masonry shall conform to the provisions of "Specification for the Design and Construction of Load-Bearing Concrete Masonry, NCMA-1970."

SECTION 1412 — REINFORCED GYPSUM CONCRETE

1412.1 — GENERAL

(a) Reinforced Poured Gypsum Concrete shall conform to the requirements of "Specifications for Gypsum Concrete, ASTM C317-68."

(b) The design and application of reinforced gypsum concrete shall be in accordance with the requirements of "Specifications for Reinforced Gypsum Concrete, ANSI A59.1-1968."

(c) A competent inspector, satisfactory to the Building Official, shall be present on the work at all times when cast-in-place gypsum concrete is being mixed or deposited.

SECTION 1413 — STRUCTURAL GLASS BLOCK

1413.1 — WHERE PERMITTED

Masonry of glass blocks may be used in non-load-bearing exterior or interior walls and in openings which might otherwise be filled with windows, either isolated or in continuous bands, provided the glass block panels have a thickness of not less than 3½ inches at the mortar joint and the mortared surfaces of the blocks are satisfactorily treated for mortar bonding.

1413.2 — SIZE OF PANELS

Glass block panels for exterior walls shall not exceed 144 square feet of unsupported wall surface nor 25 feet in length nor 20 feet in height between supports. For interior walls, glass block panels shall not exceed 250 square feet of unsupported area nor 25 feet in one direction between supports.

1413.3 — REINFORCEMENT OF EXTERIOR PANELS

(a) **Anchorage:** Exterior glass block panels shall be held in place in the wall opening to resist both external and internal pressures due to wind. Panels shall be set in recesses at the jambs and, for panels exceeding 10 feet in horizontal dimension between supports, at the head as well, so as to provide a bearing surface at least one inch wide along the panel edges; except that when approved by the Building Official for panels exceeding neither 100 square feet in area nor 10 feet in either horizontal or vertical dimension, and situated 4 stories or less, anchorage may be provided by means of non-corrodible perforated metal strips.

(b) **Placing Reinforcement:** Glass block panels shall have reinforcement in the horizontal mortar joints, extending from end to end of mortar joints, but not across expansion joints, with any unavoidable joints spliced by lapping the reinforcement not less than 6 inches. The reinforcement shall be spaced not more than 2 feet apart vertically. In addition reinforcements shall be placed in the joint immediately below and above any openings within a panel. The reinforcement shall consist of 2 parallel, longitudinal, galvanized steel wires, No. 9 gage or larger, spaced 2 inches apart, and having welded thereto No. 14 or heavier gage cross wires at intervals not exceeding 8 inches, or the equivalent approved by the Building Official.

1413.4 — MORTAR

Glass block shall be laid in Type M, S or N mortar. Both vertical and horizontal mortar joints shall be at least ¼ inch and not more than ⅜ inch thick and shall be completely filled.

1413.5 — EXPANSION JOINTS

Every exterior glass block panel shall be provided with expansion joints at the sides and top. Expansion joints shall be entirely free of mortar, and shall be filled with resilient material.

SECTION 1414 – VENEERED WALLS

1414.1 — GENERAL

(a) Veneer as used in this section refers to a facing of brick, tile, concrete masonry units, metal, including metal coated with porcelain enamel, glass, wood or similar material securely attached to a wall for the purpose of providing ornamentation, protection, or insulation but not so bonded as to exert a common reaction under load.

(b) Veneer shall not be assumed as supporting any load other than its own weight, neither shall it be assumed to add to the strength of the wall.

(c) Exterior veneer shall not be attached to wood at any point more than 20 feet above the adjacent ground elevation. See Section 1414.5.

(d) See Section 608.4 for exterior wood veneer.

1414.2 — VENEER OF STRUCTURAL MASONRY UNITS

Veneer of structural masonry units shall apply to all veneer composed of units meeting the physical requirements of this Chapter, set in mortar and not less than 1⅝ inches in actual thickness for solid masonry units, and not less than 3 inches in actual thickness for hollow masonry units.

1414.3 — SUPPORT OF VENEER OF MASONRY UNITS

The weight of all masonry veneer 1⅝ inches or greater in actual thickness shall be supported upon footings, foundation walls or other approved non-combustible structural supports.

1414.4 — ANCHORAGE OF VENEER OF MASONRY UNITS

(a) All masonry veneer 1⅝ inches or greater in actual thickness shall be attached to the supporting wall either by headers as provided in Section 1407.2, or with corrosion-resistant metal ties or other approved methods.

(b) Veneer ties, if of strand wire, shall be not less than No. 6 W. & M. gage wire with the end of the wire bent to a 90-degree angle to form a hook not less than 2 inches long embedded in the mortar joint. Veneer ties, if of corrugated sheet metal, shall be not less than 22 Manufacturer's gage. Each metal tie shall support not more than 3 square feet of wall area and shall not be spaced farther apart than 16 inches vertically and 32 inches horizontally.

1414.5 — HEIGHT OF VENEER OF MASONRY UNITS

Masonry veneer 1⅝ inches or greater in thickness, unless otherwise designed, shall not be attached to wood framing at any point more than 20 feet above the foundation, except in gables. Such veneer attached to masonry walls shall not exceed 35 feet in height above approved supports.

1414.6 — VENEER OF NON-STRUCTURAL UNITS

Veneer of non-structural units shall apply to all veneer less than 1⅝ inches in thickness and the units shall not be assumed to support any superimposed loads.

1414.7 — ANCHORAGE GENERAL

Non-structural materials used as veneer shall be anchored to the supporting wall by corrosion-resistant metal ties not less in thickness than No. 9 W. & M. gage wire, and spaced not more than 12 inches apart both horizontally and vertically or by other approved devices or methods. Such attachments and their supports shall be capable of resisting a horizontal force equal to the wind loads specified in this Code, but in no case less than 20 pounds per square foot.

1414.8 — ADHESION TYPE ANCHORAGE

(a) **Material:** Approved units or units less than 1⅝ inches in thickness of flat tile, stone or adhesion type architectural terra cotta manufactured with keyed or scored back surface may be cemented to a masonry or concrete wall or to exterior portland cement mortar on high rib metal lath with a Type M or S mortar provided the mortar bond is sufficient to withstand a shearing stress of 50 pounds per square inch after curing for 28 days. No individual unit so attached shall exceed 30 inches in any one dimension and shall not have more than 540 square inches of superficial face area.

(b) **Installation:** Just before setting, each piece shall be soaked in clean water for one hour or more and the surface of the backing wall shall be saturated with water applied through a hose nozzle at a pressure of at least 25 pounds per square inch. A brush coat of neat portland cement and water shall then be applied both to the backing and the back side of the veneer. The mortar shall average ¾ inch in thickness. One-half of the mortar (⅜ inch) shall be applied to the veneer unit and half to backing just prior to setting. Pieces disturbed after having been tapped into place shall be removed immediately, after which additional mortar shall be applied as required above and the piece reset.

1414.9 — METAL VENEERS

(a) **Material:** Metal veneers that are exposed to the weather shall be of corrosion-resistant metal, or metal covered front and back with porcelain enamel or given other approved treatment or coating to render them corrosion-resistant.

(b) **Attachment:** Exterior metal veneer shall be securely attached to the supporting masonry or framing members with corrosion-resistant fastenings, metal ties or by other approved devices or methods. The spacing of the fastenings or ties shall not exceed 24 inches either vertically or horizontally, except where wider spacing is deemed adequate and specifically approved by the Building Official; but where units exceed 4 square feet in area there shall be not less than four attachments per unit. The metal attachments shall have a cross-sectional area not less than provided by No. 9 W. & M. gage wire (0.0173 square inch). Such attachments and their supports shall be capable of resisting a horizontal force equal to the wind loads specified in this Code, but in no case less than 20 pounds per square foot.

(c) **Supports:** Metal supports for exterior metal veneer shall be protected by painting, galvanizing, or by other approved equivalent coating or treatment. Wood studs, furring strips, or other wood supports for exterior metal veneer shall be pressure-treated with an approved preservative or otherwise protected against decay in an approved manner.

(d) **Protection:** All joints and edges in metal veneer that are exposed to the weather shall be caulked or painted with approved durable waterproofing material, or shall be protected by other approved means to prevent penetration of moisture.

No masonry backup shall be required for metal veneer except as is necessary to meet the fire-resistance requirements of this Code.

(e) **Grounding Metal Veneers:** Metal veneers fastened to supporting elements which are not a part of the grounded metal framing of a building shall be made electrically continuous by contact or interconnection of individual units and shall be effectively grounded. The conductor used to ground the veneer shall have no greater resistance than the conductor used to ground the electrical system within the building. Where a metal veneer is applied to a building with no electrical wiring system, grounding shall be required only if determined to be necessary by the Building Official.

1414.10 — GLASS VENEER

(a) **Area:** The area of a single section of thin exterior structural glass veneer shall not exceed 10 square feet where not more than 15 feet above the level of the sidewalk or grade level directly below, and shall not exceed 6 square feet where more than 15 feet above that level.

(b) **Length or Height:** The length or height of any section of thin exterior structural glass veneer shall not exceed 48 inches.

(c) **Thickness:** The thickness of thin exterior structural glass veneer shall be not less than 11/32 inch.

1414.11 — ATTACHMENT

Thin exterior structural glass veneer shall be set only after backing is thoroughly dry and after application of an approved bond coat applied uniformly over the entire surface of the backing so as to effectively seal the surface. Glass shall be set in place with an approved mastic cement in sufficient quantity so that at least 50 per cent of the area of each glass unit is directly bonded to the backing by mastic not less than ¼ inch thick and not more than ⅝ inch thick. Bond coat and mastic shall preferably be from the same manufacturer and shall bond firmly together.

1414.12 — GLASS AT SIDEWALK LINE

Where glass extends to sidewalk surface, each section shall rest in an approved metal moulding, and set at least ¼ inch above the highest point of the sidewalk. The space between the moulding and the sidewalk shall be thoroughly caulked and made watertight.

1414.13 — JOINTS

(a) **Abutting Edges:** Unless otherwise specifically approved by the Building Official, all abutting edges of thin exterior structural glass veneer shall be ground square. Mitered joints shall not be used except when specifically approved for wide angles.

(b) **Treatment of Joints:** All joints shall be uniformly buttered with an approved jointing compound and all horizontal joints shall be held to not less than 1/16 inch by an approved non-rigid substance or device.

(c) **Expansion Joints:** Where thin exterior structural glass veneer abuts non-resilient material at sides or top, expansion joints not less than ¼ inch wide shall be provided.

1414.14 — SHELF ANGLES

When thin exterior structural glass veneer is installed above the level of the top of a bulkhead facing, or at a level more than 36 inches above the sidewalk level, the mastic cement binding shall be supplemented with approved non-ferrous metal shelf angles located in the horizontal joints in every course. Such shelf angles shall be of not less than 18 Manufacturer's gauge and not less than 2 inches in length, and shall be spaced at approved intervals, with not less than 2 angles for each glass unit. Shelf angles shall be secured to the wall or backing with expansion bolts, toggle bolts, or by other approved methods.

1414.15 — MECHANICAL FASTENINGS

(a) **Where Required:** All thin exterior structural glass veneer installed above the level of the heads of show windows and all such veneer installed more than 12 feet above sidewalk level, shall, in addition to the mastic cement and shelf angles, be held in place by the use of approved fastenings at each vertical or horizontal edge, or at the 4 corners of each glass unit.

(b) **Attachment to Backing:** Fastenings shall be secured to the wall or backing with expansion bolts, toggle bolts, or by other approved method.

(c) **Type and Design:** Fastenings shall be of approved type and be so designed as to hold the glass veneer in a vertical plane independently of the mastic cement. Shelf angles providing both support and fastenings may be used.

1414.16 — FLASHING

Exposed edges of thin exterior structural glass veneer shall be flashed with over-lapping corrosion-resistant metal flashing and caulked with a waterproof compound in an approved manner to effectively prevent the entrance of moisture between the glass veneer and the backing.

CHAPTER XV

STEEL CONSTRUCTION

SECTION 1501 — GENERAL

The quality, design, fabrication and erection of steel and iron used structurally in buildings or structures shall conform to the provisions of this chapter.

SECTION 1502 — STRUCTURAL STEEL CONSTRUCTION

The design, fabrication and erection of structural steel for buildings shall conform to the requirements of the AISC "Specification for the Design, Fabrication and Erection of Structural Steel for Buildings", adopted February 12, 1969, and Supplement No. 1 thereto, effective November 1, 1970, and Supplement No. 2, effective December 8, 1971.

SECTION 1503 — COLD-FORMED CARBON AND LOW ALLOY STEEL CONSTRUCTION

The design of cold-formed steel construction shall conform to the "Specification for the Design of Cold-Formed Steel Structural Members", AISI—1968 Edition.

All individual structural members and assembled panels of cold-formed steel construction, except where fabricated of approved corrosion-resistive steel or of steel having a corrosion-resistive metallic or other approved coating, shall be protected against corrosion with an acceptable shop coat of paint, enamel, or other approved protection.

SECTION 1504 — LIGHT GAGE COLD-FORMED STAINLESS STEEL CONSTRUCTION

The design of light gage cold-formed stainless steel construction shall conform to the "Specification for the Design of Light Gage Cold-Formed Stainless Steel Structural Members", AISI—1968 Edition.

SECTION 1505 — OPEN WEB STEEL JOIST CONSTRUCTION

The design, fabrication and erection of open web steel joist construction shall comply with the following specifications:

"Standard Specifications for Open Web Steel Joist, J- & H-Series", AISC-SJI, adopted November 1, 1972.

"Standard Specifications for Longspan Steel Joists, LJ- & LH-Series", AISC-SJI, adopted November 1, 1972.

"Standard Specifications for Deep Long-span Steel Joists, DLJ & DLH—Series, AISC-SJI, adopted November 1, 1972.

SECTION 1506 — WELDING

Details of design, workmanship and technique for welding, inspection of welding and qualification of welding operators shall conform to the recommendations of the "Code for Welding in Building Construction," AWS D1.0-72.

CHAPTER XVI

CONCRETE CONSTRUCTION

SECTION 1601 — GENERAL

All structures of reinforced concrete, including prestressed concrete, shall be designed and constructed in accordance with the provisions of "Building Code Requirements for Reinforced Concrete," ACI 318-71.

SECTION 1602 — CONCRETE QUALITY

1602.1 — GENERAL

(a) Concrete shall be proportioned and produced to provide an average compressive strength sufficiently high to minimize the frequency of strength tests below the value of the specified compressive strength of the concrete, f'_c. See Section 1602.2 (b) (1).

(b) Requirements for f'_c shall be based on tests of cylinders made and tested in accordance with ASTM methods as prescribed in Section 1602.2 and 1602.3.

(c) Unless otherwise specified, f'_c shall be based on 28-day tests. For high-early strength concrete, the test age for f'_c shall be as indicated in the plans or specifications.

1602.2 — SELECTION OF CONCRETE PROPORTIONS

(a) Proportions of ingredients for concrete shall be established on the basis of paragraphs (b) through (h) of this section to provide:

(1) Conformance with the strength test requirements of Section 1602.3.

(2) Adequate workability and proper consistency to permit the concrete to be worked readily into the forms and around reinforcement under the conditions of placement to be employed, without excessive segregation or bleeding.

(3) Resistance to freezing and thawing and other aggressive actions, where required.

The criteria of paragraphs (b) through (d) of this Section are solely for the purpose of establishing required mixture proportions and do not constitute a basis for conforming the adequacy of concrete strength, which is covered in Section 1602.3.

(b) Except as permitted in Paragraph (d) or required by Paragraphs (e), (f) or (g) of this Section, proportions, including water-cement ratio, shall be established on the basis either of laboratory trial batches or of field experience with the materials to be employed. The proportions shall be selected to produce an average strength at the designated test age exceeding f'c by the amount indicated below, when both air content and slump are the maximums permitted by the specifications.

(1) Where the concrete production facility has a record, based at least 30 consecutive strength tests representing similar materials and conditions to those expected, the strength used as the basis for selecting proportions shall exceed the required f'c by at least:

400 psi if the standard deviation is less than 300 psi
550 psi if the standard deviation is 300 to 400 psi
700 psi if the standard deviation is 400 to 500 psi
900 psi is the standard deviation is 500 to 600 psi

TABLE 1602.2A — MAXIMUM PERMISSIBLE WATER-CEMENT RATIOS AND MINIMUM CEMENT CONTENTS FOR CONCRETE (WHEN STRENGTH DATA FROM TRIAL BATCHES OR FIELD EXPERIENCE ARE NOT AVAILABLE)

Specified compressive strength f'c psi*	Minimum sks. cement per cu. yd. concrete	Maximum permissible water-cement ratio			
		Non-air-entrained concrete		Air-entrained concrete	
		Absolute ratio by weight	U.S. gal. per 94-lb. bag of cement	Absolute ratio by weight	U.S. gal. per 94-lb. bag of cement
2500	5	0.65	7.3	0.54	6.1
3000	5½	0.58	6.6	0.46	5.2
3500	6	0.51	5.8	0.40	4.5
4000	6½	0.44	5.0	0.35	4.0
4500	7	0.38	4.3	0.30	3.4
5000	7½	0.31	3.5	**	**

*28-day strengths for cements meeting strength limits of ASTM C150 Type I, IA, II or IIA and 7-day strengths for Type III and IIIA; with most materials, the water-cement ratios and cement contents shown will provide average strengths greater than indicated in Section 1602.2 (b) as being required.

**For strengths above 4500 psi with air-entrained concrete, proportions should be selected by the methods of Sections 1602.2 (b) and (c).

Strength data for determining standard deviation shall be considered to comply with the foregoing stipulations if they represent either a group of at least 30 consecutive tests or the statistical average for two groups totaling 30 or more tests. The tests used to establish standard deviation shall represent concrete produced to meet a specified strength or strengths within 1000 psi of that specified for the proposed work. Changes in materials and proportions within the population of background tests shall not have been more closely restricted than they will be for the proposed work.

(2) If the standard deviation exceeds 600 psi or if a suitable record of strength test performance is not available, proportions shall be selected to produce an average strength at least 1200 psi greater than the required f'_c.

Using the methods of "Recommended Practice for Evaluation of Compression Test Results of Field Concrete (ACI 214-65)," the amount by which the average strength must exceed f'_c may be reduced to an appropriate level below 1200 psi after sufficient test data become available from the job to indicate that, at the lower average strength, the probable frequency of tests more than 500 psi below f'_c will not exceed 1 in 100 and that the probable frequency of an average of three consecutive tests below f'_c will not exceed 1 in 100.

(c) When laboratory trial batches are used as the basis for selecting concrete proportions, strength tests shall be made in accordance with "Method of Test for Compressive Strength of Molded Concrete Cylinders" (ASTM C39) on specimens prepared in accordance with "Method of Making and Curing Test Specimens in the Laboratory" (ASTM C192). A curve shall be established showing the relationship between water-cement ratio (or cement content) and compressive strength. The curve shall be based on at least three points representing batches which produce strengths above and below that required. Each point shall represent the average of at least three specimens tested at 28 days or the earlier age designated.

The maximum permissible water-cement ratio (or minimum cement content) for the concrete to be used in the structure shall be that shown by the curve to produce the average strength indicated in Paragraph (b) unless a lower water-cement ratio or higher strength is required by Paragraphs (e), (f), or (g) of this Section.

(d) If suitable data from trial batches or field experience cannot be obtained, permission may be granted to base concrete proportions on the water-cement ratios and cement factors shown in Table 1602.2A. This table shall be used only for concrete to be made with cements meeting the strength requirements for Type I, Type II, or Type III of "Specification for Portland Cement" (ASTM C150) or, for air-entrained concrete only, Type IA, Type IIA or Type IIIA of "Specification for Air-Entraining Portland Cement" (ASTM C175), and shall not be applied to concrete containing lightweight aggregates or admixtures other than those for entraining air. Application of this method for estimating proportions does not remove the requirement to meet compressive strength test criteria of Section 1602.3 and the water-cement ratio limits of Paragraphs (e), (f) and (g) of Section 1602.2.

(e) Concrete that, after curing, will be subject to freezing temperatures while wet shall contain entrained air within the limits of Table 1602.2B. For such concrete made with normal weight aggregate, the water-cement ratio shall not exceed 0.53 by weight. When the concrete is made with lightweight aggregate, the specified compressive strength $f'c$ shall be at least 3000 psi.

TABLE 1602.2B — CONCRETE AIR CONTENT FOR VARIOUS SIZES OF CONCRETE AGGREGATE

Nominal maximum size of coarse aggregate, in.	Total air content, percent by volume
3/8	6 to 10
1/2	5 to 9
3/4	4 to 8
1	3.5 to 6.5
1 1/2	3 to 6
2	2.5 to 5.5
3	1.5 to 4.5

(f) When made with normal weight aggregate, concrete that is intended to be watertight shall have a maximum water-cement ratio of 0.48 for exposure to fresh water and 0.44 for exposure to sea water. With lightweight aggregate, the specified compressive strength $f'c$ shall be at least 3750 psi for exposure to fresh water and 4000 psi for exposure to sea water.

(g) Concrete that will be exposed to injurious concentrations of sulfate-containing solutions shall conform to Paragraph (f) above and be made with sulfate-resisting cement.

(h) Where different materials are to be used for different portions of the work, each combination shall be evaluated separately.

1602.3 — EVALUATION AND ACCEPTANCE OF CONCRETE

(a) Samples for strength tests of each class of concrete shall be taken not less than once a day nor less than once for each 150 cu. yd. of concrete or for each 5000 sq. ft. of surface area placed. The samples for strength tests shall be taken in accordance with "Method of Sampling Fresh Concrete" (ASTM C172). Cylinders for acceptance tests shall be molded and laboratory-cured in accordance with "Method of Making and Curing Concrete Compressive and Flexural Strength Test Specimens in the Field" (ASTM C31) and tested in accordance with "Method of Test for Compressive Strength of Molded Concrete

Cylinders" (ASTM C39). Each strength test result shall be the average of two cylinders from the same sample tested at 28 days or the specified earlier age.

(b) When the frequency of testing of Paragraph (a) above will provide less than five tests for a given class of concrete, tests shall be made from at least five randomly selected batches or from each batch if fewer than five are used. When the total quantity of a given class of concrete is less than 50 cu. yd., the strength tests may be waived by the Building Official if, in his judgment, adequate evidence of satisfactory strength is provided.

(c) The strength level of the concrete will be considered satisfactory if the averages of all sets of three consecutive strength test results equal or exceed the required $f'c$ and no individual strength test result falls below the required $f'c$ by more than 500 psi.

(d) Strength tests of specimens cured under field conditions in accordance with Section 9(c) of "Method of Making and Curing Concrete Compressive and Flexural Strength Test Specimens in the Field" (ASTM C31) may be required by the Building Official to check the adequacy of curing and protection of the concrete in the structure. Such specimens shall be molded at the same time and from the same samples as the laboratory cured acceptance test specimens. Procedures for protecting and curing the concrete shall be improved when the strength of field-cured cylinders at the test age designated for measuring $f'c$ is less than 85 percent of that of the companion laboratory-cured cylinders. When the laboratory-cured cylinder strengths are appreciably higher than $f'c$ the field-cured cylinder strengths need not exceed $f'c$ by more than 500 psi even though the 85 percent criterion is not met.

(e) If individual tests of laboratory-cured specimens produce strengths more than 50 psi below $f'c$ or if tests of field-cured cylinders indicate deficiencies in protection and curing, steps shall be taken to assure that load-carrying capacity of the structure is not jeopardized. If the likelihood of low-strength concrete is confirmed and computations indicate that the load carrying capacity may have been significantly reduced, tests of cores drilled from the area in question may be required in accordance with "Method of Obtaining and Testing Drilled Cores and Sawed Beams of Concrete" (ASTM C42). Three cores shall be taken for each case of a cylinder test more than 500 psi below $f'c$. If the concrete in the structure will be dry under service conditions, the cores shall be air dried (temperature 60 to 80 F, relative humidity less than 60 percent) for 7 days before test and shall be tested dry. If the concrete in the structure will be more than superficially wet under service conditions, the cores shall be immersed in water for at least 48 hours and tested wet.

(1) Concrete in the area represented by the core tests will be considered structurally adequate if the average of the three cores is equal to at least 85 percent of $f'c$ and if no single core is less than 75 percent of $f'c$. To check testing accuracy, locations represented by

erratic core strengths may be retested. If these strength acceptance criteria are not met by the core tests, and if structural adequacy remains in doubt, the responsible authority may order load tests as outlined in "Building Code Requirements for Reinforced Concrete" ACI 318-71, Chapter 20, for the questionable portion of the structure, or take other action appropriate to the circumstances.

SECTION 1603 – MINIMUM SLAB THICKNESS

The minimum thickness of concrete floor slabs supported directly on the ground shall not be less than three and one-half inches (3½") unless designed by an architect or engineer. An approved vapor barrier shall be installed underneath the slab.

SECTION 1604 – VERMICULITE CONCRETE

Vermiculite concrete, when used in roof systems and slabs-on-grade, shall comply with the "Specifications for Vermiculite Concrete Roofs and Slabs-on-Grade, ANSI A122.1-1965."

CHAPTER XVII

WOOD CONSTRUCTION

SECTION 1700 — GENERAL

1700.1 — GENERAL

(a) The quality and design of wood members and their fastenings used for load supporting purposes shall conform to good engineering practice.

(b) All members shall be framed, anchored, tied and braced so as to develop the strength and rigidity necessary for the purposes for which they are used.

(c) Preparation, fabrication and installation of wood members and the glues, connectors, and mechanical devices for the fastening thereof shall conform to good engineering practices.

(d) The detailed requirements contained in this Chapter are based on sound engineering principles such as those in the Standards hereunder and are intended for light frame construction in general use for structures having light loads and closely spaced framing. Where additional structural requirements should be applied because of the nature of the structure, the Standards hereunder shall be accepted as good engineering practice.

(e) For heavily loaded or engineered timber construction, structural design based on the recommendations of the Standards hereunder shall be accepted as conformance with good engineering practice. Other sections of this Chapter which are applicable, shall apply to heavily loaded or engineered timber construction as well as light frame construction.

(f) For engineered plywood structural diaphragm design, the Supplement to Chapter XVII shall apply.

National Forest Products Association:

National Design Specification for Stress Grade Lumber and Its Fastenings—1973 Edition and Supplement thereto.

Wood Construction Data No. 5—Heavy Timber Construction Details—1960.

American Institute of Timber Construction:

Timber Construction Standards — AITC 100-72.
Inspection Manual — AITC 200-73
Timber Construction Manual — First Edition 1966

Standard Specifications for Structural Glued Laminated Timber of Douglas Fir, Western Larch, Southern Pine and California Redwood AITC 117-71, and Supplement No. 1, March, 1973, and Supplement No. 2, March, 1973.

Standard Specifications for Hardwood Glued Laminated Timber AITC 119-71.

Standard Specifications for Structural Glued Laminated Timber Using "E" Rated and Visually Graded Lumber of Douglas Fir, Southern Pine, Hem-Fir, and Lodgepole Pine, AITC 120-71. Standard Specifications for Structural Glued Laminated Timber of Douglas Fir, Western Larch, Southern Pine and California Redwood—AITC 203-70.

American Plywood Association:

Design Methods for Plywood-Lumber Components (V815) APA 1972

Fabrication of Plywood-Lumber Components (V820) APA 1972

Plywood Construction Systems for Commercial and Industrial Buildings (P310) 1971

Plywood Construction Guide for Residential Building (V450) 1971

Plywood Gusset Truss Designs 64-650A through 64-660

Truss Plate Institute, Inc.:

Design Specifications For Light Metal Plate Connected Wood Roof Trusses—TPI-70

"Span Tables for Light Metal Plate Connected Wood Trusses"—1972

American Wood Preservers Institute:
Pole Building Design—1969

National Particleboard Association:
How To Install Particleboard Underlayment—1969

1700.2 — DETERMINATION OF REQUIRED SIZES

(a) All wood structural members shall be of sufficient size to carry the dead and required live loads without exceeding the allowable working stresses as contained in the Standards listed in Section 1700.1 (f).

(b) Where applicable as determined by end use, allowable working stresses may be determined by "Machine Stress Rating" as approved by the American Lumber Standards Committee.

(c) Where minimum sizes of lumber members are shown herein, they shall be construed as meaning nominal sizes. Minimum dressed sizes corresponding to nominal sizes shall conform with the provisions of the American Lumber Standards.

(d) For convenience, nominal sizes may be shown on the plans. If rough sizes or finished sizes greater or smaller than American Lumber Standard dressed sizes are to be used, computations shall be predicated upon such actual sizes, provided they are specified on the plans or in a statement appended thereto.

1700.3 — QUALITY OF MATERIALS

(a) All lumber, including end-jointed lumber, used for load supporting purposes shall be identified by the Grade Mark of an approved Lumber Grading or Inspection Bureau or Agency. The glued joints in end-jointed lumber when used for load supporting purposes shall be certified to be in accordance with the appropriate provisions of the "Voluntary Product Standard for Structural Glued Laminated Timber", PS-56-73.

(b) Structural glued laminated timber shall be manufactured and identified as required in the "Voluntary Product Standard for Structural Glued Laminated Timber," PS-56-73.

(c) All plywood when used structurally (including among others, use for siding, roof and wall sheathing, subflooring, diaphragms and built-up members) shall conform to the performance standards for its type in Product Standard PS 1-66 for Softwood Plywood, Construction and Industrial. Each panel or member shall be identified for grade and glue type by the trademarks of an approved testing and grading agency. Plywood components shall be designed and fabricated in accordance with the applicable standards listed in Section 1700.1(f) and identified by the trademarks of an approved testing and inspection agency indicating conformance with the applicable standard. In addition, all plywood, when permanently exposed in outdoor applications, shall be of Exterior type.

(d) Wood Shinges and/or shakes shall be identified by the grademark of a grading or inspection bureau or agency recognized as being competent.

(e) Fiberboard for its various uses shall conform to the standards of "Insulating Board (Cellulosic fiber) Structural and Decorative", ASTM C208-72, "Standard Methods of Testing Insulating Board Structural and Decorative", ASTM C209-72 and "Standards of IB Specification No. 3-1969". Fiberboard Nail-Base sheathing shall conform to "Standards of IB Specification No. 2-1972" and "Specification for Nail-Base Sheathing", ASTM D2277-65. Fiberboard Insulating Roof Deck shall conform to the "Standards of IB Specification No. 1-1970" and "Test for Structural Insulating Roof Deck", ASTM D2164-65.

(f) Hardboard shall conform to "Industry Standard for Hardboard," I.S. 1-1971, and shall be identified as to classification. Hardboard siding shall be not less than ¼" thickness. Underlayment-Hardboard meeting strength requirements of 7/32" or ¼" Service hardboard that is planed or sanded on one side to a thickness of 0.215 plus or minus 0.005 inches.

(g) Particleboard shall conform to "Commercial Standard for Mat-Formed Wood Particleboard", CS236-66. Particleboard shall be identified by the grade mark or Certificate of Inspection issued by an approved agency.

Particleboard floor underlayment shall conform to Type 1-B-1 of the Commercial Standard. Underlayment shall be not less than one quarter (¼) inch in thickness and shall be installed in accordance with the installation instructions of the National Particleboard Association.

(h) All lumber and plywood required to be preservatively treated in Section 1702 shall bear an approved AWPI Quality Mark or that of an independent inspection agency that maintains continuing control, testing and inspection over the quality of the product as described in the Quality Control Standards listed in Appendix C.

(i) Hardwood and decorative plywood shall be manufactured and identified as required in "Voluntary Product Standard Hardwood and Decorative Plywood", PS 51-71.

(j) Wood flooring of the various types shall be manufactured and identified as required in the appropriate standard as listed.

Laminated Hardwood Block Flooring—Interim Industry Standard HPMA-LF-71

Flooring Grading Rules (Oak, Pecan, Beech, Birch, Hard Maple [Acer Saccharum])—National Oak Flooring Manufacturers Association-1968

Mosaic-Parquet Hardwood Slat Flooring PS-27-70

Hard Maple Flooring Standard Specification (Hard Maple, Beech and Birch)—Maple Flooring Manufacturers Association-1969

1700.4 — MINIMUM LUMBER GRADES

The minimum grade of lumber used for light frame construction shall be:

(a) For Joists and Rafters: Those obtained in "Working Stresses for Joists and Rafters" as published by National Forest Products Asciation.

(b) For Load Bearing Studs: No. 3 Grade, Standard Grade, or Stud Grade. Utility Grade may be used to support roof and ceiling loads of one story buildings and the top story of multi-story buildings.

(c) For Non-Load Bearing Studs: Utility Grade.

1700.5 — MOISTURE CONTENT

All lumber members 2" and less in thickness shall contain not more than 19% moisture at the time of permanent incorporation in a building or structure.

SECTION 1701 — CONSTRUCTION PRACTICES

1701.1 — PREPARATION OF BUILDING SITE

(a) All building sites shall be graded so as to provide drainage under all portions of the building not occupied by basements or cellars.

(b) All stumps and roots shall be removed from the soil to a depth of at least twelve (12) inches.

1701.2 — REMOVAL OF DEBRIS

After all work is completed, loose wood and debris shall be completely removed from all spaces under the building. All wood forms and supports shall be completey removed. Loose or casual wood shall not be stored in contact with the ground under any building.

1701.3 — FOUNDATIONS

Foundations shall be designed and constructed in accordance with the provisions of Section 1302. Where spot piers are used, unless properly designed, spacing of such piers shall not exceed eight (8) feet center to center.

1701.4 — MUD SILLS

A one-story building, except a dwelling, which does not exceed 800 square feet in area may be constructed without masonry or reinforced concrete foundation, provided such building is placed on a sill of approved wood of natural decay resistance or pressure treated wood. No mud sills shall be less than nominal two by six inches (2x6) or three by four inches (3x4) in cross-section.

SECTION 1702 – PROTECTION AGAINST DECAY AND TERMITES

1702.1 — WOOD SUPPORTS EMBEDDED IN GROUND

Where wood is embedded in the ground for support of permanent structures, it shall have an approved pressure preservative treatment, except where continuously below the ground-water line or continuously submerged in fresh water.

1702.2 — UNEXCAVATED SPACES

When wood joists or the bottom of wood structural floors without joists are closer than eighteen (18") inches, or wood girders are closer than twelve (12") inches to exposed ground located within the periphery of the building over crawl space or unexcavated areas, they shall be of approved wood of natural decay resistance, or pressure treated wood.

1702.3 — SILLS ON EXTERIOR WALLS

All sills which rest on concrete or masonry exterior walls and are less than eight (8") inches from exposed earth shall be of approved wood of natural decay resistance or pressure treated wood.

1702.4 — SLEEPERS AND SILLS ON CONCRETE SLAB

Sleepers and sills on concrete or masonry slabs which are in direct contact with the earth shall be of approved wood of natural decay resistance or pressure treated wood.

1702.5 — BASEMENT POSTS

Wood posts or columns in basements shall be supported by piers projecting at least two (2) inches above the finish floor and separated therefrom by an approved impervious barrier except when approved wood of natural decay resistance or pressure treated wood is used. Posts or columns used in damp locations below grade shall be of approved wood of natural decay resistance or pressure treated wood.

1702.6 — GIRDERS ENTERING MASONRY WALLS

Ends of wood girders entering masonry or concrete walls shall be provided with a ½-inch air space on tops, sides and ends unless approved wood of natural decay resistance or pressure treated wood is used.

1702.7 — CLEARANCE BETWEEN SIDING AND EARTH

Clearance between wood siding and earth on the exterior of a building shall be not less than six (6) inches.

1702.8 — CRAWL SPACE VENTILATION

Crawl spaces under buildings without basements shall be ventilated in accordance with Section 1302.5(d).

1702.9 — APPROVED WOOD OF NATURAL RESISTANCE

(a) Approved wood for natural resistance to decay shall be all heartwood of bald cypress, black locust, black walnut, catalpa, and cedars, chestnut, osage orange, red mulberry, redwood and white oak.

(b) Approved wood for natural resistance to termites shall be all heartwood of bald cypress, redwood or Eastern red cedar.

1702.10 — APPROVED PRESSURE PRESERVATIVE TREATMENT

The Standards of the American Wood Preservers Bureau and the American Wood Preservers Institute shall be deemed as approved in respect to pressure treated lumber.

1702.11 — APPROVED PRE-CONSTRUCTION SOIL TREATMENT

The Standards of the National Pest Control Association shall be deemed as approved in respect to pre-construction soil treatment for protection against termites.

1702.12 — SPECIAL TERMITE PROTECTION

In territories where hazard of termite damage is known to be very heavy the building official may require floor framing of termite resistant wood, pressure treated wood, soil treatment or other approved methods of termite protection.

SECTION 1703 — FIRE PROTECTION

1703.1 — FIRESTOPPING

(a) Firestopping shall be provided to cut off all vertical and horizontal concealed draft openings. Firestopping shall be as indicated in this Section and as provided in Section 705.

(b) Firestopping, when of wood, shall be of not less than two (2) inch thickness and shall effectively fill all spaces for the entire width or depth of the framing or structural member.

(c) Firestopping, when of other materials as provided in Section 705, shall be securely and tightly fitted into place. In the case of spaces between chimneys and wood framing, such spaces shall be solidly filled with mortar or loose incombustible matter supported on non-combustible supports.

(d) Firestopping shall be installed in the following locations:

(1) In all stud walls, partitions and furred spaces at ceiling and floor levels.

(2) Around the top, bottom and side of sliding door pockets.

(3) Between stair stringers at the top and bottom, and between studs along and in line with the adjoining run of stairs.

(4) Between chimneys, fireplaces and wood framing except in the case of approved metal chimney installations as set forth in Section 802.2—Factory-Built Chimneys.

(5) In concealed spaces created by an assembly of floor or roof joists, firestopping shall be provided for the full depth of the joists at the ends and over the supports.

(6) Concealed attic spaces shall be divided into horizontal areas in accordance with Section 705.

1703.2 — FIRE RESISTANCE RATINGS

When fire resistance ratings are specified by this Code, it shall be provided in conformance with the requirements of Chapter X, and Appendix B, Fire Resistance Ratings.

1703.3 — FIRE CUTS

Where joists, beams, or girders enter and terminate in a masonry wall, they shall be provided with a fire cut of three (3) inches or provided with wall plate boxes of self-releasing type or approved hangers, and if located in a required fire resistance wall shall be separated from the opposite side of the wall by at least four (4) inches of solid masonry.

SECTION 1704 — FASTENINGS

1704.1 — NAILING AND STAPLING REQUIREMENTS

The number and size of nails or staples connecting wood members shall not be less than those specified in Table 1704.1—Nailing

Schedule. Where nails of a type other than those shown in the Table are used, the number and spacing shall be in accordance with the manufacturers instructions.

1704.2 — OTHER FASTENINGS

Where framing anchors, clips, staples, glues, or other methods of fastening are used, they shall be installed in accordance with the manufacturers instructions.

SECTION 1705 — FLOOR FRAMING

1705.1 — SILLS ON FOUNDATIONS

Sills on continuous foundation walls shall be not less than two (2) inches in thickness and shall be anchored thereto by ½-inch bolts spaced not more than six (6) feet apart and which are embedded at least six (6) inches in concrete or 15 inches in masonry units. Girders supported on piers shall be provided a true and even bearing surface. Except where wood of natural decay resistance or pressure treated wood is used an approved moisture barrier shall be provided between the sill and foundation wall.

1705.2 — BEAMS AND GIRDERS

Beams and Girders shall be designed in accordance with Section 1700.1 (d) or 1700.2 (a). Where two or more pieces of 2-inch lumber are nailed together to provide girders, the wide faces shall be vertical and the end joints shall occur over supports, provided that for a girder continuous over three or more supports the end joints may be staggered in adjacent pieces at one-fourth the distance from intermediate supports. Where a girder is spliced over the support an adequate tie shall be provided.

1705.2.1 — END-JOINTED LUMBER

End-jointed lumber may be used interchangeably with solid sawn lumber of the same grade and species. Such uses shall include, but not limited to light framing, studs, joists, planks and decking.

1705.3 — FLOOR JOISTS

(a) Maximum spans for floor joists shall be in accordance with the "Span Tables for Joists and Rafters-1970", as published by the National Forest Products Association; or may be designed in accordance with Section 1700.1 (d) or 1700.2 (a).

Spans for field-glued plywood-lumber floor systems using approved adhesives shall be as set forth in "APA Glued Floor System 1970", as published by the American Plywood Association. Approved adhesives for the APA Glued Floor System shall be those meeting the requirements of AFG-01, October, 1969, as published by the American Plywood Association.

(b) Except where supported on a 1-by-4-inch ribbon strip and nailed to the adjoining stud, the ends of each joist shall have not less than 1½ inches of bearing on wood or metal nor less than three (3) inches on masonry.

(c) Except in Dwellings and Multiple Dwellings, floor joists, having a depth to thickness ratio exceeding six (6) and/or the design live load is in excess of forty (40) pounds per square foot, shall be supported laterally by bridging or blocking installed at intervals not exceeding eight (8) feet.

(d) Joists shall be supported laterally at the ends by solid blocks or diagonal struts except where the ends of joists are nailed to a beam (wood or steel with an attached nailer), header, band joists or to an adjoining stud.

(e) Notches on the ends of joists shall not exceed one-fourth the depth. Holes bored for pipes or cable shall not be within two (2) inches of the top or bottom of the joist and the diameter of any such hole shall not exceed one-third the depth of the joist. Notches for pipes in the top or bottom of joists shall not exceed one-sixth the depth and shall not be located in the middle third of the span.

(f) Joists framing from opposite sides of a beam, girder or partition shall be lapped at least four (4) inches and fastened, or the opposing joists shall be tied together in an approved manner.

(g) Joists framing into the side of a wood girder shall be supported by framing anchors, on ledger strips not less than 2 by 2 inches, or by other approved methods.

1705.4 — FRAMING AROUND OPENINGS

Trimmer and header joists shall be doubled when the span of the header exceeds four (4) feet. The ends of header joists more than six (6) feet long shall be supported by framing anchors, joist hangers, or other approved methods unless bearing on a beam, partition or wall. Tail joists over twelve (12) feet long shall be supported at header by framing anchors or on ledger strips not less than 2 by 2 inches.

1705.5 — JOISTS SUPPORTING PARTITIONS

Bearing partitions parallel to joists shall be supported on beams, girders, walls, or other bearing partitions. Bearing partitions perpendicular to joists shall not be offset from supporting girders, walls or partitions more than the joist depth, unless such joists are of sufficient size to carry the additional load.

1705.6 — SUBFLOORS

(a) Except as provided in Section 1705.6(b), all floor joists shall be covered with subflooring of any of the following types:

Lumber

	Minimum Net Thickness For Lumber Placed	
Joist Spacing*	Perpendicular to Joists	Diagonally to Joists
24"	11/16"	¾"
16"	⅝"	⅝"
12"	⅝"	⅝"

*Joints in subflooring shall occur over supports unless end-matched lumber is used, in which case each piece shall bear on at least two joists.

Plywood

Plywood applied in accordance with the provisions of Table 1705.6(A) and fastened in accordance with Table 1704.1.

(b) Subflooring may be omitted when joist spacing does not exceed 16 inches and nominal 1-inch tongued and grooved wood strip flooring is applied perpendicular to the joists.

(c) When resilient flooring is applied directly to plywood subfloor, it shall be applied in accordance with the provisions of Table 1705.6B (pg. 17-19) and fastened in accordance with Table 1704.1 (pg. 17-16)

1705.7 — PLANK-AND-BEAM FRAMING

(a) Beams supporting plank floors shall not exceed the limitations set forth in Section 1705.2.

(b) The allowable span for 2" planks shall be determined in accordance with Section 1700.1 (d) or 1700.2 (a).

1705.8 — ANCHORAGE OF THE FLOOR FRAMING TO MASONRY WALLS

Wood Floor Construction which rests on masonry walls shall be anchored thereto in accordance with Section 1408.3.

1705.9 — STAIR FRAMING

(a) Stair framing shall be supported adequately on floor framing or on walls or partitions.

(b) Except in public stairs where the number and size of stringers shall be determined by engineering analysis, two (2) rough stringers shall be provided for each set of stairs, cut to receive finish treads and risers of uniform width and height.

(c) Unless stringers are supported on walls or partitions, the minimum effective depth at each notch shall be not less than three and one-half (3½) inches.

SECTION 1706 – VERTICAL FRAMING

1706.1 — EXTERIOR WALL FRAMING

(a) Stud Size and Spacing

Studs in one-and-two story buildings shall be not less than 2 by 4 inches with the wide face perpendicular to wall. In three-story buildings, studs in first story shall be not less than 3 by 4 inches or 2 by 6 inches.

(b) Studs supporting floors shall be spaced not more than 16 inches, those supporting ceilings and roofs shall be spaced not more than 24 inches.

1706.2 — BRACING OF EXTERIOR STUD WALLS

(a) Not less than three studs shall be installed at every corner of an exterior wall.

(b) Stud walls shall be braced by one of the following methods:

(1) Nominal 1 by 4 inch continuous diagonal strips set into the face of the studs and top and bottom plates at each corner of building.

(2) Wood boards of ⅝" (net) minimum thickness, applied diagonally.

(3) Wood sheathing panels 2 by 8 feet of ⅝" minimum thickness applied horizontally.

(4) Plywood sheathing panels not less than 48" wide and 96" long applied either vertically or horizontally, and fastened in accordance with Table 1704.1.

(5) Fiberboard sheathing, gypsum sheathing and particleboard sheathing as specified in Section 1706.2(d) when applied vertically in panels 4 feet wide and no less than 8 feet long and properly installed in accordance with Table 1704.1.

(c) Sheathing shall be applied on the exterior walls of all Type VI buildings more than one-story in height, except when back-plastered stucco construction is used.

(d) Sheathing, where required for exterior walls, shall be applied solidly over the wall surface and shall be one or more of the following materials:

(1) Wood Boards and sheathing panels five-eighths inch (⅝") minimum thickness.

(2) Plywood not less than five-sixteenths inch (5/16") thick for 16" stud spacing or not less than three-eighths (⅜") for 24" stud spacing. Plywood of exterior type complying with Section 1706.7(c), including Table 1705.6(c), may also serve as siding. Sheath-

ing paper may be omitted over plywood except where exterior stucco finish, brick veneer, and exterior wall coverings which permit passage of water are used.

(3) Fiberboard not less than seven-sixteenths (7/16") inches thick.

(4) Gypsum not less than one-half inch (½") thick.

(5) Particleboard when it conforms to Type 2-B-1 of the Commercial Standard CS236-66. Particleboard sheathing shall not be less than ⅜-inch thick on studs spaced not more than 16 inches on center.

(e) Studs shall be capped with double top plates installed to provide overlapping at corners and at intersections with bearing partitions. End joints in double top plates shall be offset at least 24 inches. In lieu of double top plates, a continuous header may be used.

(f) For Platform Frame Construction, studs shall rest on a single bottom plate.

1706.3 — OPENINGS IN EXTERIOR WALLS

(a) Double studs shall be provided on each side of openings exceeding three (3) feet in width, and triple studs shall be provided on each side of openings exceeding six (6) feet in width.

(b) Headers shall be provided over each opening in exterior bearing walls. Headers shall not exceed the spans shown in Table 1706.3, or may be of solid lumber of equivalent cross-section. Where the opening does not exceed three (3) feet, each end of the header shall be supported on a stud or framing anchor.

(c) Where the openings exceed three (3) feet in width, each end of the header shall be supported on one stud and where the opening exceeds six (6) feet, each end shall be supported on two studs.

1706.4 — POST AND BEAM FRAMING

(a) Where post and beam framing is used in lieu of stud and joist construction, the posts shall be located to support the beams above and shall be designed in accordance with Section 1700.1 (d) or 1700.2 (a).

(b) Intermediate framing shall be attached to the posts and braced in the manner specified in Section 1704.2.

1706.5 — INTERIOR BEARING PARTITIONS

(a) Studs in one- and two-story buildings shall be not less than 2 x 4 inches with the wide face perpendicular to the partitions. In three-story buildings, studs in the first story shall be not less than 3 by 4 inches or 2 by 6 inches.

(b) Studs supporting floors shall be spaced not more than 16 inches, those supporting ceilings and roofs shall be spaced not more than 24 inches.

(c) Double studs shall be provided on each side of openings exceeding three (3) feet in width, and triple studs shall be provided on each side of openings exceeding six (6) feet in width.

(d) Headers shall be provided over each opening in bearing partitions. Headers shall not exceed the spans shown in Table 1706.3, or may be of solid lumber of equivalent cross-section. Where the opening does not exceed three (3) feet, each end of the header shall be supported on a stud or framing anchor. Where the opening exceeds three (3) feet in width each end of the header shall be supported on one stud and where the opening exceeds six (6) feet, each end shall be supported by two studs.

(e) Studs shall be capped with double top plates installed to provide overlapping at corners and at intersections with exterior walls. End joints in double top plates shall be offset at least 24 inches. For platform frame construction, studs shall rest on a single bottom plate.

1706.6 — INTERIOR NON-BEARING PARTITIONS

(a) Framing for non-bearing partitions shall be of adequate size and spacing to support the finish applied thereto in accordance with the manufacturers recommendations. In non-bearing walls and partitions, studs may be spaced not more than 28" o.c. and may be set with the long dimension parallel to the wall.

(b) Openings in the non-bearing partitions may be framed with single studs and headers.

1706.7 — EXTERIOR WALL COVERINGS

Exterior wall coverings of other than the following shall be of a material approved for exterior use and shall be applied in accordance with the manufacturers recommendations when not otherwise covered by the Code.

(a) Weatherboarding. Wood siding when in place shall have an average thickness of not less than five-eighths inch (⅝"). Siding of less than these dimensions may be applied, provided the outside face of the stud is covered with sheathing, as provided in Section 1706.2(c).

(b) Wood Shingles or Shakes. Wood shingles or shakes attached to sheathing other than wood or plywood shall be secured with approved mechanically-bonding nails or by corrosive resisting common nails on shingle nailing boards securely nailed to each stud with two (2) 8d nails. Wood shingles or shakes may be applied over fiberboard shingle backer and fiberboard sheathing with approved non-corrosion annular grooved nails or may be nailed directly to fiberboard nailbase sheathing with non-corrosion annular grooved nails. The minimum thickness of wood shingles or shakes between nailing boards shall be not less than three-eighths inches (⅜").

(c) Plywood. Plywood shall be of the exterior type and shall have a thickness of three-eighths (⅜") inches, except as provided in

Table 1705.6(c). All Plywood joints shall be backed solidly with nailing pieces not less than two (2) inches in width, unless wood or plywood sheathing is used, or joints are lapped horizontally, or otherwise made waterproof.

(d) Fiberboard Siding shall be medium density not less than one-half (½") nominal thickness.

(e) Hardboard Siding shall conform with the requirements of Section 1700.3(f).

(f) Particleboard siding shall be of the Exterior Type 2-B-1 conforming to the Commercial Standard CS 236-66, not less than ⅜-inch thick when applied over acceptable sheathing on framing spaced 16-inches on centers; or not less than ⅝-inch thick when applied directly to framing spaced 16-inches on centers; or not less than ¾-inch thick when applied directly to framing spaced 24-inches on centers. Unless applied over 1-inch wood sheathing or ½-inch plywood sheathing joints shall occur over framing members and shall be covered with a continuous wood batt; or joints shall be lapped horizontally or otherwise made waterproof to the satisfaction of the Building Official.

(g) Asbestos Shingles attached to sheathing other than wood or plywood shall be secured with approved mechanically-bonding nails or by corrosion-resistive common nails on shingle nailing boards securely nailed to each stud with two (2) 8d nails, except that asbestos shingles may be attached directly to fiberboard nail-base sheathing with corrosion-resistive annular grooved nails. Asbestos shingles shall have a minimum thickness of five thirty-second (5/32") inches.

(h) Masonry Veneer. Masonry veneer shall conform to the requirements of Chapter XIV—Masonry Construction. Brick or other unit veneers shall be backed with solid sheathing covered with waterproof building paper, except where the sheathing is water-repellant. Brick veneer shall be securely attached to the structure as required in Section 1414.4.

(i) Stucco. Stucco or exterior plaster shall conform to requirements of Section 1804.

(j) Metal. Exterior wall coverings may be of formed metal not less than twenty-eight (28) Manufacturer's Standard Gauge.

(k) Solid Masonry. Solid masonry walls shall conform to requirements of Chapter XIV—Masonry Construction.

(l) Flashing shall be provided as necessary to prevent the entrance of water at openings in, or projections through, exterior walls; at intersections of exterior wall coverings of different materials, unless such materials provide a self-flashing joint; at other points subject to the entrance of water.

SECTION 1707 — ROOF AND CEILING FRAMING

1707.1 — CEILING JOIST AND RAFTER FRAMING

(a) Maximum spans for ceiling joists and rafters shall be in accordance with "Span Tables for Joists and Rafters-1970" as published by the National Forest Products Association; or may be designed in accordance with Section 1700.1 (d) or 1700.2 (a).

(b) Where rafters meet to form a ridge, they shall be placed directly opposite each other and nailed to a ridge board of not less than 1-inch in thickness, and not less in depth than the cut end of the rafters.

(c) Ceiling joists and rafters shall be nailed to each other where possible and the assembly shall be nailed to the top wall plate in an adequate manner to secure the roof framing to the walls.

(d) Ceiling joists shall be continuous or where they meet over interior partitions, shall be securely joined to provide a continuous tie across the building.

(e) Where ceiling joists are not parallel to rafters, sub-flooring or metal straps attached to the ends of the rafters shall be installed in a manner to provide a continuous tie across the building.

(f) Valley rafters shall be doubled. Hip rafters may be single members. Valley and hip rafters shall be two (2) inches deeper than jack rafters.

(g) Collar beams of 1 x 6 inch boards shall be installed in the upper third of the roof height to every third pair of rafters.

(h) Notches on the ends of joists shall not exceed one-fourth the depth. Holes bored for pipes or cable shall not be within 2 inches of the top or bottom of the joist and the diameter of any such hole shall not exceed one-third the depth of the joist. Notches for pipes in the top or bottom of joists shall not exceed one-sixth the depth and shall not be located in the middle third of the span.

1707.2 — TRUSSED RAFTERS

(a) Trussed rafters shall be designed in accordance with accepted engineering practice. Members may be joined by nails, glue, bolts, timber connectors or other approved framing devices.

(b) Where trusses are to support a finished ceiling, the deflection under live load shall not exceed 1/360.

(c) The design of metal plate connected wood roof trusses shall comply with the "Design Specifications For Light Metal Plate Connected Wood Roof Trusses," TPI-70.

1707.3 — ROOF JOISTS

(a) Maximum spans for roof joists shall be in accordance with "Span Tables for Joists and Rafters, 1970" as published by the

National Forest Products Association; or may be designed in accordance with Section 1700.1 (d) or Section 1700.2 (a).

(b) Joists shall be supported laterally at the ends by solid blocks or diagonal struts. Such bridging may be omitted where ends of joists are nailed to a header, band joist or to an adjoining stud.

(c) Notches on the ends of joists shall not exceed one-fourth the depth. Holes bored for pipes or cable shall not be within 2 inches of the top or bottom of the joist and the diameter of any such hole shall not exceed one-third the depth of the joist. Notches for pipes in the top or bottom of joists shall not exceed one-sixth the depth and shall not be located in the middle third of the span.

1707.4 — ROOF SHEATHING

(a) All rafters and roof joists shall be covered with sheathing as follows:

Lumber

Solid sheathing—Wood boards of ⅝ inch (net) minimum thickness

Spaced sheathing—Wood boards of ¾ inch (net) minimum thickness

Plywood

Applied in accordance with the provisions of Table 1705.6 (A) and nailed in accordance with Table 1704.1.

Insulating Roof Deck.

Fiberboard insulating roof deck not less than 1-inch nominal thickness.

(b) Joints in lumber sheathing shall occur over supports unless end-matched lumber or approved clips are used in which case each piece shall bear on at least two rafters or joists.

1707.5 — PLANK AND BEAM ROOFS

Beams shall be supported on posts, piers or other beams and shall conform to Section 1705.2. Roof planks shall conform to Section 1705.7.

1707.6 — ANCHORAGE OF ROOF FRAMING TO MASONRY WALLS

Wood roof construction which rests on masonry walls shall be anchored thereto in a manner equivalent to that specified in Section 1408.2.

1707.7 — ACCESS TO ATTIC SPACE

Attic spaces shall be provided with an interior access opening not less than twenty-two inches by thirty-six inches (22" x 36"). Access opening shall be readily accessible and provided with a lid or device that may be easily removed or operated.

1707.8 — VENTILATION OF ATTIC SPACE

Attic spaces shall be ventilated in accordance with the following:

(a) For gabled and hipped roofs ventilation shall be provided to furnish cross ventilation of each separate space with weather protected vents. The ratio of total net free ventilating area to the area of the ceiling shall be not less than 1/150. That ratio may be reduced to 1/300 provided: (1) A vapor barrier having a transmission rate not exceeding one perm is installed on the warm side of the ceiling, or (2) at least 50% of the required ventilating area is provided by ventilators located in the upper portion of the space to be ventilated (at least 3'-0" above eave or cornice vents) with the balance of the required ventilation provided by eave or cornice vents.

(b) For flat roofs, blocking and bridging shall be arranged so as not to interfere with the movement of air. Such roofs shall be ventilated along the overhanging eaves, with the net area of opening being not less than 1/250 of the area of the ceiling below.

1707.9 — ROOF COVERING

(a) Any roof covering permitted in this Code may be applied to dwellings. (See Chapter VII—Fire Protection Requirements, Section 706, Roof Coverings). Whenever composition roofing is used, solid sheathing shall be applied.

(b) Flashing shall be placed around openings and extensions of mechanical appliances or equipment through the roof and otherwise as necessary to provide adequate drainage.

SECTION 1708 — FRAMING AT CHIMNEYS AND FIREPLACES

1708.1 — FRAMING MEMBERS

All headers, beams, joists and studs shall be kept at least two (2) inches from the outside face of chimney or fireplace masonry.

1708.2 — FIREPLACE TRIM

All wood mantles and similar trim shall be kept at least six inches from fireplace opening. Parts of the mantle which project more than 1½ inches from the face of the fireplace shall have additional clearance equal to the projection.

TABLE 1704.1—FASTENING SCHEDULE	Fastener	Number or Spacing
Joist to sill or girder, toe nail	8d common	3
Bridging to joist, toe nail each end	8d common	2
Ledger strip	16d common	3 at each joist
1"x6" subfloor or less to each joist, face nail	8d common	2
Over 1"x6" subfloor to each joist, face nail	8d common	3
2" subfloor to joist or girder, blind and face nail	16d common	2
Sole plate to joist or blocking, face nail	16d common	16" o.c.
Top or sole plate to stud, end nailed	16d common	2
Stud to sole plate, toe nail	8d common	4
Doubled studs, face nail	16d common	24" o.c.
Doubled top plates, face nail	16d common	16" o.c.
Top plates, laps and intersections face nail	16d common	2
Continuous header, two pieces	16d common	16" o.c. along each edge
Ceiling joists to plate, toe nail	8d common	3
Continuous header to stud, toe nail	8d common	3
Ceiling joists, laps over partitions, face nail	16d common	3
Ceiling joists to parallel rafters, face nail	16d common	3
Rafter to plate, toe nail	8d common	3
1-inch brace to each stud and plate, face nail	8d common	2
1"x8" sheathing or less to each bearing, face nail	8d common	2
Over 1"x8" sheathing to each bearing, face nail	8d common	3
Built-up corner studs	16d common	o.c.
Built-up girders and beams	20d common	32" o.c. at top and bottom and staggered 2 ends and at each splice.
2-inch Planks	16d common	2 each bearing

TABLE 1704.1—FASTENING SCHEDULE (Continued)

Plywood Subflooring		
1/2"	6d Common, annular or spiral thread	6" o.c. edges and 10" o.c. intermediate
5/8", 3/4"	8d Common or 6d annular or spiral thread	6" o.c. edges and 10" o.c. intermediate
1", 1-1/8"	10d Common or 8d annular or spiral thread	6" o.c. edges and 6" o.c. intermediate
1/2"	16 ga. galvanized wire staples, 3/8" minimum crown	4" o.c. edges and 7" o.c. intermediate
5/8"	1-5/8" length	2½" o.c. edges and 4" o.c. intermediate
Plywood Roof & Wall Sheathing		
1/2" or less	6d Common	6" o.c. edges and 12" o.c. intermediate
5/8" or greater	8d Common	6" o.c. edges and 12" o.c. intermediate
5/16", 3/8", 1/2"	16 ga. galvanized wire staples, 3/8" min. crown. Length of 1" plus plywood thickness	4" o.c. edges and 8" o.c. intermediate
5/8", 3/4"		2" o.c. edges and 5" o.c. intermediate
1/2" Fiberboard Sheathing	1-1/2" Galvanized roofing nail 6d common nail	3" o.c. at edges 6" at other bearings
25/32" Fiberboard Sheathing	1-3/4" Galvanized roofing nail 8d common nail	3" o.c. at edges 6" o.c. at other bearings
1/2" Gypsum Sheathing	12 gage 1-1/4" Large Head Corrosion-Resistive	4" o.c. at edges 8" o.c. at other bearings

TABLE 1704.1—FASTENING SCHEDULE (Continued)

Particleboard Siding	
3/8"-1/2"[1]	6d[2]
5/8"[3]	8d[2]
3/4"[4]	8d[2]
Particleboard Wall Sheathing	
3/8"-1/2"	6d[5]
5/8"-3/4"	8d[5]

1. Siding applies to 1-inch wood sheathing or ½-inch plywood or ½-inch particleboard sheathing.
2. Corrosion resistant nails spaced 6-inches on center at edge and 8-inches on center at intermediate supports. Nails shall have a minimum edge distance of ⅜-inch.
3. Siding applied to studs spaced 16-inch on center maximum.
4. Siding applied directly to studs spaced 24-inches on center maximum.
5. Nails spaced 6-inches on center at edges and 12-inches at intermediate supports. Nails shall have a minimum edge distance of ⅜".

*Fiberboard sheathing may be stapled using 16 gage galvanized staples 1-1/8" long for 1/2" sheathing and 1-1/2" long for 25/32" sheathing. Staples to have minimum crown of 7/16" and spaced 3" o.c. at edges and 6" o.c. at other bearings.

TABLE 1705.6A – ALLOWABLE SPANS FOR PLYWOOD FLOOR AND ROOF SHEATHING CONTINUOUS OVER TWO OR MORE SPANS AND FACE GRAIN PERPENDICULAR TO SUPPORTS[1]

Panel Identification Index[2]	Roof				Floor
	Maximum Span (Inches)		Load (psf)		Maximum Span[4] (Inches)
	Edges Blocked or other Support	Edges Unblocked	Total Load	Live Load	
12/0	12	12	130	100	0
16/0	16	16	75	55	0
20/0	20	20	55	45	0
24/0	24(5)	24	60	45	0
30/12	30	26	55	40	12(6)
32/16	32	28	50(3)	40	16(7)
36/16	36	30	50(3)	35(3)	16(7)
42/20	42	32	45(3)	35(3)	20(7)
48/24	48	36	40(3)	40	24

(1) These values apply for Structural I and II, Standard Sheathing and C-C grades only. Spans shall be limited to values shown because of possible effect of concentrated loads.

Edges may be blocked with lumber or other approved type of edge support.

(2) Identification Index appears on all panels in the construction grades listed in footnote (1).

(3) For roof live load of 40 psf or total load of 55 psf, decrease spans by 13 percent or use panel with next greater identification index.

(4) Plywood edges shall have approved tongue and groove joints or shall be supported with blocking, unless one-fourth (¼) inch minimum thickness underlayment is installed, or finished floor is 25/32" wood strip. Allowable uniform load based on deflection of 1/360 of span is 100 psf.

(5) ½" Structural I, when continuous over one support, may be laid with face grain parallel to supports provided all panel edges are blocked or other approved type edge support is provided, the spacing of the supports does not exceed twenty-four inches (24") on center, and the live load does not exceed 30 pounds per square foot. For other grades, a minimum thickness of five-eighths inch (⅝") 5 ply is required.

(6) May be 16" if 25/32" wood strip flooring is installed at right angles to joists.

(7) May be 24" if 25/32" wood strip flooring is installed at right angles to joists.

TABLE 1705.6B — ALLOWABLE SPANS FOR PLYWOOD COMBINATION SUBFLOOR-UNDERLAYMENT[1]

Plywood Continuous over Two or More Spans and Face Grain Perpendicular to Supports

Species Groups	Maximum Spacing of Joists (inches)		
	16"	20"	24"
1	½"	⅝"	¾"
2,3	⅝"	¾"	⅞"
4	¾"	⅞"	1"

(1) Applicable to Underlayment grade, C-C (Plugged) and all grades of sanded Exterior type plywood. Spans limited to values shown because of possible effect of concentrated loads. Allowable uniform load based on deflection of 1/360 of span is 100 psf. Plywood edges shall have approved tongue and groove joints or shall be supported with blocking, unless one-fourth (¼) inch minimum thickness underlayment is installed, or finish floor is 25/32" wood strip. If wood strips are perpendicular to supports, thicknesses shown for 16" and 20" spans may be used on 24" spans.

TABLE 1705.6C — PLYWOOD EXTERIOR WALL COVERINGS

Panel thickness shall be not less than ⅜" nominal except for "303 Specialty Siding" panels which are grooved, striated, brushed, or have other surface texture. Average thickness of "303 Specialty Siding" panels after texturing shall be not less than 5/16".

Nailing of plywood shall be as indicated:

	Plywood (1) Thickness	Nail Size	Nail Type	Panel Edges	Nail-Spacing on Supports (2) Intermediate
Panel Siding	⅜", ½" ⅝" & Thicker	6d 8d	Non-corrosive, box or casing nails	6" 6"	12" on each stud 12" on each stud
Lap Siding	⅜" ½" & Thicker	6d 8d	Non-corrosive, box or casing nails	4"	One nail per stud on widths 12" or less. 8" for widths greater than 12"
303 Specialty Sidings(3)	⅜", ½" ⅝" & Thicker	6d 8d	Non-corrosive, box or casing nails	6" 6"	12" on each stud 12" on each stud
Texture 1-11(3)	⅝"	8d	Non-corrosive, box or casing nails	6"	12" on each stud

(1) Minimum edge distance of ⅜".

(2) Special Requirement: Nails on ship-lap edges ⅜" from exposed edge and slant driven towards it; do not set.

(3) In direct-to-stud applisations 5-ply panels of ½"nominal thickness or more may be used over studs 24" o.c. if texturing does not penetrate through the face veneer. All other panels must be used over studs spaced not more than 16" o.c.

TABLE 1706.3—MAXIMUM SPANS FOR HEADERS*

In this table, headers consist of two pieces of nominal 2-inch framing lumber set on edge and nailed together. The span for the two pieces is expressed as a percentage of the maximum allowable span for floor joists of the same species and grade spaced 16 inches on centers and subjected to a live load of 40 pounds per square foot. Spans for floor joists are included in Section 1705.3(a).

Exterior Wall Openings			Bearing Partition Openings	
Rafters with Bearing Partition (Slope of 3 in 12 or less) No attic storage 	Trussed Rafters (Slope of 3 in 12 or less) No attic storage  Rafters with Bearing Partition (Slope over 3 in 12) Attic storage 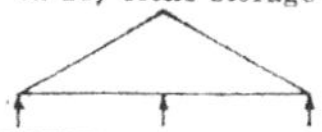	Trussed Rafters (Slope over 3 in 12) Attic storage 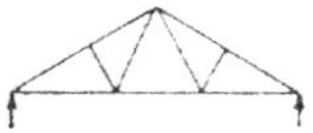	Rafters with Bearing Partition (Slope of 3 in 12 or less) No attic storage 	Rafters with Bearing Partition (Slope over 3 in 12) Attic storage 
Buildings up to 26 ft. wide—1 story or 2nd story of 2-story buildings				
60%	45%	40%	50%	35%
Buildings up to 26 ft. wide—1st story of 1½ or 2 story buildings				
40%	35%	35%	35%	30%
Buildings 27 to 32 ft. wide—1 story or 2nd story of 2 story buildings				
55%	40%	35%	45%	35%
Buildings 27 to 32 ft. wide—1st story of 1½ or 2 story buildings				
35%	35%	35%	35%	30%

Note 1—Span for a header of two 2x4's should not exceed 2'-6" in bearing partitions under attic storage nor 3'-0" elsewhere.
Note 2—Example—Section 1705.3(a)—for the species and grade in question, may show a span of 14 feet for 2x8 floor joists spaced 16 inches on center under a 40-pound live load. If the factor from the table is 35%, the allowable span for a header consisting of 2 pieces of 2x8 of this grade and species would be 35% of 14 feet or 4 feet, 11 inches.

*Headers may be designed in accordance with Section 1700.1(d).

SUPPLEMENT

TO

CHAPTER XVII

PLYWOOD DIAPHRAGMS

GENERAL

Plywood diaphragms may be used to resist horizontal forces in horizontal and vertical distributing or resisting elements, provided the deflection in the plan of the diaphragm, as determined by calculations, tests, or analogies drawn therefrom, does not exceed the permissible deflection of attached distributing or resisting elements.

Permissible deflection shall be that deflection up to which the diaphragm and any attached distributing or resisting element will maintain its structural integrity under assumed load conditions, i.e., continue to support assumed loads without danger to occupants of the structure.

Connections and anchorages capable of resisting the design forces shall be provided between the diaphragms and the resisting elements. Openings in diaphragms which materially affect their strength shall be fully detailed on the plans, and shall have their edges adequately reinforced to transfer all shearing stresses.

Size and shape of diaphragms shall be limited as set forth in Table 1. In buildings of wood construction where rotation is provided for, transverse shear resisting elements normal to the longitudinal element shall be provided at spacings not exceeding 2 times the width for plywood diaphragms. In masonry or concrete buildings plywood diaphragms shall not be considered as transmitting lateral forces by rotation.

TABLE 1

Maximum Diaphragm Dimension Ratios

Type	Horizontal Diaphragms Maximum Span-Width Ratios	Vertical Diaphragms Maximum Height-Width Ratios
Plywood, nailed all edges	4:1	3½:1
Plywood, blocking omitted at intermediate joints	4:1	2:1

Design

Horizontal and vertical diaphragms sheathed with plywood may be used to resist horizontal forces not exceeding those set forth in

Table 2 and 3, may be calculated by principles of mechanics without limitations by using values for nail strength and plywood shear strength given elsewhere in this code. Plywood thickness for horizontal diaphragms shall not be less than set forth in Tables 1705.6 for corresponding joist spacing and loads, except that one-fourth inch (¼") may be used where perpendicular loads permit.

All boundary members shall be proportioned and spliced where necessary to transmit direct stresses. Framing members shall be at least two inches nominal in width. In general panel edges shall bear on the framing members and butt along their center lines. Nails shall be placed not less than three-eighths inch (⅜") in from the panel edge, not more than twelve inches (12") apart along intermediate supports, and six inches (6") along panel edge-bearings, and shall be firmly driven into the framing members. No unblocked panel less than twelve inches (12") wide shall be used.

TABLE 2—ALLOWABLE SHEAR IN POUNDS PER FOOT FOR HORIZONTAL PLYWOOD DIAPHRAGMS[1]

PLYWOOD GRADE	Common Nail Size	Minimum Nominal Penetration in Framing (in Inches)	Minimum Nominal Plywood Thickness (in Inches)	Minimum Nominal Width of Framing Member (in Inches)	BLOCKED DIAPHRAGMS: Nail Spacing at diaphragm boundaries (all cases) and continuous panel edges parallel to load (cases 3 & 4)				UNBLOCKED DIAPHRAGM: Nails spaced 6" max. at supported end	
					6	4	2½	2	Load perpendicular to unblocked edges and continuous panel joints (case 1)	All other configurations (cases 2, 3 & 4)
					Nail spacing at other plywood panel edges: 6	6	4	3		
STRUCTURAL I	6d	1¼	5/16 or ¼	2	185	250	375	420	165	125
				3	210	280	420	475	185	140
	8d	1½	⅜	2	270	360	530	600	240	180
				3	300	400	600	675	265	200
	10d	1⅝	½	2	320	425	640[2]	730[2]	285	215
				3	360	480	720	820	320	240
STRUCTURAL II, C-C Exterior, Standard Sheathing and other grades covered in PSI-66.	6d	1¼	5/16 or ¼	2	170	225	335	380	150	110
				3	190	250	380	430	170	125
			⅜	2	185	250	375	420	165	125
				3	210	280	420	475	185	140
	8d	1½	⅜	2	240	320	480	545	215	160
				3	270	360	540	610	240	180
			½	2	270	360	530	600	240	180
				3	300	400	600	675	265	200
	10d	1⅝	½	2	290	385	575[2]	655[2]	255	190
				3	325	430	650	735	290	215
			⅝	2	320	425	640[2]	730[2]	285	215
				3	360	480	720	820	320	240

[1]These values are for short time loads due to wind or earthquake and must be reduced 25 per cent for normal loading. *Space nails 12 inches on center along intermediate framing members.*
[2]Reduce tabulated allowable shears 10 per cent when boundary members provide less than 3-inch nominal nailing surface.

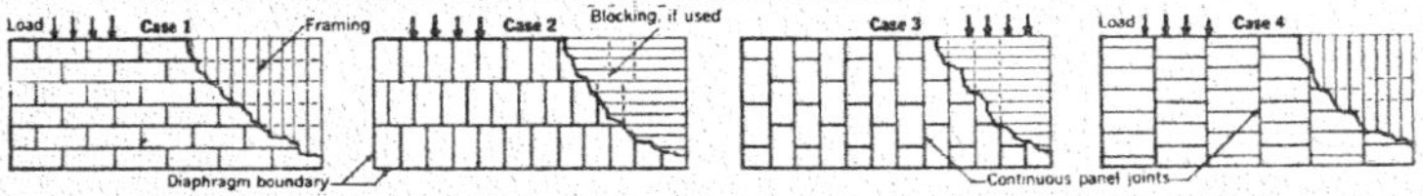

TABLE 3—ALLOWABLE SHEAR FOR WIND OR SEISMIC FORCES IN POUNDS PER FOOT FOR PLYWOOD SHEAR WALLS[1]

PLYWOOD GRADE	NAIL SIZE (Common or Galvanized Box)	MINIMUM NAIL PENETRATION IN FRAMING (Inches)	MINIMUM NOMINAL PLYWOOD THICKNESS (Inches)	PLYWOOD APPLIED DIRECT TO FRAMING — Nail Spacing at Plywood Panel Edges: 6	4	2½	2	NAIL SIZE (Common or Galvanized Box)	PLYWOOD APPLIED OVER ½-INCH GYPSUM SHEATHING — Nail Spacing at Plywood Panel Edges: 6	4	2½	2
STRUCTURAL I	6d	1¼	5/16	200	300	450	510	8d	200	300	450	510
	8d	1½	⅜	280	430	640	730	10d	280	430	610	730
	10d	1⅝	½	340	510	770	870	—	—	—	—	—
STRUCTURAL II C-C Exterior Standard Sheathing, Panel Siding Plywood and other grades Covered in PSI-66.	6d	1¼	5/16	180	270	400	450	8d	180	270	400	450
	8d	1½	⅜	260	380	570	640	10d	260	380	570	640
	10d	1⅝	½	310	460	690	770	—	—	—	—	—
	Nail SIZE (Galvanized Casing)							NAIL SIZE (Galvanized Casing)				
Plywood Panel Siding in Grades Covered in PSI-66.	6d	1¼	5/16	140	210	320	360	8d	140	210	320	360
	8d	1½	⅜	160	240	360	410	10d	160	240	360	410

[1]All panel edges backed with two-inch (2") nominal or wider framing. Plywood installed either horizontally or vertically. Space nails at twelve inches (12") on center aong intermediate framing members. These values are for short time loads due to wind or earthquake and must be reduced 25 per cent for normal loading.

CHAPTER XVIII

LATHING, PLASTERING AND GYPSUM WALLBOARD

SECTION 1800 – GENERAL

1801 — SCOPE

(a) Lathing, plastering and gypsum wallboard application shall be done in the manner and with the materials specified in this Chapter, and when required for fire protection shall also comply with the provisions of Chapter X.

(b) No plaster shall be applied until the lathing has been inspected and approved by the Building Official.

(c) The Building Official may require that test holes be made in the wall for the purpose of determining the thickness and/or proportioning of the plaster, provided the permit holder has been notified 24 hours in advance of the time of making such tests.

(d) Joint treatment of gypsum wallboard shall not be applied until the wallboard application has been inspected and approved by the Building Official.

SECTION 1802 – MATERIALS

Materials	Designation
Aggregate	
Sand—Shall be washed and when used with portland cement for scratch coat plastering the amount of sand retained on a No. 16 sieve shall not be less than 10% or more than 40%	ASTM C 35-70
Perlite	ASTM C 35-70
Vermiculite	ASTM C 35-70
Gypsum Plasters	ASTM C 28-68
Gypsum Veneer Plaster	ASTM C 587-68
Gypsum Veneer Base	ASTM C 588-68
Water Resistant Gypsum Backing Board	ASTM C 630-70
Bonding Compounds for Interior Plastering	ASTM C 631-70
Lime-Special Finishing Hydrated Lime Type "S"	ASTM C 206-68

Materials	Designation
Quicklime for structural purposes (Lime putty shall be made from quicklime or hydrated lime and shall be prepared in an approved manner.)	ASTM C 5-59
Keene's Cement	ASTM C 61-70
Portland Cement Type I, II, or III	ASTM C 150-67
Type I-A, II-A, or III-A	ASTM C 175-67
EXCEPTION: Approved types of plasticizing agents may be added to portland cement Type I or Type II in the manufacturing process, but not in excess of 12 percent of the total volume. Plastic or water-proofed cements so manufactured shall meet the requirements for portland cement as specified in ASTM C 150-64 except in respect to the limitation on insoluble residue, air-entrainment and additions subsequent to calcination	
Masonry Cement Type II	ASTM C 91-68
Portland Blast Furnace Slag Cement	ASTM C 205-64 T Type IS Type IS-A
Gypsum Lath	ASTM C 37-69
Metal and Wire Lath, Metal Accessories and Channels	ANSI A42.4-67
Gypsum Wallboard	ASTM C 36-68
Gypsum Backing Board	ASTM C 442-67
Joint Reinforcing Tape and Adhesive Materials	ASTM C 474-67 ASTM C 475-70
Steel Studs (for use with Gypsum Sheet Materials)	ASTM C 645-70
Screws for use with Steel Studs covered by ASTM-C-645-70)	C 646-70

SECTION 1803 — APPLICATION OF INTERIOR LATHING AND PLASTERING

(a) Interior lathing and furring shall be done in accordance with the procedures set forth in "Specifications for Interior Lathing and Furring, ANSI A42.4-1967."

(b) Interior gypsum plastering shall be done in accordance with the procedures set forth in "Specifications for Gypsum Plastering, ANSI A42.1-1964," or the Recommended Specifications, "Gypsum Plastering" as published by the Gypsum Association. Portland cement plaster shall be in accordance with the provisions of "Specifications for Portland Cement Plastering, ANSI A42.3-1946."

SECTION 1804 – APPLICATION OF EXTERIOR LATHING AND PLASTERING

(a) Exterior use of gypsum plaster shall be in strict conformance with the applicable requirements of "Recommended Specifications—Gypsum Plastering" published by the Gypsum Association and the provisions of "Interior Lathing and Furring, ANSI A42.4-1967."

(b) Exterior use of portland cement plaster shall be in conformance with the applicable requirements of "Specifications for Portland Cement Stucco, ANSI A42.2-1946."

SECTION 1805 – PNEUMATICALLY PLACED PORTLAND CEMENT PLASTER

(a) Pneumatically placed portland cement plaster shall be a mixture of portland cement and aggregate conveyed by air through a pipe or flexible tube, and deposited by air pressure in its final position.

(b) Rebound material may be screened and reused as aggregate in an amount not greater than 25 percent of the total sand in any batch.

(c) Pneumatically placed portland cement plaster shall consist of a mixture of one part cement to not more than five parts of aggregate. Plasticity agents may be used as specified elsewhere in this Chapter. Except when applied to concrete or masonry, such plaster shall be applied in not less than two coats to a minimum total thickness of seven-eighths inch (⅞").

SECTION 1806 – APPLICATION OF GYPSUM WALLBOARD

Interior and exterior application of gypsum wallboard shall be done in accordance with the procedures set forth in "Specifications for the Application and Finishing of Wallboard, ANSI A97.1-1965."

1807 — APPLICATION OF LIGHT GAUGE STEEL STUDS

When screw type steel framing members are used in non-load bearing and/or non-combustible fire resistive assemblies, they shall conform to the "Gypsum Association Specification for the Installation of Screw Type Steel Framing Members" to receive gypsumboard.

1808 — ALLOWABLE PARTITION HEIGHTS

TABLE NO. 1 — ALLOWABLE PARTITION HEIGHTS BASED ON WALLBOARD AND NO. 25 GAUGE STUDS[1] ACTING AS A COMPOSITE SECTION[2]

STUD SPACING (In Inches)	FACING ON EACH SIDE	STUD DEPTH (In Inches)					
		1⅝	2½	3¼	3⅝	4	6
		HEIGHT IN FEET AND INCHES					
16	½"-one ply	11'0"	14'8"	17'10"	19'5"	20'3"	18'10"
24	½"-one ply	10'0"	13'5"	16'0"	17'3"	18'5"	17'8"
24	½"-two ply	12'4"	15'10"	18'3"	19'5"	20'8"	19'0"

[1] The tabulated stud heights are based on 25 gauge steel studs and installed in conformance with Gypsum Association specifications for installation of screw type steel framing members to receive gypsumboard.

[2] Gypsumboard product must have a minimum thickness of ½" and may be applied vertically or horizontally.

CHAPTER XIX

RAT-PROOF CONSTRUCTION

SECTION 1900 — RAT-PROOFING

This Chapter provides regulations governing the rat-proofing of occupancies that provide conditions favorable to rodent infestation and propagation.

SECTION 1901 — OCCUPANCY

Every building in which feed, foodstuff or food is stored, prepared, processed, served or sold, shall be made, maintained or built of rat-proof construction by the owner, tenant, or occupant in accordance with the provisions of this section. This Chapter applies particularly to the following occupancies:

Group B—Cafes or Restaurants
(Curb Markets see Section 1902.3.)
Group E—Large Restaurants
Group F—Storage
Group G—Slaughter Houses

1901.1 — EXISTING OR NEW BUILDINGS

(a) **Foundation Wall Ventilation Openings.** All foundation wall ventilator openings shall be covered for their entire height and width with perforated sheet metal plates of a thickness not less than fourteen (14) Manufacturer's Standard Gauge or with expanded sheet metal of a thickness not less than eighteen (18) Manufacturer's Standard Gauge or with cast iron grills or gratings, or with hardware cloth of nineteen (19) B. & S. Gauge or heavier. The openings therein shall not exceed one-half (½") inch.

(b) **Miscellaneous Foundation and Exterior Wall Openings**—All foundation and exterior wall openings (except those used for doors and screened windows), such as those openings around pipes, electric cables, conduits, openings due to deteriorated walls, broken masonry or concrete, shall be protected against the passage of rats by closing such openings with cement mortar, concrete masonry or non-corrodible metal.

(c) **Doors.** All exposed edges of the lower ten inches of wooden doors, door sills and jambs serving as rear or side entrances into business buildings from the ground, basement, or cellar floors, and other doors accessible to rats shall be protected against gnawing by covering doors, door sills and jambs with solid sheet metal of not less than twenty-four (24) Manufacturer's Standard Gauge thickness.

All doors on which metal protection has been applied shall be hinged so as to be freeswinging. When closed, doors shall fit snugly so that the maximum clearance between any door, door jambs and sills shall not be greater than three-eighths (3/8) inches.

Door jambs and sills constructed of metal, concrete, masonry, stone or cement mortar will be acceptable without metal protection as specified under paragraphs above.

(d) **Windows**—All windows and other openings for the purpose of light or ventilation located in exterior walls within two (2) feet above the existing ground level immediately below such opening shall be covered for their entire height and width, including frame, with wire cloth of at least nineteen (19) B. & S. Gauge having a mesh not larger than one-half (½) inch.

All windows and other openings for the purpose of light and ventilation in the exterior walls not covered in paragraph above, accessible to rats by way of exposed pipes, wires, conduits, and other appurtenances shall be covered with wire cloth of at least nineteen (19) B. & S. Gauge or heavier having a mesh not larger than one-half (½) inch or in lieu of wire cloth covering, said pipes, wires, conduits and other appurtenances shall be blocked from rat usage by installing solid sheet metal guards of twenty-four (24) Manufacturer's Standard Gauge or heavier. Guards shall be fitted snugly around pipes, wires, conduit or other appurtenances. In addition, they shall be fastened securely to and shall extend perpendicularly from the exterior wall for a minimum distance of twelve (12) inches beyond and on either side of pipe, wire, conduit or appurtenance.

SECTION 1902 – CONSTRUCTION

1902.1 — MASONRY CONSTRUCTION AND CONCRETE FLOORS

All new buildings with concrete floors shall be constructed with no intervening space between the edge of the floor slab and building walls. All openings in the floor slab shall be properly protected against the passage of rats.

1902.2 — PIER AND WOOD CONSTRUCTION

(a) **Less Than Twelve Inches Above Ground.** All new buildings constructed on piers having wooden floor sills less than twelve (12) inches above the surface of the ground shall have the intervening space between the floor sills and ground protected against the passage of rats by installing a solid masonry, concrete curtain wall around the entire perimeter of the building and extending said curtain wall to a depth of not less than twenty-four (24) inches below the surface of the ground level and fastened securely to the exterior wall of the building.

Where curtain walls are not desirable, all ground floors of wood construction may be replaced with concrete of not less than four (4) inches in thickness with the exterior wall protected for a height of twenty-four (24) inches above the concrete floor with masonry, concrete or solid sheet metal of twenty-four (24) Manufacturer's Standard Gauge or heavier. Exterior wall protection shall be securely tied into the concrete floor at all points.

(b) **Over Twelve Inches Above Ground**—Buildings constructed on piers having wooden floor sills more than twelve (12) inches above the ground level, shall have the intervening spaces between floor sills and ground protected against the passage of rats by installing curtain walls in accordance with Section 1902.2 (a), or protecting said building against the passage of rats by installing solid sheet metal collars of twenty-four (24) Manufacturer's Standard Gauge or heavier at top of each pier and snugly around each pipe, cable, wire, conduit or other utility service passing through wooden ground floor. Metal collars shall be not less than eight (8") inches greater in diameter than the pier, pipe, cable, wire, conduit or other utility service and shall be securely fastened underneath the wooden floor. All other openings in wooden ground floors through which rats may gain entrance into double walls or the interior of business buildings, such as openings that may exist in floors at double walls above floor sills, shall be closed with twenty-four (24) Manufacturer's Standard Gauge or heavier solid sheet metal or sixteen (16) B. & S. Gauge or heavier wire cloth of one-half (½) inch mesh or with concrete or masonry.

1902.3 — CURB OR FARMER'S MARKET

The floors of Curb or Farmer's Markets in which fruit or vegetables are exposed and offered for sale shall be paved with four inches of concrete for the entire surface area of the market. The floor shall be protected by a curtain wall of concrete or masonry of not less than four (4) inches in thickness, hemetically sealed to the surface pavement, and extending not less than twenty-four inches below the ground surface. Curtain wall shall be extended around the entire perimeter of the floor pavement. Display racks, stands, or platforms on which fruit or vegetables are stored or offered for sale shall be of sufficient height that all such fruit or vegetables shall be kept at a distance of not less than eighteen (18) inches above the floor pavement and so constructed that rats cannot harbor in or under such racks.

CHAPTER XX

LIGHT, VENTILATION AND SANITATION

SECTION 2001 – LIGHT AND VENTILATION

2001.1 — MINIMUM REQUIREMENTS

(a) Every habitable room of buildings hereafter erected shall have one or more windows, unless otherwise, specifically provided herein, to afford adequate light and ventilation. The requirements specified in this chapter shall be considered as minimum requirements supplementary to all State laws regulating light and ventilation.

(b) Where windows are required such windows shall open on a street, public space, yard, or approved open space that will afford adequate air and light. Required windows shall be so constructed that when fully opened, the total open space shall not be less than one-half (½) the required window area.

(c) Attics not used for habitational purposes shall have provisions for the emission of excess heat.

(d) Skylights, vents, louvers or mechanical ventilation may be substituted for windows when approved by the Building Official, provided adequate light and ventilation is provided to meet the requirements of this chapter.

(e) Except as otherwise provided herein, required windows shall have glazed openings of clear glass of area not less than one-tenth (1/10) of the floor area of the room served by them with the following exceptions:

Basements and cellars not used for habitational purposes—windows shall have an area not less than one-fiftieth (1/50) of the floor area served.

Storage rooms—windows shall have an area not less than one-twentieth (1/20) of the floor area served.

Obscure glass, glass blocks and similar glazed panels that admit less light than clear glass shall have area increased to admit amount of light equivalent to the above requirement.

Year around mechanically ventilating conditioned air systems may be substituted for windows, as required herein, in rooms other than rooms used for sleeping purposes. Window type air conditioning units are not included in this exception.

(f) Rooms where by reason of use or occupancy, dust fumes, gases, vapors, odors or other hazardous, obnoxious, or injurious impurities exist shall be provided with adequate additional ventilation to insure safe and healthful conditions.

2001.2 — ALCOVES

An alcove opening off a habitable room may be included as part of that room in determining the window area required, provided that a portion of the common wall between the habitable room and the alcove, equal to fifty per cent (50%) of the floor area of the alcove, is open and unobstructed.

2001.3 — TOILETS

(a) Toilet rooms shall not open directly into a kitchen or room used for the preparation of food.

(b) Every toilet room shall have windows as specified for habitable rooms providing in no case less than three (3) sq. ft. of open space, or shall have approved, equivalent mechanical ventilation.

2001.4 — CLASSROOMS

Classrooms in School occupancies shall have at least unilateral light. The windows shall be located on the long axis of the room.

EXCEPTION: When mechanical ventilation is provided in accordance with Section 2001.5, artificial lighting, with foot candle values at desk top or work table level not less than set forth below, may be substituted for unilateral lighting.

Type Room	Minimum Values
Cooking	50 foot candles
Lecture and Demonstration	150 foot candles
Sewing	150 foot candles
Drafting	100 foot candles
Others	70 foot candles

2001.5 — MECHANICAL VENTILATION

(a) Where ventilation is provided by mechanical means, fresh air in sufficient quantity to maintain healthful conditions shall be provided to meet the requirements of all State laws. In the absence of such requirements, ventilation at least equivalent to the requirements of this code governing natural ventilation shall be provided.

(b) Lavatories, toilets, bathrooms, and rest rooms shall be provided with at least two (2) cubic feet of fresh air per minute per square foot of floor area.

SECTION 2002 – SANITATION

2002.1 — TOILET FACILITIES

Every building or portion thereof where persons are employed shall be provided with at least one toilet. Every building and each subdivision thereof where both sexes are employed shall be provided with access to at least two toilets, except where there are five or less employees on any one shift, located either in such building or conveniently in a building adjacent thereto on the same property.

2002.2 — SURROUNDING MATERIALS

The walls and floors of all public rest rooms shall be lined with non-absorbent material to a height of 4 feet above the floor.

(See Southern Standard Plumbing Code for regulations governing installation and facilities.)

CHAPTER XXI

SAFEGUARDS DURING CONSTRUCTION

SECTION 2101 — GENERAL

The temporary use of streets or public property for the storage or handling of materials or of equipment required for construction or demolition, and the protection provided to the public shall be in accordance with the provisions of this chapter.

2101.1 — ALLOWABLE USE OF PUBLIC PROPERTY DURING CONSTRUCTION

The amount of space and conditions under which public property may be used for construction or demolition purposes shall be as set forth below:

(a) One-third (⅓) of the width of street that is adjacent to the curb in front of the building being erected and for which a permit has been issued. If street in front of property adjoining such building is to be used for similarly limited storage, a due waiver of claim against the applicable governing authority for damages on account of such use, issued by the owner of such property, must be filed with the Building Official before such use shall be allowed.

(b) Provisions under which street or sidewalk space may be used:

(1) That such one-third (⅓) allocated space or any portion thereof shall not come within five (5) feet of a rail or railway track.

(2) That a walkway be constructed in the outer portion of the permissible occupied street space, conforming to the requirements of Section 2101.3.

(3) That no building material, fence, shed or any obstruction of any kind shall be placed so as to obstruct free approach to any fire hydrant, lamp post, manhole, fire alarm box, or catch basin, or so as to interfere with the passage of water in the gutter. Protection against damage shall be provided to such utility fixtures during the progress of the work, but sight of them shall not be obstructed.

(4) That a ten (10) foot clear roadway be maintained through any alley located along the building site.

(5) That proper precaution shall be made during construction to prevent concrete, mortar washings, or any other material from entering a sewer.

(6) The person or persons to whom a permit is issued for such purposes as stated above, shall post with the appli-

cable governing authority a bond of such type and amount as may be deemed advisable by the applicable governing authority as protection from any and all liability.

2101.2 — WHERE COVERED WALKWAYS ARE REQUIRED

(a) During the erection or demolition of any building exceeding one (1) story in height that is located at a distance less than ten (10) feet or less than one-quarter (¼) of the height of the building from any street or alley property line, or when required by the Building Official, a roof covering for the entire length of the project shall be provided over the temporary or permanent sidewalk, from the time the construction or demolition extends above the second floor level until materials are no longer being used or handled on the front above such walk.

(b) Buildings having their exteriors altered or repaired in an extensive manner involving any hazard shall be provided with a covered walk as required for new structures during erection.

Exception:

Where, in the opinion of the Building Official, a covered walk is not necessary, a permit may be issued to block off part of the sidewalk and have a temporary walk constructed as provided in Section 2101.3.

2101.3 — CONSTRUCTION OF WALKWAYS FENCES AND PROTECTIVE COVERINGS

Before any construction work is commenced the owner or his agent shall construct a temporary walkway in conformity with this section.

(a) All fences, barriers, or temporary structures of any kind located on public highways, shall be so constructed as not to obstruct vision at the intersection of streets.

(b) Walkways shall be not less than four (4) feet wide in the clear except that in congested districts the Building Official may require a walkway as wide as, in his opinion, is necessary. Walks shall be built in safe and substantial manner and be maintained in that condition at all times. A smooth handrail of substantial construction, not less than three (3) feet high, shall be provided on the traffic or street side of the walkway, and also on the building side when considered necessary by the Building Official.

(c) Where the distance from building to street or alley property line is less than half the height of the building, a fence of substantial solid construction at least eight (8) feet high shall be provided on the building side of the walkway.

(d) Roof coverings over walkways, as required by Section 2101.2, shall be constructed of not less than one layer of two (2) inch nominal dimension wood plank spanning not over three (3) feet between supports, or equivalent decking. The framework supporting the walkway covering shall be well braced and designed to support at least one hundred fifty (150) Lbs. per Sq. Ft. but the top deck shall be designed to carry not less than two hundred fifty (250) Lbs. per Sq. Ft. The roof covering shall be of width sufficient to cover the entire walkway or sidewalk, and shall be made watertight. Suitable provision shall be made for adequate lighting of the walk under the covering, at all times. A minimum clearance of eight foot six inches (8'-6") shall be maintained above walkways.

2101.4 — WALKWAYS OVER EXCAVATED AREAS

When the area occupied by the sidewalk or temporary walkway is to be excavated, such walk shall be made of boards not less than two (2) inches nominal dimension designed to support a load of not less than one hundred and fifty (150) pounds per Sq. Ft., provided with suitable ramps at each end. Such walkways shall be provided with a fence and handrails on each side.

2101.5 — STORAGE OF MATERIALS OVER WALKWAYS

Whenever roofs of walkways are used for the storing of materials, it shall be designed for the load to which it is to be subjected and a railing and footboard shall be installed so as to prevent the materials from spilling into the streets. The posts and/or other supporting members on the street side shall be protected so as to insure against failure due to impact from street traffic.

2101.6 — WALKWAYS TO BE KEPT IN REPAIR

The street side of any barricade or fence, handrails and sidewalks shall be kept reasonably smooth and in good repair while construction work is in progress, or while such barricades, fences, or walkways are placed on or over public property.

2101.7 — CLEANING OF SIDEWALKS AND STREETS

The owner or his agent, upon the completion of the building, shall immediately remove all walkways, debris or any other obstructions and leave such public property in as good a condition as it was before such work was commenced.

2101.8 — RED LIGHTS REQUIRED

Every walkway shall be kept well lighted continuously between sunset and sunrise and the outer edge of the occupied space of the street or sidewalk shall have placed thereon "red lights" which shall burn continuously between sunset and sunrise.

2101.9 — SAFETY REQUIREMENTS DURING CONSTRUCTION

(a) **Construction Scaffolds, Hoists, Equipment, Etc.**—All equipment such as temporary stairs, ladders, ramps, scaffolds, hoists, runways, barricades, chutes, elevators, etc., as required for the execution of any construction work shall be substantially constructed and erected to insure the safety of the workmen using them or passing under, on or near them. Where a large amount of scaffolding is used the Building Official may require the use of non-combustible material or fire retardant treated wood. The flame proofing of tarpaulins may also be required by the Building Official where, in his opinion, the fire hazard warrants such precaution.

(b) **Sanitation**—Adequate sanitary facilities for the convenience of all workmen shall be provided. These facilities shall be kept in a clean and sanitary condition throughout the duration of the work. The temporary workman's convenience shall be enclosed, screened, and weatherproofed and shall be connected to a sewer. Upon removal of the temporary facilities the sewer connection shall be removed and the sewer capped. In lieu of connecting to a sewer, the temporary facility may be a portable, enclosed, chemically treated, tank-tight unit.

(c) **Standpipes**—In buildings under construction over three (3) stories in height, standpipes where required by the Code shall be installed as the work progresses. Connection outlets shall be provided at each floor level and all such connections and fittings from ground level up shall be designed to fit the fire department equipment. All standpipes shall be not less than two and one half inches (2½") in diameter and so located that 100 feet of one and one half (1½) inch hose will reach within 40 feet of the most remote part of each floor area.

2101.10 — PROTECTION OF ROOFS AND SKYLIGHTS OF ADJOINING BUILDINGS

When a building or structure is to be carried above the roof of an adjoining building, protection for the skylights and roof of such adjoining building shall be provided, at his own expense, by the person constructing or causing the construction of such building or structure; provided that if the owner, lessee or tenant of the adjoining building should refuse permission to have the roofs and skylights protected, the responsibility and expense for the necessary protection shall devolve on the person refusing such permission.

CHAPTER XXII

USE OF PUBLIC PROPERTY

SECTION 2201 — GENERAL

The use of public property or any portion thereof, shall be in accordance with the provisions of this chapter, except signs which shall conform to the requirements of Chapter XXIII, Signs and Outdoor Displays, and allowable use of public property during construction, Section 2101.1.

2201.1 — DOORS AND WINDOWS

No door in the city limits shall open or project upon public property. Exit doors, as specified in Chapter XI, which are required to open in the direction of exit travel, shall be set back from the property line by means of vestibules or similar enclosures. Windows which swing over public property shall have a clearance of not less than eight feet above the sidewalk or ground level.

2201.2 — MARQUEES, CANOPIES OR FIXED AWNINGS

(a) Fixed awnings, marquees or canopies shall be entirely supported from the building.

(b) All combustible materials used in the construction of such canopies, awnings or marquees, shall be protected with not less than one hour fire-resistance protection as specified in Chapter X. All glazing in marquees, canopies, or fixed awnings shall be of wired glass.

(c) Every fixed awning, canopy or marquee shall be at least nine feet in the clear, between the lowest point or projection and a sidewalk immediately below. (See 2303.5—Marquee Signs.)

(d) No fixed awning, canopy or marquee shall extend or occupy more than two-thirds (⅔) of the width of sidewalk measured from the building, except that such fixed awning, canopy or marquee may occupy the entire width of the sidewalk, provided it is fourteen feet in the clear above the sidewalk. The overall height of any marquee or canopy including signs, shall not exceed eight (8) feet, measured from the roof of such marquee.

(e) Marquees or canopies constructed with a roof live load of not less than forty (40) pounds per square foot shall be permitted the full width of the building (front) and full depth of the building, corner location.

2201.3 — MOVABLE AWNINGS (METAL OR CANVAS)

(a) Metal or canvas awnings may extend over public property for a distance of not more than five (5) feet, provided such awnings or any part thereof maintain a clear height of eight (8) feet above the sidewalk.

(b) All such movable awnings shall be supported on metal frames attached to the building.

2201.4 — PROHIBITIVE LOCATIONS

(a) Every awning, marquee or canopy shall be so located as not to interfere with the operation of any exterior standpipe, stairway or exit from the building.

(b) Awnings, marquees or canopies shall not be used as a landing for any fire escape or exterior stair.

2201.5 — CONSTRUCTION REQUIREMENTS

Marquees, canopies and awnings shall be so constructed and anchored to the building so as to support all live and dead loads as specified in Chapter XII.

2201.6 — ROOF DRAINAGE REQUIRED

The roof of every fixed awning, marquee or canopy shall be sloped to down spouts at the building, which shall conduct all drainage under the sidewalk to the curb.

SECTION 2202 – OTHER PROJECTIONS

2202.1 — GENERAL

(a) Every projection of any character over or upon public property shall maintain a clear height above the sidewalk or ground level of not less than eight feet. The allowable projection over public property shall not exceed the following measurements from the building.

(b) Bay windows, porches, balconies, fire escapes—three (3) feet.

(c) Cornices, Belt Courses, sills, pilasters, water tables or any decorative features—six (6) inches.

(d) See Section 710 for fire protective requirements.

2202.2 — SIDEWALK OR STREET OBSTRUCTIONS

Public property shall be maintained clear of any and all obstructions, including among others, posts, columns, display of wares or merchandise and sidewalk signs.

SECTION 2203 – SPACE UNDER PUBLIC PROPERTY

2203.1 — SPACE UNDER SIDEWALK

Where space under the sidewalk is used for any purpose a special permit shall be required.

2203.2 — SIDEWALK LIGHTS

When glass is set in the sidewalk to provide light for spaces underneath, the glass shall be supported by metal or reinforced concrete frames and such glass shall be not less than one-half (½) inch in thickness. Where such glass is over twelve (12) square inches in area, it shall have wire mesh embedded in the glass. All portions of sidewalk lights shall be of not less strength than required in Section 1203.3 for sidewalks.

SECTION 2204 — MOVING OF BUILDINGS

2204.1 — GENERAL

No building or part of any building shall be moved through or across any sidewalk, street, alley or highway within the governmental limits without first obtaining a permit from the Building Official.

2204.2 — WRITTEN APPLICATION

Any person desiring to move a building shall first file with the Building Official a written application setting forth the following information.

(a) Type and kind of building to be moved.

(b) The original cost of such building.

(c) The extreme dimensions of the length, height and width of the building.

(d) Its present location and proposed new location by lot, block, subdivision and street numbers.

(e) The approximate time such building will be upon the streets, and the contemplated route that will be taken from present to new location.

2204.3 — BUILDING OFFICIAL SHALL REJECT WHEN

(a) If in the opinion of the Building Official, the moving of any building will cause serious injury to persons or property or serious injury to the streets or other public improvements, or the building to be moved has deteriorated more than fifty per cent of its original value by fire or other element, or the moving of the building will violate any of the requirements of this Code or of the Zoning Regulations, the permit shall not be issued and the building shall not be moved over the streets.

(b) Any building being moved for which permit was granted shall not be allowed to remain in or on the streets for more than forty-eight hours.

2204.4 — BOND REQUIRED

The Building Official, as a condition precedent to the issuance of such permit, shall require a bond to be executed by person desiring such removal permit, with corporate surety to his satisfaction. Such bond shall be made payable to applicable governing body and for such amount as he prescribes. It shall indemnify the applicable governing body against any damage caused by the moving of such building to streets, curbs, sidewalks, shade trees, highways and any other property which may be affected by the moving of a building. Such surety bond shall also be conditioned upon and liable for strict compliance with the terms of said permit, as to route to be taken and limit of time in which to effect such removal and to repair or compensate for the repair and to pay said applicable governing body as liquidated damages an amount not exceeding fifty dollars ($50.00) to be prescribed by the Building Official for each and every day's delay in completing such removal or in repairing any damage to property or public improvement or in clearing all public streets, alleys or highways of all debris occasioned thereby.

2204.5 — NOTICES TO BE GIVEN BY BUILDING OFFICIAL

Upon the issuance of said moving permit, the Building Official shall cause notice to be given to the Superintendent of Fire Alarm, Chief of Fire Department, telephone or light companies, or others whose property may be affected by such removal. The Building Department shall set forth in all notices the route that will be taken, time started, and approximate time of completion.

2204.6 — PUBLIC SAFETY REQUIREMENTS

(a) **Lights required**

Every building which occupies any portion of public property after sundown, shall have sufficient lights continuously burning between sunset and sunrise for the protection of the public.

(b) **Number and location**

There shall be a minimum of five red lights placed on each street side of the building; such red lights shall be attached to the building in such a fashion as to indicate extreme width, height, and size.

(c) **Flares required**

There shall be placed in addition to the red lights on the building, flares at regular intervals for a distance of two hundred feet up the street on each side of the building.

(d) **Flagmen required**

When more than fifty per cent of the street, measured between curbs, is occupied at night by the building, or when in the opinion of the Building Official, flagmen are necessary to divert or caution traffic, the owner or person moving such building shall employ at their expense, two flagmen, one at each street intersection beyond the building; such flagmen shall remain at these intersections, diverting or cautioning traffic from sunset to sunrise. Red lights shall be employed in flagging traffic at night.

CHAPTER XXIII

SIGNS AND OUTDOOR DISPLAYS

SECTION 2301 — GENERAL

2301.1 — OUTDOOR ADVERTISING DISPLAYS

Outdoor advertising displays, means any letter, figure, character, mark, plane, point, marquee sign, design, poster, pictorial, picture, stroke, stripe, line, trademark, reading matter or illuminated service, which shall be so constructed, placed, attached, painted, erected, fastened or manufactured in any manner whatsoever, so that the same shall be used for the attraction of the public to any place, subject, person, firm, corporation, public performance, article, machine or merchandise, whatsoever, which are displayed in any manner whatsoever out of doors. Every outdoor display shall be classified and conform to the requirements of that classification as set forth in this chapter.

2301.2 — CLASSIFICATIONS

For the purpose of this chapter and the regulations and provisions thereof, outdoor advertising displays shall be classified into one of the following type signs:

(a) **Spectacular Signs**—means an "Outdoor Advertising Display Sign" advertising copy usually animated, constructed of metal, wired for lights or luminous tubing, or both, with copy action controlled by the flashed circuit breakers or matographs and attached on an open face steel structure built especially for the purpose.

Spectacular signs may be built upon the ground, attached to a wall, or above the roof, or projecting from a wall, provided that such spectacular sign meets the requirements of the provisions of this Code governing ground, roof, wall, projection or marquee sign, depending upon where such sign is built, as set forth below.

Spectacular signs shall be illuminated with electricity only.

(b) **Ground Sign**—means an "Outdoor Advertising Display Sign" when such sign is supported by uprights or braces in or upon the ground.

(c) **Roof Sign**—means an "Outdoor Advertising Display Sign" erected, constructed, or maintained above the roof of any building.

(d) **Wall Sign**—means an "Outdoor Advertising Display Sign" that shall be affixed to the wall of any building, when such sign shall project not more than twelve (12) inches from the building.

(e) **Projection Sign**—means an "Outdoor Advertising Display Sign" which is affixed to any building wall or structure and extends beyond the building wall, structure, building line or property line more than twelve (12) inches.

(f) **Marquee Sign**—means a projecting sign attached to or hung from a marquee and said marquee shall be known to mean a canopy or covered structure projecting from and supported by a building, when such canopy or covered structure extends beyond the building, building line or property line.

(g) **Shingle Sign**—means a projection or wall sign not over six (6) square feet in area, constructed of metal or other non-combustible material attached securely to a building and not projecting more than twenty-four (24) inches over public property.

2301.3 — PERMITS REQUIRED

(a) No "Outdoor Advertising Display Sign" shall hereafter be erected, constructed, altered or maintained except as provided in this Code, until after permit for the same has been issued by the Building Official as required in Sections 105 and 106 and the fee paid as specified in Section 107, Fees.

2301.3 (b) — EXCEPTION

No permit fee shall be required for a shingle sign over a show window or door of a store or business establishment, announcing without display or elaboration, only the name of the proprietor and nature of the business; nor shall a permit be required for a ground sign advertising for sale or rent property, providing such sign is not over fifteen (15) square feet in area.

2301.4 — IDENTIFICATION OF SIGNS

Every Outdoor Advertising Display Sign hereafter erected, constructed or maintained, for which a permit is required shall be plainly marked with the name of the person, firm or corporation erecting and maintaining such sign and shall have affixed on the front thereof the number of permit issued for said sign by the Building Official.

2301.5 — ANNUAL INSPECTION

It shall be the duty of the Building Official or his authorized agent to inspect every ground sign, roof sign, wall sign, and projection sign at least once annually.

2301.6 — UNSAFE SIGNS

Should any sign become insecure or in danger of falling or otherwise unsafe in the opinion of the Building Official, the owner thereof, or the person or firm maintaining the same, shall upon written notice from the Building Official, forthwith in the case of immediate danger and in any case within ten (10) days, secure the same in a manner to be approved by the Building Official, in conformity with the provisions of this Code or remove such sign. If such order is not complied with in ten (10) days the Building Official shall remove such sign at the expense of the owner or lessee thereof.

2301.7 — MAINTENANCE

All signs for which a permit is required, together with all their supports, braces, guys and anchors shall be kept in repair and unless of galvanized or non-corroding metal shall be thoroughly painted at least once every two years. The Building Official may order the removal of any sign that is not maintained in accordance with the provisions of this section. Such removal shall be at the expense of the owner or lessee.

2301.8 — UNLAWFUL SIGNS

In case any sign shall be installed, erected, or constructed in violation of any of the terms of this Code the Building Official shall notify by registered mail or written notice served personally, the owner or lessee thereof to alter such sign so as to comply with this Code or of the Zoning Regulations and to secure the necessary permit therefor, or to remove the sign. If such order is not complied with in ten (10) days the Building Official shall remove such sign at the expense of the owner or lessee thereof.

2301.9 — LOCATION RESTRICTIONS

No Outdoor Advertising Display Sign shall be erected, constructed or maintained so as to obstruct any fire escape or any window or door or opening used as a means of egress or so as to prevent free passage from one part of a roof to any other part thereof. No sign shall be attached in any form, shape or manner to a fire escape, nor be placed in such manner as to interfere with any opening required for legal ventilation.

2301.10 — SIGNS PROJECTING OVER PUBLIC PROPERTY

Signs projecting from a building or extending over public property shall maintain a clear height of nine (9) feet above the sidewalk and all such signs shall not extend more than eighteen (18) inches of the curbline.

SECTION 2302 – STRUCTURAL REQUIREMENTS

2302.1 — DESIGN AND STRESS DIAGRAMS REQUIRED

Before a permit shall be granted the erector of every Outdoor Advertising Sign with the exception of shingle signs and light cloth temporary signs, shall submit to the Building Official a design and stress diagram or plan, containing the necessary information to enable the Building Official to determine that such sign complies with all the regulations of this Code.

2302.2 — WIND PRESSURE

In the design and erection of all Outdoor Advertising Display Signs, the effect of wind shall be carefully considered. All signs shall be so constructed as to withstand the wind pressure as specified in Section 1205.

2302.3 — WORKING STRESSES

(a) In all Outdoor Advertising Display Signs, the allowable working stresses shall conform with the requirements of Section 1205 of this Code, except as specified below.

(b) The allowable working stresses for steel and wood shall be in accordance with the provisions of Chapter XV—"Steel Construction" and Chapter XVII—"Wood Construction."

(c) The working strength of chains, cables, guys or steel rods shall not exceed one-fifth (1/5) of the ultimate strength of such chains, cables, guys or steel rods.

SECTION 2303 — CONSTRUCTION

2303.1 — GROUND SIGNS

(a) No ground sign constructed entirely of wood material shall be at any point over twenty-four (24) feet above the ground level or located in the Fire District, but when the facing of a ground sign is constructed entirely of sheet metal or other non-combustible material such ground sign may be erected within or without the Fire District and the supports, braces, battens, ornamental moulding, platform and decorative trim may be of wood material.

(b) Lighting reflectors may project beyond the face of the sign.

(c) The bottom coping of every ground sign shall be at least three (3) feet above the ground or street level, which space may be filled with platform decorative trim or light wooden construction.

(d) Every ground sign shall provide rigid construction to withstand wind action in all directions.

(e) Any person or persons, partnership, firm or corporation occupying any vacant lot or premises by means of a ground sign, shall be subject to the same duties and responsibilities as the owner of the lot or premises, with respect to keeping the same clean, sanitary, inoffensive, free and clear of all obnoxious substances and unsightly conditions on the ground in the vicinity of such ground sign on said premises for which they may be responsible.

(f) Wherever anchors or supports consist of wood embedded in the soil, the wood shall be pressure-treated with an approved preservative.

2303.2 — ROOF SIGNS

(a) All roof signs shall be so constructed as to leave a clear space of not less than six (6) feet between the roof level and the lowest part of the sign and shall have at least five (5) feet clearance between the vertical supports thereof; no portion of any roof sign structure shall project beyond an exterior wall.

(b) Every roof sign shall be constructed entirely of steel construction, including the upright supports and braces, except that only the ornamental moulding and battens behind the steel facing and the decorative lattice work may be of wood construction.

(c) The bearing plates of all roof signs, shall distribute the load directly to or upon masonry walls, steel roof girders, columns or beams. The building shall be designed to avoid overstress of these members.

(d) No roof sign having a tight or solid surface shall be at any point over twenty-four (24) feet above the roof level.

(e) Open roof signs in which the uniform open area is not less than forty (40) per cent of total gross area may be erected to a height of seventy-five (75) feet on buildings of Type I or Type II Construction and on other type buildings to a height of forty (40) feet, all such signs shall be thoroughly secured to the building upon which they are installed, erected or constructed by iron, metal anchors, bolts, supports, chains, stranded cables, steel rods or braces and they shall be maintained in good condition as set forth in Section 2301.7.

2303.3 — WALL SIGNS

(a) Wall signs attached to exterior walls of solid masonry, concrete or stone, shall be safely and securely attached to the same by means of metal anchors, bolts or expansion screws of not less than three-eighths (⅜) inch in diameter and shall be embedded at least five (5) inches. No wood blocks shall be used for anchorage, except in the case of wall signs attached to buildings with walls of wood. No wall sign shall be supported by anchorages secured to an unbraced parapet wall.

(b) The surface face of all wall bulletins must be of sheet metal but the ornamental moulding surrounding same may be of wood construction.

(c) Temporary cloth signs with wood frames may be kept in place for a period not exceeding thirty (30) days.

2303.4 — PROJECTING SIGNS

(a) All projecting signs shall be constructed entirely of metal or other non-combustible material and securely attached to a building or structure by metal supports such as bolts, anchors, supports, chains, guys or steel rods. No staples or nails shall be used to secure any projecting sign to any building or structure.

(b) The dead load of projecting signs, not parallel to the building or structure and the load due to wind pressure shall be supported with chains, guys, or steel rods having net cross sectional dimension of not less than three-eighths (⅜) inch in diameter. Such supports shall be erected or maintained at angle of at least 45 degrees with the horizontal to resist the dead load and at an angle of 45 degrees or more with the face of the sign to resist the specified wind pressure. If such projecting sign exceeds thirty (30) square feet in one facial area, there shall be provided at least two such supports on each side not more than eight feet apart to resist the wind pressure.

(c) **Anchorage**—All supports shall be secured to a bolt or expansion screw that will develop the strength of the supporting chain, guys or steel rod, with a minimum five-eighths (⅝) inch bolt or lag screw, by an expansion shield. Turn buckles shall be placed in all chains, guys or steel rods supporting projecting signs.

(d) Chains, cables, guys, or steel rods used to support the live or dead load of projecting signs may be fastened to solid masonry walls with expansion bolts or by machine screws in iron supports, but no such supports shall be attached to an unbraced parapet wall. Where the supports must be fastened to walls made of wood, the supporting anchor bolts must go through the wall and be plated or fastened on the inside in a secure manner.

(e) No projecting sign shall be erected on the wall of any building so as to project above the roof or cornice wall or above the roof level where there is no cornice wall; except that a sign erected at a right angle to the building, the horizontal width of which sign perpendicular to such wall does not exceed eighteen (18) inches may be erected to a height not exceeding two (2) feet above the roof or cornice wall or above the roof level where there is no cornice wall. A sign attached to a corner of a building and parallel to the vertical line of such corner, shall be deemed to be erected at a right angle to the building wall.

2303.5 — MARQUEE SIGNS

Marquee signs shall be constructed entirely of metal or non-combustible material and may be attached to, or hung from a marquee, and such signs when hung from a marquee shall be at least eight (8) feet at its lowest level above the sidewalk or ground level, and further, no such sign shall extend outside the line of such marquee. Marquee signs may be attached to the sides and front of a marquee, and such sign may extend the entire length and width of said marquee, provided such sign does not extend more than six (6) feet above, nor one (1) foot below such marquee, but under no circumstances, shall the sign or signs have a vertical dimension greater than eight (8) feet.

2303.6 — SPECTACULAR SIGNS

All permits for spectacular signs shall be issued by the Building Official upon application therefor, after approval by the department having jurisdiction over electricity, upon payment of the required permit fee. The permit to erect or maintain a spectacular sign shall be good for one (1) year. The Building Official may issue a renewal of the permit issued hereunder upon the expiration thereof or within thirty (30) days thereafter upon the payment by the applicant of a renewal fee of two dollars ($2.00) and by surrendering the old permit, accompanied by satisfactory proof in the form of an affidavit that the sign is as safe as when originally licensed and that the wiring or piping of the same is in good condition. All spectacular signs shall be constructed of non-combustible materials.

SECTION 2304 – USE OF PLASTIC MATERIALS

(a) Notwithstanding any other provisions of this Code, plastic materials which burn at a rate no faster than 2.5" per minute when tested in accordance with "Test for Flammability of Rigid Plastics Over 0.050 Inches in Thickness, ASTM D635-63," shall be deemed approved plastics and may be used as the display surface material and for the letters, decorations and facings on signs and outdoor display structures, provided, that in the Fire District the structure of the sign in which the plastic is mounted or installed is non-combustible.

(b) Individual plastic facings of electric signs shall not exceed 200 square feet in area. If the area of a display surface exceeds 200 square feet, the area occupied or covered by approved plastics shall be limited to 200 square feet plus 50 percent of the difference between 200 square feet and 2,000 square feet.

(c) In no case shall the area of plastic on a display surface exceed 1,100 square feet.

(d) Letters and decorations mounted upon an approved plastic facing or display surface may be made of approved plastics.

CHAPTER XXIV

ELEVATORS AND ESCALATORS

2401.1

Elevators, Dumbwaiters, Escalators and Moving Walks shall be constructed, installed and maintained to provide proper safety in accordance with the requirements of this Code. Construction, installation and maintenance shall conform with the "Standard Safety Code for Elevators, Dumbwaiters, Escalators and Moving Walks, ANSI A17.1-1971."

2401.2

(a) All openings in elevator shafts shall be protected. Elevators and escalators shall be enclosed as required in Section 701.

(b) The safe capacity of every elevator shall be conspicuously posted.

(c) Elevators and Escalators shall be provided with safety devices conforming to the standard specified in Section 2401.1.

(d) Every passenger elevator shall have an emergency exit as required by "Standard Safety Code for Elevators, Dumbwaiters, Escalators and Moving Walks, ANSI A17.1-1971."

2401.3 — MANLIFTS

All Manlifts shall be installed in accordance with the "American Standard Safety Code for Manlifts, ANSI A90.1-1969."

CHAPTER XXV

PREFABRICATED CONSTRUCTION AND MOBILE HOMES OVER EIGHT (8) FEET WIDE

2501 — GENERAL

Prefabricated construction and mobile homes over eight (8) feet wide shall conform to the requirements of this Code, except as otherwise provided for in this Chapter.

2502 — LOADS

Live, dead and wind load requirements shall conform to the requirements, as set forth in Chapter XII.

2503 — STRUCTURAL DESIGN

Where the size and spacing of framing members on materials are in conflict with, or not covered by this Code, they will be acceptable if the assembly meets the load requirements of this Code, provided they conform to the requirements as set forth for tests under Section 2504. The Building Official may waive load tests for plywood components and other structural elements, designed in accordance with applicable standards listed elsewhere in this code.

2504 — TESTS

(a) Every manufacturer of prefabricated construction shall file with the Building Official, duplicate copies of a certificate from a recognized testing laboratory, which states that tests have been made on this particular type of prefabricated construction, and showing the live, dead and wind load capacities in pounds, per square foot, uniformly distributed, together with a detailed physical description of the panels tested.

(b) Panels and other elements tested for loads shall sustain, without failure, for a period of 24 hours, a superimposed load equal to 2½ times the live load. Recovery within 24 hours, after removal of the full test load, shall be not less than seventy-five (75) percent of the observed deflection. The measured deflection of any panel or element under full live load shall be not over one three-hundred-and-sixtieth (1/360) of the clear span.

Exception: The measured deflection of any wall or roof panel or element under full live load shall be not over one two-hundred-and-fortieth (1/240) of the clear span, providing they are not plastered.

(c) When it is definitely ascertained by the Building Official that the requirements of this Code have been met, a permit shall be issued. A copy of all testing laboratory certifications shall be filed as a permanent record in the office of the Building Official.

2505 — PLYWOOD

Plywood shall conform to U. S. Product Standard PS 1-66, Softwood Plywood, Construction and Industrial. Plywood of species not covered in PS 1-66, when used structurally, shall be identified as to veneer grade and glue type by an approved agency and shall meet the performance standards in U. S. Product Standard PS1-66 for its type.

2506 — MOBILE HOMES

See Appendix I for Mobile Home Standards.

CHAPTER XXVI

PLASTICS

SECTION 2601 — GENERAL

2601.1 — SCOPE AND USES

(a) Plastic materials regulated by this chapter shall be those materials made wholly or principally from standardized plastics listed and described in "Technical Data on Plastics," 1957 edition, published by the Manufacturing Chemists' Association.

(b) This Chapter pertains to the use of plastics in the following applications only.

Wall panels.
Glazing.
Roof Panels.
Skylights.
Light-transmitting panels in monitors and sawtooth roofs.
Light-diffusing systems in ceilings.
Partitions.
Exterior veneers.
Awnings and similar shelters.
Greenhouses.
Signs and outdoor displays.

2601.2 — APPROVED PLASTIC MATERIAL DEFINED

(a) For the purposes of this Chapter, an approved plastic shall be the material found to be marked or labeled or to carry an identification by the manufacturer of the material denoting its class of plastics; or as approved by the Building Official.

(b) The basis of approval of such approved plastic material shall take into consideration the materials when tested by a recognized laboratory in accordance with "Test for Flammability of Rigid Plastics Over 0.050 Inches in Thickness, ASTM D635-63," under which the material shall not burn faster than 2.5 inches per minute, or "Test for Flammability of Plastics 0.050 Inch and Under in Thickness, ASTM D568-61, under which the material shall not be consumed in less than 2 minutes, and "Test for Thickness of Solid Electrical Insulation, ASTM D374-57T. The Building Official may accept the classes of plastics as approved by the Committee on Compliance of the Southern Building Code Congress.

2601.3 — CLASSES OF PLASTICS

(a) Class A plastic materials shall be those reinforced or unreinforced approved plastic materials which are self-extinguishing when tested in accordance with the test procedures described in 2601.2 above.

(b) Class B plastic materials shall be those approved plastic materials which are reinforced with glass fiber or other non-combustible material mounting to not less than 1.5 ounces per square foot and not less than 20 per cent by weight of the plastic panel or sheet.

(c) Class C plastic materials shall be those approved plastic materials which are reinforced with glass fiber or other non-combustible material amounting to not less than 10 per cent by weight of the plastic panel or sheet.

(d) Class D plastic materials shall be approved plastic materials other than Class A, B or C, which meet the requirements of Section 2601.2.

2601.4 — DESIGN LOADS

Design Loads shall conform to the provisions of Chapter XII of this Code.

SECTION 2602 – WALL PANELS

2602.1 — GENERAL

Wall panels shall mean plastic sheets used as a light-transmitting medium in exterior walls. Such panels may be installed only in wall areas in which openings are not required to be fire protected, and access panels shall be provided as required by the Code for structure and occupancy.

2602.2 — CLASS A AND B PLASTICS

Class A and Class B plastic sheets may be used in wall panels in Types III, IV, V and VI buildings except in occupancy classifications E, D and H provided:

(a) The total area of plastic panels does not exceed 20 per cent of the wall area in any one story of the structure.

(b) No section of plastic panels shall exceed one hundred feet (100') in length horizontally and no section shall exceed twelve feet (12') in height.

(c) In Types III, IV and V buildings, sections up to forty feet (40') in length shall be separated longitudinally by a section of conforming materials equal in width to 10 per cent of the length of the section or four feet (4'), whichever is greater. Sections over forty feet (40') in length shall be separated by a section of approved non-combustible siding at least eight feet (8') in width.

(d) In Types III, IV and V buildings, parallel sections shall be separated vertically by a section of conforming materials at least eight feet (8') in height.

2602.3 — CLASS C AND D PLASTICS

Class C and D plastics may be used as wall panels in locations and subject to the conditions specified for Class A and B plastics provided the area of such panels does not exceed 20 per cent of the wall area in any one story of the structure and no section of such panels is over fifty feet (50') in length or eight feet (8') in height.

SECTION 2603 — GLAZING OF UNPROTECTED OPENINGS

Doors, sash and framed openings may be glazed or equipped with transparent or translucent approved plastic materials, where such openings are not required to be fire protected, provided that the area so glazed shall not exceed 20 per cent of the wall area nor be located at a height greater than twenty feet (20') above grade level, unless greater areas and heights are approved by the Building Official. (See Table 2706 for application where subject to impact loads.)

SECTION 2604 — ROOF PANELS

2604.1 — GENERAL

Transparent or translucent plastic panels may be used in roofs not required to have a fire-resistive rating and in all roofs where sprinkler protection is provided, except for Occupancy Classifications D, E and H, provided:

(a) That on structures or over occupancies required to have fire-retardant or non-combustible roofing, the panels conform to the slope of the roof which shall be at least four inches in twelve inches (4" in 12") or steeper, and each area of plastic panels shall be separated from every other area of plastic panels by at least eight feet (8') laterally and ten feet (10') along the slope of the roof.

(b) All plastic roof panels shall be attached directly to the building framework or shall be mounted individually in steel or other approved metal frames.

(c) Corrugated panels shall be pitched in the direction of the corrugations.

2604.2 — AREA LIMITATIONS

Plastic sections installed on roofs required to have a fire-retardant or non-combustible roofing shall conform to the following area limitations:

(a) **Class A Plastics.** No section shall exceed three hundred square feet (300 sq. ft.) in area, and the aggregate area of such sections shall not exceed 20 per cent of the floor area of the room or occupancy sheltered by the roof.

(b) **Class B Plastics.** No section shall exceed three hundred square feet (300 sq. ft.) in area, and the aggregate area of such sections shall not exceed 12½ per cent of the floor area of the room or occupancy sheltered by the roof.

(c) **Class C and D Plastics.** No section shall exceed one hundred square feet (100 sq. ft.) in area, and the aggregate area of such sections shall not exceed 7½ per cent of the floor area of the room or occupancy sheltered by the roof.

SECTION 2605 — SKYLIGHTS

2605.1 — GENERAL

Approved plastics may be used in skylights provided that:

(a) The skylight is not installed over a shaft or over occupancy classification H—Special Hazardous.

(b) The plastic is mounted at least four inches (4") above the roof on a non-combustible or metal clad curb of at least twelve inches (12") for industrial and commercial structures, and six inches (6") for residential structures.

(c) The plastic units are installed at least five feet (5') apart and not less than five feet (5') from any exterior wall, and in no case shall such units be installed within the distance from an exposure within which openings in walls are required to be fire protected.

(d) Flat or corrugated panels shall slope from the horizontal at least three inches in twelve inches (3" in 12") and the panel shall not exceed ten feet (10') from the bottom to the top of the inclined plane. Corrugations shall run with the inclined plane.

(e) Dome-shaped or curved units shall rise above the mounting flange a minimum distance equal to 10 per cent of its maximum span or five inches (5"), whichever is the greater.

(f) The edges of the plastic material are enclosed in metal.

2605.2 — CLASS A PLASTICS

Class A plastics may be used in skylights provided:

(a) The maximum area enclosed within the curb of units shall not exceed three hundred square feet (300 sq. ft.).

(b) The aggregate area of all such units shall not exceed 33⅓ per cent of the floor area of the room sheltered by the roof in which the units are installed.

2605.3 — CLASS B PLASTICS

Class B plastics may be used in skylights under the same conditions as allowed in subsection 2605.2 except that the aggregate area shall not exceed 25 per cent of the floor area sheltered by the roof upon which it is erected.

2605.4 — CLASS C AND D PLASTICS

Class C and D plastics may be used in skylights provided:

(a) The maximum area enclosed within the curb of units equipped with flat or corrugated plastic sheets shall not exceed one hundred square feet (100 sq. ft.).

(b) The aggregate area of all such units shall not exceed 20 per cent of the floor area of the room sheltered by the roof in which the units are installed.

(c) Flat or corrugated units shall slope from the horizontal at least four inches in twelve inches (4" in 12") and shall not exceed eight feet (8') from the bottom to the top of the inclined plane.

SECTION 2606 – LIGHT-TRANSMITTING PANELS IN MONITORS AND SAWTOOTH ROOFS

2606.1 — GENERAL

Approved plastics may be used as light-transmitting panels in monitors and sawtooth roofs with or without sash, provided:

(a) The lower edge of the plastic material shall be at least six inches (6") above the horizontal surface of the roof.

(b) The areas of such plastic panels shall be separated from each other by a section of non-combustible material or by a section of the roofing material of the structure, said section to be equal in length to one tenth (1/10) of the length of the plastic section or five feet (5'), whichever is greater.

2606.2 — CLASS A AND B PLASTICS

When Class A and B plastics are used the maximum length of a section of plastic panels shall not exceed one hundred feet (100'). and the distance between the upper and lower edges shall not exceed ten feet (10').

2606.3 — CLASS C AND D PLASTICS

When Class C and D plastics are used, the maximum length of a section of plastic panels shall not exceed fifty feet (50') and the distance between the upper and lower edges shall not exceed eight feet (8').

SECTION 2607 – LIGHT-DIFFUSING SYSTEMS IN CEILINGS

2607.1 — GENERAL

(a) Plastic light-diffusing systems in ceilings shall mean installations of plastic panels suspended below lighting fixtures for the purpose of diffusing light throughout a room or space and supported directly or indirectly from floor or roof construction. Plastic diffusers installed in surface mounted or recessed fixtures shall not be subject to the requirements of this section unless the aggregate area of the diffusers exceeds 15 per cent of the area of the ceiling.

(b) No plastic light-diffusing system shall be installed in areas required to be equipped with automatic sprinklers unless appropriate tests by a recognized laboratory have shown that such system does not prevent effective operation of the sprinklers or unless sprinklers are located both above and below the light-diffusing system to give effective sprinkler protection.

(c) Plastic diffusers in systems in which the aggregate plastic area exceeds 30 per cent of the ceiling area shall comply with Section 704.3 unless the plastic panels have heat distortion temperature of 225° F. or less (in accordance with "Test for Deflection Temperature of Plastics Under Load, ASTM D648-56"), and will fall from their mountings at an ambient temperature of at least 200° F. below the ignition temperature of the plastic material as shown in appropriate tests by a recognized testing laboratory. Such panels which exceed two (2) ounces per sq. ft. in weight shall not exceed 10 ft. in length, and such panels which weigh less than two (2) ounces per sq. ft. shall not exceed 25 ft. in length. In no case shall the weight of such panels be more than eight (8) ounces per sq. ft.

(d) In Type I, II, III and IV buildings all hangers and fasteners shall be of non-combustible material. Hangers shall be at least No. 12 Manufacturer's Standard gauge galvanized wire or equivalent.

(e) The maximum temperature in the space between the panel and the ceiling shall not exceed the manufacturer's recommended maximum service temperature for the plastics employed in the panel.

2607.2 — CLASS A PLASTICS

Class A plastics may be used for light-diffusing systems in ceilings in all types of buildings and occupancies, subject to approval of the Building Official.

2607.3 — CLASS B, C AND D PLASTICS

Class B, C, and D plastics shall not be installed in required fire exits or corridors or in occupancies D, E and H. In other rooms or spaces, panels of Class B, C and D plastics may be installed as light-diffusing systems provided the aggregate plastic area does not exceed 15 per cent of the ceiling area.

SECTION 2608 — PARTITIONS

2608.1 — CLASS A PLASTICS

Where partitions may be of combustible construction, Class A plastics may be used for the construction of the entire partition.

2608.2 — APPROVED PLASTICS IN LIEU OF PLAIN GLASS

Plastics may be used to provide the light-transmitting medium in the upper half of partitions where plain glass is permitted, provided the area of plastic so installed does not exceed in the aggregate one-third (1/3) of the area of the partition in which installed.

2608.3 — APPROVED PLASTICS IN MOVABLE PARTITIONS

Approved plastics may also be installed in openings in movable partitions made of metal or other non-combustible material, provided the area of plastic so installed does not exceed in the aggregate one-half (½) of the area of the partition in which installed.

SECTION 2609 — EXTERIOR VENEER

Class A plastics may be used as exterior veneer in accordance with general requirements of Chapter XIV—Masonry Construction.

SECTION 2610 — AWNINGS AND SIMILAR SHELTERS

Class A and B plastics may be used on awnings and similar structures in conformance with general provisions of Chapter XXII—Use of Public Property.

SECTION 2611 — GREENHOUSES

Class A and B plastics may be substituted for wire glass in the construction of greenhouses and Class A, B, C and D plastics may be used in lieu of plain glass.

SECTION 2612 — SIGNS AND OUTDOOR DISPLAYS

The use of plastics in signs, outdoor displays and similar structures shall be governed by provisions of Chapter XXIII—Signs and Outdoor Displays.

CHAPTER XXVII – GLASS

SECTION 2701 – GLASS

2701.1 — LABELING

Each light shall be labeled showing Type, Thickness and Manufacturer except where permanent marking is required. To identify glass with special impact resistance characteristics, laminated and fully-tempered glass shall be identified permanently by the manufacturer, except labeling of heat-strengthened Spandrel glass may be omitted.

SECTION 2702 – GLASS DIMENSIONAL TOLERANCES

There is established glass dimensional tolerances. Where minimum dimensions are required, the nominal values stated are subject to the tolerances shown in "Federal Specification DDG-00451b."

SECTION 2703 – MAXIMUM AREAS – MINIMUM THICKNESS

2703.1 — IMPACT LOADS

(a) For safety, glazing in hazardous locations such as: frameless glass doors, glass in exit and entrance doors, storm doors, sliding doors, fixed glazed panels, shower doors and tub enclosures, being subjective to accidental human impact, shall meet the requirements set forth in Table 2706, or by comparative tests shall be proven to produce equivalent performance. Fixed panels or glazed openings located not less than eighteen (18") inches above the finished floor or walking surface, or where the least dimension is not greater than eighteen (18") inches are exempt from these requirements.

(b) Where the risk of accidental breakage may involve extreme hazard to human safety (as in explosive process areas, in exposed skylights over public use areas, in underwater view windows, etc.), detailed shop drawings, specifications and rational analysis and/or test data assuring safe performance for the specific installation shall be prepared by engineers experienced in this work and shall be submitted for and receive, if warranted, formal approval by the Building Official.

2703.2 — WIND LOADS

(a) For safety, glass or glass areas in exterior walls in screens, in partitions and in other openings subject to wind loading shall be capable of safely withstanding the wind loads as shown in Section 1205.1 acting either inward or outward. In the case of regular plate, float or sheet glass supported on four sides, the design factor shall be not less than 2.5.

(b) Adjustment Factors for other types of glass are given in Table 2707.

SECTION 2704 – GLASS SUPPORTS

(a) Glass supports such as sash members, glazing stops or glazing clips shall be considered firm when deflection of the support at design load does not exceed $\frac{1}{175}$ of the span.

(b) Where other than firm support on all sides is provided, (3 sides, 2 sides, cantilever, or highly flexible, for example), detailed shop drawings, specifications and rational analysis and/or test data assuring safe performance for the specific installation shall be prepared by engineers experienced in this work and shall be submitted for and receive, if warranted, formal approval by the Building Official.

SECTION 2705 – JALOUSIES (IMPACT, AND/OR WIND LOAD)

Thickness shall be not less than 7/32-inch. Length shall be not more than 48 inches. Edges shall be seamed. Other types may be considered only if detailed shop drawings, specifications and rational analysis and/or test data assuring safe performance for the specific installation shall be prepared by engineers experienced in this work and shall be submitted for and receive, if warranted, formal approval by the Building Official.

TABLE 2706 — IMPACT LOADS — GLAZING

Glazing shall conform to the following limits:

1. Annealed glass less than single strength (SS) in thickness shall not be used.
2. If short dimension is larger than twenty-four (24) inches, annealed glass must be double strength (DS) or thicker.
3.

For Specific Hazardous Locations	Size of Individual Glazed Area	Requirements
Glazing in swinging exit and entrance doors	Over 6 sq. ft.	Each glazed area shall pass the test requirements of ANSI Z97.1-1966[1] if not protected by a protective grill or push-bar[2] firmly attached to stiles on each exposed side.*
Glazing in fixed glazed panels which may be mistaken for means of egress or ingress.	Over 6 sq. ft.	Each glazed area shall pass the test requirements of ANSI Z97.1-1966[1] if not protected by a protective grill or push-bar[2] firmly attached to stiles on each exposed side.*
Glazing in patio type sliding doors (Both fixed and sliding panels)	Over 6 sq. ft.	Each glazed area shall pass the test requirements of ANSI Z97.1-1966.[1]

Table 2706 — Continued

For Specific Hazardous Locations	Size of Individual Glazed Area	Requirements
Glazing in storm doors	Over 4 sq. ft.	Each glazed area shall pass the test requirements of ANSI Z97.1-1966[1] if not protected by a protective grill firmly attached to stiles on each exposed side.*
Glazing in all unframed doors (Swinging)	All sizes	Shall be fully-tempered glass and pass the test requirements of ANSI Z97.1-1966.[1]
Glazing in shower doors and tub enclosures:	All sizes	
(a) Obscure wired glass		(a) Not less than 7/32" nominal thickness. Must be protected by towel bar or push-bar[2] firmly attached to stiles on one exposed side.*
(b) Transparent wired glass or fully-tempered glass or laminated glass or transparent rigid plastic		(b) Shall pass the test requirements of ANSI Z97.1-1966.[1]
4. For Application of Specific Glazing Materials Subject to Impact Loads		
Annealed glass (Regular plate, float, sheet, rolled or obscure)	Over 6 sq. ft.	Not less than 3/16" nominal thickness. Each glazed area must be protected by protective grill or push-bar[2] firmly attached to stiles on each exposed side.*

Table 2706 — Continued

For Specific Hazardous Locations	Size of Individual Glazed Area	Requirements
Annealed glass (Regular plate, float, sheet, rolled or obscure) surface sandblasted, etched or otherwise depreciated.	Over 6 sq. ft.	Not less than 7/32" nominal thickness. Each glazed area must be protected by protective grill or push-bar[2] firmly attached to stiles on each exposed side.*
Fully-tempered glass	All sizes	Shall pass the test requirements of ANSI Z97.1-1966.[1]
Laminated glass	All sizes	Shall pass the test requirements of ANSI Z97.1-1966.[1]
Wired glasses (a) Obscure, patterned (b) Transparent	All sizes	(a) Not less than 1/4" nominal thickness. (b) Shall pass the test requirements of ANSI Z97.1-1966.[1]
Transparent rigid plastic	All sizes	Shall pass the test requirements of ANSI Z97.1-1966.[1]

*Building owners and tenants shall maintain towel bars, push-bars or protective grills in safe condition at all times.

[1]American National Standards Institute Performance Specification and Method of Test for Transparent Safety Glazing Materials Used in Buildings.

[2]Shall be constructed and attached in such a manner so as to limit or prevent human impact from being delivered to glass surface.

TABLE 2707 — RELATIVE RESISTANCE TO WIND LOAD
(Assuming equal thickness)

Glass Type	Approximate* Relationship
Laminated	0.6
Wired Glass	0.5
Heat Strengthened	2.0
Fully Tempered	4.0
Metal Edge Double Glazing**	1.5
Rough-rolled Plate	1.0
Sandblasted	0.4
Regular Plate or Sheet	1.0

*Before using Wind Load Chart, divide the Design Wind Load from 1205.1 by the value shown here for the glass type involved.

**Use thickness of the thinner of the two lights, not thickness of unit.

WIND LOAD CHART

REQUIRED NOMINAL THICKNESS OF REGULAR PLATE, FLOAT OR SHEET GLASS

Based on Minimum Thicknesses Allowed in Federal Specification DDG-00451 b Before Weathering

Design Factor = 2.5

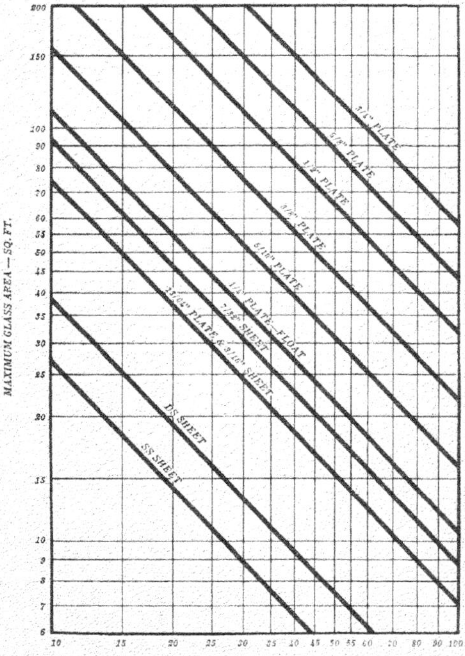

Design Wind Load From Section 1205.1—Pounds Per Square Foot

Chart Applies for Ratios of Width-to-Length from 2:10 to 10:10

CHAPTER XXVIII

ALUMINUM CONSTRUCTION

SECTION 2801 – GENERAL

The quality, design, fabrication and erection of aluminum used structurally in buildings or structures shall conform to good engineering practice, the provisions of this Chapter and other applicable requirements of this Code.

SECTION 2802–STRUCTURAL ALUMINUM CONSTRUCTION

(a) The design, fabrication and assembly of structural aluminum for buildings or structures shall conform to Specifications for Aluminum Structures, Aluminum Construction Manual, Section 1, of the Aluminum Association, Second Edition, November, 1971.

(b) The use of aluminum alloys, other than those listed in the above noted manual, shall be permitted for structural members and assemblies, provided standards of performance not less than those required by the "Specification for Aluminum Structures, Aluminum Construction Manual," AA, 1967 edition, are substantiated to the satisfaction of the Building Official.

CHAPTER XXIX

ACOUSTICAL CEILING SYSTEMS

SECTION 2901 — GENERAL

The quality, design, fabrication and erection of metal suspension systems for acoustical tile and lay-in panel ceilings in buildings or structures shall conform to good engineering practice, the provisions of this Chapter and other applicable requirements of this Code.

SECTION 2902 — ACOUSTICAL MATERIALS

(a) Acoustical materials complying with the interior finish requirements of Section 704.3 shall be installed in accordance with the manufacturer's recommendations and applicable provisions for applying interior finish.

(b) Suspended acoustical ceiling systems shall be installed in accordance with the provisions of "Standard Specification for Metal Suspension Systems for Acoustical Tile and Lay-In Panel Ceilings" ASTM C635-69 and "Recommended Practice for Installation of Metal Ceiling Suspension Systems for Acoustical Tile and Lay-In Panels" ASTM C636-69.

(c) Acoustical ceiling systems which are part of a fire resistive construction shall be installed in the same manner used in the assembly tested and shall comply with the provisions of Section 1001 of this Code.

APPENDIX "A"

WEIGHTS OF BUILDING MATERIALS

MATERIALS	Weight Lb. Per Sq. Ft.
CEILINGS	
Gypsum ceiling block, 2" thick, unplastered	10
Plaster board, unplastered	3
Plaster, ¾", and wood lath	8
Plaster, ¾" and metal lath	8
Plaster, on tile or concrete	5
Suspended, metal lath and plaster	10
FLOORS	
Hardwood flooring, ⅞" thick	4
Sheathing, yellow pine, 1" Oregon pine, spruce or hemlock, ⅞" thick	2½
Sheathing, yellow pine, 1" thick	4
Wood block, creosoted, 3" thick	15
Cement finish, per inch thick	12
Cinder concrete, per inch thick	9
Cinder concrete fill, per inch thick	5
Terrazzo, Tile, Mastic, Linoleum, per inch thick, including base	12
Gypsum slab, per inch thick	5
ROOFS	
Corrugated metal, galvanized:	
20 Manufacturer's Standard Gauge	1.66
24 Manufacturer's Standard Gauge	1.16
28 Manufacturer's Standard Gauge	.78
Roofing felt, 3 ply and gravel	5½
Roofing felt, 5 ply and gravel	6½
Roofing felt, 3 ply and slag	4½
Roofing felt, 5 ply and slag	5½
3-ply ready roofing	1
Shingles, wood	3
Tile or slate	5-20
PARTITIONS	
Channel studs, metal lath, cement plaster, solid 2" thick	17.5
Studs, 2" x 4", wood or metal lath, ¾" plaster both sides	18
Studs, 2" x 4", plaster board ½" plaster both sides	18
Plaster, ½", on gypsum block or clay tile (one side)	4
Hollow clay tile, 2"	13
Hollow clay tile, 3"	16
Hollow clay tile, 4"	18
Hollow clay tile, 5"	20
Hollow clay tile, 6"	25
Hollow clay tile, 8"	30
Hollow clay tile, 10"	35
Hollow gypsum block, 3"	10
Hollow gypsum block, 4"	13
Hollow gypsum block, 5"	15½
Hollow gypsum block, 6"	16½
Solid gypsum block, 2"	9½
Solid gypsum block, 3"	13
Steel partitions	2

MATERIALS	Weight Lb. Per Sq. Ft.
WALLS	
Brick, 8" thick	84
Brick, 12" thick	121
Brick, 16" thick	168
Brick, 20" thick	205
Brick, 24" thick	243
Wall tile, 6" thick	30
Wall tile, 8" thick	33
Wall tile, 10" thick	40
Wall tile, 12" thick	45
Brick 4", tile backing 4"	60
Brick 4", tile backing 8"	75
Brick 9", tile backing 4"	100
Brick 9", tile backing 8"	115
Limestone 4", brick 9"	140
Limestone 4", brick 13"	175
Limestone 4", tile 8"	90
Limestone 4", tile 12"	100
Corrugated metal siding same as roofs	
Windows, glass, frame and sash	8
ASHLAR MASONRY Per Cu. Ft.	
Granite, syenite, gneiss	165
Limestone, marble	160
Sandstone, bluestone	140
MORTAR RUBBLE MASONRY	
Granite, syenite, gneiss	155
Limestone, marble	150
Sandstone, bluestone	130
DRY RUBBLE MASONRY	
Granite, syenite, gneiss	130
Limestone, marble	125
Sandstone, bluestone	110
BRICK MASONRY	
Pressed brick	140
Common brick	120
Soft brick	100

MATERIALS	Weight Lb. Per Cu. Ft.
CONCRETE MASONRY	
Cement, stone, sand	144
Cement, slag, etc.	130
Cement, cinder, etc.	100
VARIOUS BUILDING MATERIALS	
Ashes, cinders	40-45
Cement, portland, loose	94
Cement, portland, set	183
Lime, gypsum, loose	53-64
Mortar, set	103
Slags, bank slag	67-72
Slags, bank screenings	98-117
Slags, machine slag	96
Slags, slag sand	49-55
EARTH, ETC., EXCAVATED	
Clay, dry	63
Clay, damp, plastic	110
Clay, and gravel, dry	100
Earth, dry, loose	76
Earth, dry, packed	95
Earth, moist, loose	78
Earth, moist, packed	96
Earth, mud, flowing	108
Earth, mud packed	115
Riprap, limestone	80-115
Riprap, sandstone	90
Riprap, shale	105
Sand, gravel, dry, loose	90-105
Sand, gravel, dry, packed	100-120
Sand, gravel, dry, wet	118-120
MINERALS	
Asbestos	153
Barytes	281
Basalt	184
Bauxite	159
Borax	109

MATERIALS	Weight Lb. Per Cu. Ft.
Chalk	137
Clay, marl	137
Dolomite	181
Feldspar, orthoclase	159
Gneiss, serpentine	159
Granite, syenite	175
Greenstone, trap	187
Gypsum, alabaster	159
Hornblende	187
Limestone, marble	165
Magnesite	187
Phosphate rock, apatite	200
Pumice natural	40
Porphyry	172
Quartz, flint	165
Sandstone, bluestone	147
Shale, slate	175
Soapstone, talc.	169
STONE, QUARRIED, PILED	
Basalt, granite, gneiss	96
Limestone, marble, quartz	95
Sandstone	82
Shale	92
Greenstone, hornblende	107
BITUMINOUS SUBSTANCES	
Asphaltum	81
Coal, anthracite	97

MATERIALS	Weight Lb. Per Cu. Ft.
Coal, bituminous	84
Coal, lignite	78
Coal, peat, turf, dry	47
Coal, charcoal, pine	23
Coal, charcoal, oak	33
Coal, coke	75
Graphite	131
Paraffine	56
Petroleum	54
Petroleum, refined	50
Petroleum, benzine	46
Petroleum gasoline	42
Pitch	69
Tar, bituminous	75
EXCAVATIONS IN WATER	
Sand or gravel	60
Sand or gravel and clay	65
Clay	80
River mud	90
Soil	70
Stone riprap	65
COAL AND COKE, PILED	
Coal, anthracite	47-58
Coal, bituminous lignite	40-54
Coal, peat, turf	20-26
Coal, charcoal	10-14
Coal, coke	23-32

APPENDIX "B"

FIRE RESISTANCE RATINGS

FOR MATERIALS AND CONSTRUCTION

(a) In various Chapters and Sections of this Code there are specific requirements of fire protection as expressed in terms of fire-resistance. This Appendix B provides a guide in the form of Tables in terms of minimum requirements for the stated fire resistance related to materials and assemblies of construction materials.

(b) The fire resistance ratings as stipulated in the Tables in this Appendix are not intended to be all inclusive but rather those which are acceptable as the result of having successfully performed under accepted tests as prescribed in Section 1001 of this Code. Other fire resistance ratings may be accepted by the Building Official as fire protection on compliance with Section 1001 of this Code.

TABLE 1

FIRE RESISTANCE RATINGS for NON-LOAD BEARING WALLS and PARTITIONS

WALL OR PARTITION ASSEMBLY	MINIMUM NOMINAL THICKNESS For Fire Ratings Indicated (Inches)			
	4-Hr.	3-Hr.	2-Hr.	1-Hr.
BRICK—Clay, Shale, Concrete or Sand-Lime.				
Solid-Unplastered	8	8	8	4
—½" Plaster on both sides	8	8	4	
BRICK				
Hollow—Plastered or Unplastered	12		8	
—Plastered on both sides	8			
—Hollow Rolok-Bak; Plastered	8			
Hollow—Cells filled with perlite loose fill insulation	8			
CONCRETE—Plastered and Unplastered 0.2% reinforcement in each direction; ¾" max. size aggregate	6	5	4	3
—Plastered both sides	6	5	3	3

TABLE 1 (Continued)

FIRE RESISTANCE RATINGS for NON-LOAD BEARING WALLS and PARTITIONS

WALL OR PARTITION MINIMUM NOMINAL THICKNESS	MINIMUM NOMINAL THICKNESS For Fire Ratings Indicated (Inches)			
	4-Hr.	3-Hr.	2-Hr.	1-Hr.
HOLLOW CLAY TILE				
—Plastered or unplastered 2-units:				
3 cells in wall thickness	12			
2-units; 4 cells in wall thickness		8		
3-cells; in wall thickness		12	8	
3-interior cells and double shells			8	
T-shaped 2 cells in wall thickness			8	
—Plastered				
2-cells in wall thickness, both sides	8		6	
2-cells in wall thickness, one side			8	
Plastered both sides				4
HOLLOW CLAY TILE, BRICK FACING **CONCRETE TILE, BRICK FACING** Tile units bond to 4 inch brick facing, having ⅝" 1:3 plaster applied to tile, total masonry thickness	12			
STONE MASONRY-SOLID WALLS Plastered or Unplastered	12			8

MINIMUM EQUIVALENT THICKNESS (IN INCHES) FOR FIRE-RESISTANCE RATINGS INDICATED

WALL OR PARTITION ASSEMBLY	Members Framed Into Wall or Partition None or Noncombustible			
	4-Hr.	3-Hr.	2-Hr.	1-Hr.
CONCRETE MASONRY UNITS Expanded Slag or Pumice Aggregates	4.7	4.0	3.2	2.1
Expanded Clay or Shale Aggregates	5.7	4.8	3.8	2.6
Expanded Shale, Clay or Slate Aggregates (Rotary Kiln Process)	5.35	4.50	3.95	2.6
Limestone, Cinders, or Unexpanded Slag Aggregates	5.9	5.0	4.0	2.7
Calcareous Gravel Aggregates	6.2	5.3	4.2	2.8
Siliceous Gravel Aggregates	6.7	5.7	4.5	3.0

Note (a) Where combustible members are framed into the wall, the wall must be of such thickness or be so constructed that the thickness of solid material between the end of each member and the opposite face of the wall, or between members set in from opposite sides, will not be less than 93% of the thickness shown in table.

Note (b) Equivalent thickness is the average thickness of the solid material in the wall. It may be found by taking the total volume of a wall unit, subtracting the volume of core spaces, dividing this by the area of the face of the unit. Where walls are plastered or faced with brick the thickness of plaster or brick may be included in determining the equivalent thickness. The minimum nominal thickness of non-load bearing concrete masonry walls (not including thickness of plaster) shall be not less than 3" for single wythe walls and not less than 10" (two 4" wythes plus 2" air space) for cavity walls.

TABLE 2

FIRE RESISTANCE RATINGS for NON-LOAD BEARING WALLS and PARTITIONS[4]

WALL OR PARTITION ASSEMBLY			Fire Resistance Ratings
SOLID PLASTER PARTITIONS		Total thickness of partition: 2" unsanded gypsum plaster.	2-Hr.
Solid Plaster Steel-Framed Partitions	Plaster applied to metal lath (except as noted) supported on ¾" steel channel studs.	2½" vermiculite-gypsum or perlite-gypsum plaster.	2-Hr.
		2" sanded gypsum plaster (mix 1:½).	1½-Hr.
		2" vermiculite-gypsum or perlite-gypsum plaster	1-Hr.
		2" sanded gypsum plaster (mix 1:2).	1-Hr.
		2½" sanded gypsum plaster (mix 1:2, 1:3) on metal lath or paper-backed wire fabric.	1-Hr.
Solid Plaster Studless Partitions	1" plaster applied to each side of ½" long length plain gypsum lath inserted at top and bottom in steel or other non-combustible runners.	Total thickness of partition: 2½" vermiculite-gypsum or perlite-gypsum plaster.	2-Hr.
	½" long length gypsum lath, ¾" gypsum plaster applied each side. Lath inserted at top and bottom in steel runners.	Total thickness of partition: 2" gypsum-perlite or vermiculite (mix 1:2, 1:2).	1½-Hr.
	¾" plaster applied to each side of ½" long-length plain gypsum lath inserted at top and bottom in steel or other non-combustible runners.	Total thickness of partition: 2" vermiculite-gypsum or perlite-gypsum plaster.	1-Hr.
	Plaster applied to each side of diamond mesh or rib metal lath fastened top and bottom in steel or other non-combustible runners.	Total thickness of partition: 2" sanded gypsum-plaster (mix 1:2).	1-Hr.
	13/16" plaster applied to each side of ⅝" or ¾" plaster applied to each side of ½" long length plain gypsum lath inserted at top and bottom in steel or other non-combustible runners.	Total thickness of partition: 2" sanded gypsum plaster (mix 1:1, 1:2).	1-Hr.

TABLE 2 (Continued)

FIRE RESISTANCE RATINGS for NON-LOAD BEARING WALLS and PARTITIONS[4]

WALL OR PARTITION ASSEMBLY			Fire Resistance Ratings
WALLBOARD PARTITIONS		Total thickness of partition:	
Solid Wallboard Partitions	1" gypsum board core (solid or laminated) ½" face panels, metal runners.	2"	2-Hr.
Hollow Wallboard Partitions	1" T & G gypsum coreboard attached to 1⅝" steel or wood runner by gypsum screws. ½" gypsum wallboard laminated and screw attached.	Total thickness of partition: 4⅝"	2-Hr.
	⅝" Type "X" face panels laminated to both sides 1" x 6" gypsum ribs 24" o.c., wood or metal runners.	2¼"	1-Hr.
HOLLOW STEEL FRAMED CONSTRUCTION			
Steel Stud Hollow Partitions	Steel studs to both faces of which are attached resilient clips, ¼" diam. rods, metal lath and plaster. Minimum stud size 2".	Each face — 1" vermicuite-gypsum or perlite-gypsum plaster (mix 2 cu. ft. of aggregate per 100 lbs. gypsum for scratch coat and 3 cu. ft. aggregate per 100 lbs. gypsum for brown coat).	3-Hr.
	Steel studs to both faces of which is attached plaster applied over face of metal lath with not less than 2" air space between inside faces of lath.	Each face — ⅞" unsanded gypsum plaster.	2½-Hr.
		⅞" vermiculite gypsum or perlite gypsum plaster.	2-Hr.
		¾" unsanded gypsum plaster or ⅞" sanded gypsum plaster (mix 1:½).	2-Hr.
		¾" sanded gypsum plaster (mix 1:½) or ⅝" unsanded gypsum plaster.	1½-Hr.
		⅝" sanded gypsum plaster (mix 1:2).	1-Hr.
		¾" sanded gypsum plaster (mix 1:2, 1:3)	1-Hr.
		¾" sanded portland cement-asbestos fiber plaster (mix 1:2, 1:3—3 lb. asbestos fiber per bag of cement).	1-Hr.

TABLE 2 (Continued)

FIRE RESISTANCE RATINGS for NON-LOAD BEARING WALLS and PARTITIONS[4]

WALL OR PARTITION ASSEMBLY			Fire Resistance Ratings
Steel Stud Hollow Partitions	Steel studs to both faces of which is attached plaster applied to ⅜" perforated gypsum lath (the lath shall be fastened to the studs by approved wire clips supporting the lath fully across its face), with not less than 2" air space between inside faces of lath.	Each face—½" vermiculite-gypsum or perlite-gypsum plaster (mix 2½ cu. ft. of aggregate per 100 lbs. gypsum).	1-Hr.
		½" sanded gypsum plaster (mix 1:2).	1-Hr.
		¾" perlite gypsum plaster (1:2 - 1:2).	2-Hr.
	Steel studs to both faces of which is attached gypsum wallboard with not less than 1⅝" air space between wallboard. Studs 24" o.c. Wallboard Screw applied.	Each face— ⅝" type X wallboard	1-Hr.
		Two layers of ⅝" type X wallboard with face layer laminated.	2-Hr.
	⅜" perforated gypsum lath, clip attached to steel studs.	½" gypsum sanded plaster (mix 1:2).	1½-Hr.
HOLLOW WOOD FRAMED CONSTRUCTION		Each face —	
Wood Studs Hollow Partitions	⅜" perforated gypsum lath nailed.	½" perlite or vermiculite gypsum plaster (mix 1:2).	1½-Hr.
	Fire Retardant Treated Wood studs 2" x 4" (or larger) spaced not more than 24" o.c.	Each face — ⅝" Type X Gypsum Wallboard	1-Hr.
MASONRY WALLS AND PARTITIONS		Thickness of Partition:	
Hollow Structural Clay Tile Partitions	Shale tile or dense hard-burned clay tile unplastered.	4 inches (Units not less than 40% solid)[2]	1-Hr.
		6 inches (Units not less than 30% solid)[2]	2-Hr.
	⅝" (1:3) sanded gypsum plaster on each side.	Thickness of Partition: 4 inches[1] (Units not less than 50% solid.)	1-Hr.
		6 inches[1] (Units not less 30% solid).	1-Hr.
		4 inches[1] (Units not less than 40% solid)[2].	2-Hr.
		6 inches[1] (Units not less than 30% solid)[2].	3-Hr.

TABLE 2 (Continued)

FIRE RESISTANCE RATINGS for NON-LOAD BEARING WALLS and PARTITIONS[4]

WALL OR PARTITION ASSEMBLY			Fire Resistance Ratings
Hollow Structural Clay Tile Partitions	Medium-burned clay tile unplastered.	6 inches (Units not less than 45% solid). (Two cells in wall thickness).	1-Hr.
		6 inches (Units not less than 30% solid)[2].	2-Hr.
	⅝" (1:3) sanded gypsum plaster on each side.	3 inches[1] (Units not less than 50% solid).	1-Hr.
		6 inches[1] (Units not less than 30% solid).	1-Hr.
		4 inches[1] (Units not less than 60% solid). (Two cells in wall thickness).	2-Hr.
		6 inches[1] (Units not less than 30% solid)[2].	3-Hr.
Gypsum Block Partitions	Solid block unplastered.	Thickness of Partition: 2 inches (Units not less than 100% solid).	1-Hr.
		3 inches (Units not less than 100% solid).	3-Hr.
		5 inches (Units not less than 100% solid).	4-Hr.
	Hollow block unplastered.	Thickness of Partition: 3 inches (Units not less than 70% solid).	1-Hr.
	Hollow block ½" (1:3) sanded gypsum plaster on either side.	4 inches[1] (Units not less than 70% solid).	3-Hr.
	Hollow block ½" (1:3) sanded gypsum plaster on each side.	3 inches[1] (Units not less than 70% solid).	3-Hr.
		4 inches[1] (Units not less than 70% solid).	4-Hr.
	Hollow block ⅝" sanded gypsum plaster on unexposed side (1:3).	3 inches (Units not less than 70% solid).	2-Hr.

TABLE 2 (Continued)

FIRE RESISTANCE RATINGS for NON-LOAD BEARING WALLS and PARTITIONS[4]

WALL OR PARTITION ASSEMBLY			Fire Resistance Ratings
Perlite Concrete Masonry Walls	Hollow perlite concrete masonry units with cores filled with perlite concrete mortar unplastered.	Thickness of Wall: 4 inches.	4-Hr.
Lightweight Concrete Panel or Curtain Wall	**Exterior Section**—4" perlite-portland cement concrete applied on paper-backed wire fabric encasing steel framing members and **Interior Section**—Channel furred section comprising a 1" air space and 1" perlite-gypsum plaster applied on paper-backed wire fabric.	Thickness of Wall: 6 inches[3].	4-Hr.
	Vermiculite-portland cement concrete applied to paper-backed wire fabric spaced 1½" from (1½" x 1½") steel channels, 2' on centers. Concrete encasing channels.	Thickness of Wall: 4 inches.	4-Hr.
	3⅛" thickness of perlite-portland cement spray-applied to paper-backed wire lath attached to 4" steel studs 16" on center.	Thickness of Wall: 4 inches.	2-Hr.

Footnotes: [1]Thickness of partition (not including plaster).

[2]Cells filled with broken tile, stone, slag, cinders or sand mixed with mortar.

[3]Total thickness of partition, including plaster.

[4]Appropriate fire resistance ratings as listed in Design Data—Fire Resistance, 1973-74 Edition, as published by the Gypsum Association, may be accepted as if herein listed.

TABLE 3

FIRE RESISTANCE RATINGS for LOAD BEARING WALLS and PARTITIONS[17]

MINIMUM NOMINAL THICKNESS (In Inches) FOR FIRE-RESISTANCE RATINGS INDICATED

WALL OR PARTITION ASSEMBLY	Members Framed Into Wall or Partition							
	Combustible				None or Non-Combustible			
	4-Hr.	3-Hr.	2-Hr.	1-Hr.	4-Hr.	3-Hr.	2-Hr.	1-Hr.
SOLID BRICK WALLS								
Solid—(Clay, shale, concrete or sand-lime) —unplastered	12	12	8	8	8	8	8	
Solid (as above) ½" (1:3) sanded gypsum plastered on one side	12 (1)	8 or 12 (1) (2)	8 (1)	8 (1)	8	8	8	
Solid—(as above) ½" (1:3) sanded gypsum plastered on each side	12	8 or 12 (2)	8	8	8	8		
Cavity Wall consisting of two 3 in. (actual) brick wythes separated by 2-in. air space; masonry joint reinforced spaced 1 ft. 3 in. o.c. vertically; mortar fill (1 ft. in height) placed between wythes at top of wall.						8		
HOLLOW BRICK WALLS								
Hollow units (clay or shale) cells filled with perlite loose fill insulation.					8			
HOLLOW CAVITY WALLS								
Hollow Cavity Type (clay or shale) Unplastered or ½" (1:3) sanded gypsum on each side—¼" metal ties for each 3 sq. ft. of wall area.			9 (3)	9	9	9	9	9

TABLE 3 (Continued)

	4-Hr.	3-Hr.	2-Hr.	1-Hr.	4-Hr.	3-Hr.	2-Hr.	1-Hr.
Hollow Rolok (clay or shale) Unplastered		12	12	8	12		8	
½" (1:3) sanded gypsum plaster on one side	12			8		8		
½" (1:3) sanded gypsum plaster on each side					8			
Hollow Rolok Bak (clay or shale) Unplastered					8			
CONCRETE WALLS								
Plain Concrete Solid Walls (Unplastered)	8				8			
Solid Concrete Reinforced inch wire welded fabric near the center plane of wall			6			6		
STRUCTURAL TILE BRICK FACED WALLS								
Hollow Structural Tile (Clay or Shale) Walls Unplastered	16-12 (5) (6)	12 or 16 (6) (7)	12 (6) (8)	8 or 12 (6) (9)	12 (10)	8 or 12 (11)	8 or 12 (14)	8 (13)
⅝" (1:3) sanded gypsum plaster on one side	16 (6) (7)	12 (6) (8)	8 or 12 (6) (12)	8 (6) (13)	8 or 12 (11)	8 or 12 (14)	8 (15)	
Hollow Structural Tile (Clay or Shale)—Brick Faced Walls (Tile Bonded to 4" Brick Facing Unplastered, total Masonry thickness)			12	8	12	8		
⅝" (1:3) sanded gypsum plaster on tile side, total masonry thickness		16	12	8	8			

TABLE 3 (Continued)

LOAD BEARING WALLS AND PARTITIONS

WALL OR PARTITION ASSEMBLY	FIRE-RESISTANCE RATING	
	Plaster Side Exposed	Brick-Faced Side Exposed
STEEL-FRAMED BRICK-VENEERED WALLS		
Steel studs faced outer side with ½" wood fiberboard sheathing next to studs, ¾" air space formed with ¾" x 1⅝" wood strips placed over the fiberboard and secured to the studs; metal or wire lath nailed to such strips, 3¾" brick veneer held in place by filling ¾" space between the brick and lath with mortar. Inside facing of studs: ¾" unsanded gypsum plaster on metal (or wire) lath attached to 5/16" wood strips secured to the edges of the studs.	1½-hr.	4-hr.
Steel studs faced outer side with 1" insulation board sheathing attached to studs, 1" air space, and 3¾" brick veneer attached to steel frame with metal ties every fifth course. Inside facing of studs: ⅞" sanded gypsum plaster (1:2 mix) applied on metal (or wire) lath attached directly to the studs.	1½-hr.	4-hr.
Same as the above except use ⅞" vermiculite-gypsum plaster or 1" sanded gypsum plaster (1:2 mix) applied on metal (or wire).	2-hr.	4-hr.
Steel studs faced outer side with ½" gypsum sheathing board, attached to studs, and 3¾" brick veneer attached to the steel frame with metal ties every fifth course. Inside facing of studs: ½" sanded gypsum plaster (1:2 mix) applied on ½" perforated gypsum lath (see Sec. 1001.5) securely attached to the studs and having strips of metal lath 3 inches wide applied to all horizontal joints of gypsum lath.	2-hr.	4-hr.

TABLE 3 (Continued)

LOAD BEARING WALLS AND PARTITIONS

WALL OR PARTITION ASSEMBLY			Fire Resistance Ratings
WOOD-FRAMED WALLS AND PARTITIONS			
	Metal or wire lath (except as noted) and plaster on both sides of 2" x 4" (or larger) wood stud framing, effectively fire-stopped.	⅝" sanded gypsum plaster (mix 1:2) or ¾" sanded gypsum plaster (mix 1:2, 1:3) on metal lath or paper-backed wire fabric.	1-hr.
		⅝" vermiculite-gypsum or perlite-gypsum plaster on metal lath.	1-hr.
		¾" sanded portland cement plaster (mix 1:2, 1:3 with 3 lbs. of short asbestos fiber per bag of cement) on metal or wire lath.	1-hr.
Wood Framed Plaster and Wallboard Partitions	Gypsum lath and plaster on both sides of 2" x 4" (or larger) wood stud framing, effectively fire-stopped.	½" sanded gypsum plaster (mix 1:2) applied to ⅜" plain or perforated gypsum lath	1-hr.
		½" vermiculite-gypsum or perlite-gypsum plaster apto ⅜" perforated gypsum lath.	1-hr.
		½" wood-fibered gypsum plaster applied to ⅜" plain gypsum lath.	1-hr.
		½" sanded gypsum plaster applied to ⅜" Type "X" gypsum lath (mix 1:2).	1-hr.
		1" perlite or vermiculite gypsum plaster applied over 1" wire mesh nailed over ⅜" perforated gypsum lath (mix 1:2, 1:2).	2-hr.
Wood-Framed Plaster and Wallboard Partitions	Gypsum wallboard on both sides of 2" x 4" (or larger) wood stud framing, effectively fire-stopped.	Two layers of ⅜" or ½" gypsum wallboard attached to both sides of wood studs.	1 hr.
		One layer of ½" gypsum wallboard attached to both sides of wood studs, with stud spaces filled with mineral wool insulation bats nailed in place to the studs.	1-hr

TABLE 3 (Continued)

LOAD BEARING WALLS AND PARTITIONS

WALL OR PARTITION ASSEMBLY			Fire Resistance Ratings
Wood-Framed Plaster and Wallboard Partitions	Type "X" gypsum wallboard or coreboard of the same thickness, on both sides of 2" x 4" (or larger) wood stud framing, effectively fire-stopped.	One layer ⅝" Type "X" gypsum wallboard attached to both sides of wood studs.	1-hr.
		Two layers ⅝" Type "X" gypsum wallboard attached to both sides of wood studs.	2-hr.
Wood Framed Exterior Walls	Wood studs 2" x 4" (or larger) effectively fire-stopped; outer face gypsum sheathing board and wood siding; inner face gypsum plaster on gypsum lath.	Wood or plywood siding on ½" gypsum sheathing board attached to outer face of wood studs; ½" sanded gypsum plaster (mix 1:2) on ⅜" perforated gypsum lath on inner face of studs; rating for outside fire exposure.	1-hr.
	Fire retardant treated wood studs 2" x 6" (or larger) spaced not more than 16" o.c.	Faced outer side one layer Type "X" gypsum wallboard, fire resistant vapor barrier, self furring wire mesh and two coats ½" thick cement stucco. Inside facing two layers of ⅝" Type "X" gypsum wallboard.	2-hr.
	Wood studs 2" x 4" (or larger) spaced not more than 16" o.c.	Faced outer side with ½" gypsum sheathing attached to studs, and 3¾" brick veneer attached to each stud with metal ties every sixth course. Inside facing two layers of ⅝" type "X" gypsum wallboard.	2-hr.
METAL FRAMED PARTITIONS			
Steel Stud Hollow Partitions	Steel studs to both faces of which is attached metal or wire lath, with not less than 2" airspace between inside faces of lath.	Each face — ⅞" unsanded gypsum plaster.	2½-hr.
		⅝" unsanded gypsum plaster.	1½-hr.
	Gypsum wallboard nailed to both sides 3⅝" nailable steel studs.	⅝" type "X" gypsum wallboard.	1-hr.
	Steel studs to both faces of which is attached plaster applied to ⅜" perforated gypsum lath with not less than 2" airspace between inside faces of lath.	Each face— ⅝" unsanded gypsum plaster.	1-hr.

REFERENCE NOTES TO TABLE 3 — FIRE-RESISTANCE RATINGS FOR WALLS AND PARTITIONS

(1)—Rating is applicable only when the plastered side of wall or partition is on the exposed side.

(2)—8" for sand-lime or concrete brick or 12" for clay or shale brick (3-hr. rating).

(3)—A 9" wall may be used for a 2-hr. rating if hollow spaces near combustible members are filled with fire-resistive material for the full thickness of the wall and for at least 4" above and below and between the combustible members.

(4)—Grade A and Grade B concrete shall comply with the requirements in Section 1001.2.

(5)—Two or three units, four or five cells in wall thickness. Units not less than 40% solid.

(6)—The ratings for tile walls with combustible members projecting into the walls may be increased to those given for similar walls with non-combustible members if the spaces surrounding the ends of the members are filled solidly with mortar or masonry.

(7)—16" tile (two units, four cells in wall thickness); or 12" hollow tile (one unit, three cells in wall thickness, units not less than 49% solid).

(8)—One unit, three cells in wall thickness. Unit not less than 40% solid.

(9)—12" hollow tile three cells in wall thickness (8" tile if hollow spaces are filled as required in Reference Note (3) herein); or 8" hollow tile (one unit, two cells in wall thickness, units not less than 46% solid).

(10)—Two units, three or four cells in wall thickness. Units not less than 45% solid.

(11)—12" hollow tile (two units, three cells in wall thickness); or 8" hollow tile (one unit, three or four cells in wall thickness, units not less than 53% solid).

(12)—12" hollow tile, two cells in wall thickness (8" tile if hollow spaces are filled as required in Reference Note (3) herein); or 8" hollow tile (one unit, two cells in wall thickness, units not less than 49% solid).

(13)—One unit, two cells in wall thickness. Units not less than 40% solid.

(14)—12" hollow tile (three cells in wall thickness); or 8" hollow tile (one unit, two cells in wall thickness, units not less than 49% solid).

(15)—8" hollow tile (three cells in wall thickness); or 8" hollow tile (one unit, two cells in wall thickness, units not less than 40% solid).

(16)—Equivalent thickness is the average thickness of the solid material in the wall. It may be found by taking the total volume of a wall unit, subtracting the volume of core spaces, dividing this by the area of the face of the unit. Where walls are plastered or faced with brick the thickness of plaster or brick may be included in determining the equivalent thickness.

(17)—Appropriate fire resistance ratings as listed in Design Data—Fire Resistance, 1973-74 Edition, as published by the Gypsum Association, may be accepted as if herein listed.

(18)—When core spaces of the blocks are filled with a water repellent vermiculite masonry fill insulation, the protection time of U.L. Designs 1-3 Hr. and 1-2 Hr. will be increased to provide 4 Hr. fire resistance. U.L. R3653.

GENERAL REQUIREMENTS—FIRE-RESISTANCE RATINGS FOR WALLS AND PARTITIONS—TABLE 3

PLASTER

Plaster, when applied to masonry or gypsum lath, shall be no less than ½-inch in thickness. Thickness shall be measured from the face of the plaster base.

Gypsum plaster as required for the fire-resistance ratings indicated in Appendix B may be fibered or unfibered.

THICKNESS OF WALLS

The thickness prescribed in Table 3 for masonry and tile walls and partitions of the various fire-resistance ratings do not include the thickness of plaster.

MORTAR

For the fire-resistance ratings indicated in Table 3, masonry walls shall be laid in Type M, S, N or O mortar, provided that gypsum blocks shall be laid in gypsum mortar only. Hollow structural clay tile partitions also may be laid in gypsum mortar. See Sections 1402.11 and 1402.12 for mortar proportions and types of mortar required.

CLOSURE OF ENDS

All open cells in tile or block occurring at wall ends shall be filled solid with concrete or gypsum for at least a depth of 6 inches, or solid units or closure tile set in the opposite direction shall be used. No fire-wall of hollow units and no 8 inch solid wall shall be broken into, subsequent to erection, for chases or for the insertion of structural members.

MAXIMUM ALLOWABLE HEIGHTS OF PARTITIONS

All masonry, tile, or block load-bearing partitions required to have a specific fire-resistance rating by the provisions of this Code shall be limited in height not to exceed thirty times their thickness, except when securely anchored at top and bottom, and except where fire-resistance ratings for such partitions of greater height than herein specified have been established by actual fire tests conforming with the requirements of Section 1001. For height of non-bearing partitions of masonry and those of plaster on metal or wire lath, see Section 1405.3. Two-inch solid, studless partitions of gypsum plaster on metal lath or gypsum lath shall not exceed 12 ft. in height.

FIRESTOPPING

Hollow partitions shall be firestopped with non-combustible materials at every floor.

LATH

Metal or wire lath used in fire-resistant walls or partitions shall meet the requirements of Section 1001.7. Gypsum lath, gypsum wallboard, and gypsum sheathing board used in fire-resistant walls or partitions shall meet the requirements of Section 1001.6.

COMBUSTIBLE MEMBERS FRAMED INTO WALLS

For plaster on the same side as combustible framing, and for plaster on one side of walls or partitions which support combustible members from both sides, the fire-resistance rating for such walls or partitions shall be the same as the rating for the same walls unplastered.

TABLE 4

FIRE RESISTANCE RATINGS FOR PROTECTED STEEL COLUMNS[dd]

	Minimum Nominal Thickness Outside of Column (in inches) for Rating Indicated.			
PROTECTIVE MATERIAL	4-hr.	3-hr.	2-hr.	1-hr.
Concrete (Re-entrant, space filled solid.) (a)—Grade A Concrete (r) (course aggregate other than trap rock) and except that it shall have a combined total of not more than 10% of quartz, chert, or flint.				
6" x 6" columns	2	1½	1	1
8" x 8" columns	1½	1	1	1
12" x 12" columns or larger	1	1	1	1

TABLE 4 (Continued)

PROTECTIVE MATERIAL	Minimum Nominal Thickness Outside of Column (in inches) for Rating Indicated.			
	4-hr.	3-hr.	2-hr.	1-hr.
(b)—Trap rock course aggregate, or Grade A Concrete (r) with coarse aggregate having a combined total of from 10% to 30% of quartz, chert, or flint with wire ties as prescribed in Reference Note (c): or cinder, sandstone or granite aggregate of Grade B. Concrete (r) if concrete is held in place as prescribed in Reference Note (d).				
6" x 6" columns	2½	2	1½	1
8" x 8" columns	2	1½	1	1
12" x 12" columns or larger	1½	1	1	1
(c)—Cinder, sandstone, or granite aggregate of Grade B Concrete (r) with wire ties as prescribed in Reference Note (c); or siliceous aggregate of Grade B Concrete (r) having a combined total of 60% or more of quartz, chert, or flint if concrete is held in place as prescribed in Reference Note (d).				
6" x 6" columns	3	2	2	1
8" x 8" columns	3	2	1½	1
12" x 12" columns or larger	2	1½	1	1
(d)—Siliceous aggregate of Grade B Concrete (r) having a combined total of 60% or more of quartz, chert, or flint with wire ties as prescribed in Reference Note (c).				
6" x 6" columns	4½	3½	2½	1½
8" x 8" columns	4	3	2	1
12" x 12" columns or larger	3	2	1½	1
Hollow Clay or Shale Tile (e) Unplastered	f	2 (g)		2 (h)
Plastered with ¾" sanded gypsum plaster (1:3 mix by volume).	2 (i)			
Gypsum (e) — Poured solid (re-entrant Space filled) and reinforced with 4" by 4" wire mesh reinforcement wrapped around column Unplastered	2	1½	1	1
Gypsum (Solid) Block (e) Unplastered	4 (j)		2 (k)	
Plastered with ½" sanded gypsum plaster	2 (m)			
Gypsum (Hollow) Block (e) Unplastered			3 (n)	
Plastered with ½" sanded gypsum plaster	3 (o)			
Cinder Concrete (Hollow) Block (e) Unplastered	3 (p)			
Solid Brick (Clay or Shale) (e) Re-entrant space filled with brick and mortar	3¾	3¾	3¾	2¼

TABLE 4 (Continued)

PROTECTIVE MATERIAL	Minimum Nominal Thickness Outside of Column (in inches) for Rating Indicated. 4-hr.	3-hr.	2-hr.	1-hr.
Metal or wire Lath and Plaster Sanded gypsum plaster (mix 1:3)			2¼ (q)	¾
Sanded portland cement plaster (mix 1:2½)			2¾ (q)	1
Vermiculite-gypsum or perlite-gypsum plaster on metal lath.	(a)	(b)	(s)	
Gypsum Lath and Plaster Sanded gypsum plaster		(y)	(x)	(t)
Vermiculite-gypsum or perlite-gypsum plaster	(u)	(v)	(w)	
Vermiculite Type MK direct-to-steel fireproofing	1¼" (ff) 2½"	⅞" (ff) 1⅞"	1½"	
Gypsum wallboard		(bb)	(cc) (aa)	(z)
Vermiculite portland cement plaster on paper backed wire fabric	(ee)			

Reference Notes to Table 4 — Fire Resistance Ratings for Protected Steel Columns

(a)—1¾ inch vermiculite or perlite-gypsum plaster on self-furring metal lath wrapped around column — no back-fill; 1½ inch vermiculite-gypsum or perlite-gypsum plaster on metal lath — lath furred out 1¼ inches from column flanges, with space between lath and column flanges filled with the same plaster—no back-fill; or 1 inch vermiculite-gypsum plaster on metal lath—lath spaced 1 inch from flanges of column—loose vermiculite back-fill; or 1½ " perlite-gypsum plaster on metal lath furred 7/16" from column with ¾" channels; no back-fill.

(b)—1 inch vermiculite-gypsum or perlite-gypsum plaster on metal lath — lath furred out 1¼ inches from column flanges with space between column flanges and lath filled with vermiculite-gypsum or perlite-gypsum plaster—no back-fill, or 1⅜ inch vermiculite-gypsum or perlite gypsum on self-furring metal lath wrapped around column—no back-fill.

(c)—Concrete shall be held in place with wire ties consisting of No. 5 B & S gage (0.18 inch diam.) steel wire spirally around the column on a pitch of 8 in., or equivalent ties.

(d)—Concrete shall be held in place with wire mesh or expanded metal having not larger than 4" mesh, weighing not less than 1.7 lbs. per sq. yd., or equivalent.

(e)—The ratings indicated are applicable to steel columns 6" x 6" or larger.

(f)—Structural steel columns protected with 2" hollow clay or shale tile, having wire mesh in horizontal joints, flanges covered with mortar or concrete, re-entrant space filled with concrete, shall have a fire-resistance rating of 4-hours if minimum area of solid material is not less than 225 sq. in. For columns of less size two 2" layers of hollow clay or shale tile, ½" mortar between tile and column, ⅜" metal mesh in horizontal joints, hollow clay tile fill, shall be required for 4-hour rating.

(g)—Hollow clay tile with outside wire ties (not less than No. 12 B & S gage — 0.08" diameter steel wire tied around the outside of each course of tile at the middle) or with ⅜" metal mesh in horizontal joints; limestone or trap rock concrete fill extending 1" outside column on all sides.

(h)—Hollow clay tile with outside wire ties (not less than those prescribed in Reference Note (g) herein), with or without concrete fill ¾" mortar between column and tile.

(i)—¾" mortar between column and tile; ⅜" metal mesh in horizontal joints, limestone concrete fill.

(j)—⅜" metal mesh (or equivalent) ties in horizontal joints; 1" gypsum mortar on flange; poured gypsum fill, or re-entrant space filled with gypsum block and mortar.

(k)—Same as prescribed in Reference Note (j) herein—or ⅞" x 12 gage (or equivalent) metal clamps, at horizontal joints, set in holes drilled in blocks; 1" gypsum mortar on flange at horizontal joints only; re-entrant space not filled.

(m)—Metal or wire lath, or mesh (or equivalent ties) in horizontal joints; ½" mortar between column and block; poured gypsum fill, or re-entrant space filled with gypsum block and mortar or ⅞" x 12 gage (or equivalent) metal clamps, at horizontal joints set in holes drilled in blocks; re-entrant space not filled.

(n)—⅞" x 12 gage (or equivalent) metal clamps, at horizontal joints, set in holes drilled in blocks; re-entrant space not filled.

(o)—Same as prescribed in Reference Note (n) herein except ¼" mortar is required between column flange and block.

(p)—1¼" mortar between column and block; re-entrant space filled with broken block and mortar.

(q)—Total thickness, consisting of two layers of plaster, with ¾" air space between, provides a fire-resistance rating of 2½ hours; no fill.

(r)—For definition of Grade A Concrete and Grade B Concrete see Section 1001.2.

(s)—1 inch vermiculite-perlite-gypsum plaster on metal lath — lath furred out 1¼ inch from column flanges—no back-fill; or 1 inch vermiculite-perlite-gypsum plaster on self-furring metal lath wrapped around column—no back-fill.

(t)—⅜ inch perforated gypsum lath applied vertically with double strands 18 gage tie wire spaced 2 inches from ends of lath and

15 inches center to center at intermediate points — plastered with ½ inch sanded gypsum plaster (mix 1:2½).

(u)—Two layers ½ inch long-length gypsum lath tied with double strand 18 gage tie wire spaced approximately 24 inches center to center and wrapped with 1 inch hexagonal mesh, 20 gage galvanized fabric — two ¾ inch coats perlite-gypsum or vermiculite plaster — no back-fill.

(v)—⅜" perforated gypsum lath applied vertically and tied with double strands of 18 gage tie wires spaced 2" from ends of lath and 15" center to center at intermediate points. Plastered with 1⅜" gypsum perlite or gypsum vermiculite plaster (mix 1:2, 1:3). (NBS Test No. 321).

(w)—⅜ inch perforated gypsum lath applied vertically and tied with double strands 18 gage tie wire spaced 2 inches from ends of lath and 15 inches center to center at intermediate points—plastered with 1 inch vermiculite-perlite-gypsum plaster (mix 1:2½).

(x)—⅜" perforated gypsum lath applied vertically and tied with double strands of 18 gage tie wires spaced 2" from ends of lath and 15" center to center at intermediate points. Plastered with 1⅜" gypsum sand plaster (mix 1:2, 1:3). (NBS Test No. 351 and letter from Dr. A. F. Robertson, Chief, Fire Protection Section, National Bureau of Standards, October 7, 1957).

(y)—⅜" perforated gypsum lath applied vertically and tied with double strands of 18 gage tie wires spaced 2" from ends of lath and 15" center to center at intermediate points. Plastered with 2" gypsum sand plaster (mix 1:2, 1:3). (NBS Test No. 344).

(z)—2 layers of ½" gypsum wallboard adhesively attached together and to column. 1st layer wire tied to column.

(aa)—4 layers of ½" gypsum wallboard adhesively attached to column. Third layer wire tied to column, or

(aa)—3 layers ⅝" type X wallboard. Second layer held by steel bands.

(bb)—3 layers of ⅝" type "X" gypsum wallboard screw attached to 1⅝" steel studs at each corner of column.

(cc)—3 layers of ⅝" type "X" gypsum wallboard first and second layers attached with ring shank nails 24" o.c. at corners. Third layer attached with 1" long gypsum wallboard screws spaced 12" o.c. at corners.

(dd)—Appropriate fire resistance ratings as listed in Design Data—Fire Resistance, 1973-74 Edition, as published by the Gypsum Association, may be accepted as if herein listed.

(ee)—2" of vermiculite portland cement plaster spray applied on paper backed wire fabric and reinforced with plain wire fabric between coats.

(ff)—Column size tested 14WF228.

FIRE-RESISTANCE RATINGS FOR STEEL COLUMNS PARTLY PROTECTED WITH CONCRETE

Steel columns of solid section (not latticed) unprotected outside, but having re-entrant space filled with Grade A Concrete (see Sec. 1001.2) and having minimum area of solid materials not less than 64 sq. inches, shall have their fire-resistance rated at 1 hour.

GENERAL REQUIREMENTS — FIRE-RESISTANCE RATINGS FOR PROTECTED STEEL COLUMNS — TABLE 4

THICKNESS

The thickness in Table 4 refer to thicknesses of protective material before the application of plaster except that the thickness indicated for plaster protection is total plaster thickness. Thicknesses shall be measured from the extreme outer edge of the member, except that the thickness of protective material required at the extreme edges of lugs, brackets, wind bracing and other connections shall be not less than 1 inch.

REINFORCEMENT

Except as otherwise prescribed in Table 4, poured protection shall be adequately reinforced with not less than 4" x 4" wire mesh weighing not less than 1½ lbs. per square yard, or equivalent reinforcement.

MORTAR JOINTS

Protective coverings that consist of masonry units shall be solidly bedded and laid in Type M, S, N or O mortar, except that gypsum blocks shall be laid in gypsum mortar; structural clay tile units may likewise be laid in gypsum mortar.

BONDS AND TIES

Except as otherwise prescribed in Table 4, block and tile protective coverings shall be securely anchored or bonded by wall ties or metal mesh laid in the horizontal joints, by metal clips connecting one unit to another, by outside tie wires not smaller than No. 12 B and S gage (0.08" diameter) with at least one such tie around every course, or by means of specially designed units providing positive anchorage to the member or to other units. Outside tie wires shall in all cases be protected by at least ½ inch of mortar or plaster.

METAL OR WIRE LATH

Where metal lath is prescribed in Table 4, it shall meet the minimum requirements of Section 1001.7 of this Code.

GYPSUM LATH

Where gypsum lath is prescribed in Table 4, it shall meet the minimum requirements of Section 1001.6 of this Code. Applicable to ratings prescribed in Table 4 for steel columns protected with gypsum lath and plaster, the column protection shall be provided with corner beads with expanded metal wings wired with No. 18 gage tie wires surrounding the column at intervals not to exceed 15 inches.

TABLE 5 — FIRE RESISTANCE RATINGS FOR PROTECTED STEEL BEAMS, GIRDERS AND TRUSSES(o)

Minimum Nominal Thickness Outside of Members (in inches) for Rating Indicated.

Protective Material	4-hr.	3-hr.	2½-hr.	2-hr.	1½-hr.	1-hr.
Concrete (Re-entrant space filled solid)						
Grade A Concrete (a)	2	2	1½	1½	1½	1
Grade B Concrete (b)	2½	2½	2	2	2	1½
Hollow Clay or Shale Tile or Concrete						
Block Unplastered				2 (c)	2 (c)	2
Plastered with ½" sanded portland cement or gypsum plaster	3 (c)	2 (c)	2 (c)	2		
Gypsum - Poured solid (re-entrant space filled) Unplastered	2	1½	1½	1½	1	1
Plastered with ½" sanded gypsum plaster	1½	1	1	1	½	½
Gypsum (Solid) Block Unplastered - Joints grouted			2	2	2	2
Plastered with ½" sanded gypsum plaster	2	2	2	2	2	2
Gypsum (Hollow) Block Unplastered - Joints grouted		3	3	3	3	3
gypsum plaster	3	3	3	3	3	3
Brick - Hollow or Solid (Clay, Concrete or Sand-lime) - Unplastered	3¾	3¾	3¾	2¼	2¼	2¼
Metal or Wire Lath and Plaster Sanded Portland cement plaster					(l)	(m)
Sanded gypsum plaster			(d)	(g)	(j)	(k)
Unsanded gypsum plaster		(h)	(i)			
Vermiculite - gypsum or perlite gypsum plaster	(e) (n)	(f)	(f)			
Vermiculite Type MK direct-to-steel fireproofing	½	½	½			
Gypsum Wallboard		(p)				

Reference Notes to Table 5 — Ratings for Protected Steel Beams, Girders, and Trusses

(a)—Grade A Concrete shall comply with the requirements in Section 1001.2—Metal ties bent around beam flanges and other projecting parts.

(b)—Grade B Concrete shall comply with the requirements in Section 1001.2—3 inch or smaller metal mesh placed 1 inch from surface.

(c)—All spaces between structural member and tile or block shall be filled solid.

(d)—⅞ inch sanded gypsum plaster (mix 1:2) on metal or wire lath directly attached, furred or suspended—structural member protected on top with concrete or gypsum slab not less than 2½" thick or equivalent.

(e)—⅞ inch vermiculite-gypsum plaster on metal lath directly attached, furred or suspended—or ⅝ inch vermiculite-gypsum plaster plus ½ inch vermiculite acoustical plastic on metal lath.

(f)—⅝ inch vermiculite-gypsum plaster on metal or wire lath directly attached, furred, or suspended with non-combustible construction above.

(g)—⅝ inch sanded gypsum plaster (mix 1:2, 1:3) on metal lath attached, furred, or suspended—structural member protected on top with a concrete or gypsum slab not less than 2¼ inches thick or equivalent.

(h)—⅞ inch unsanded gypsum plaster on metal or wire lath directly attached, furred, or suspended.

(i)—⅞ inch unsanded gypsum plaster on metal or wire lath attached, furred or suspended—structural member protected on top with a concrete or gypsum slab not less than two inches thick or equivalent.

(j)—¾ inch sanded gypsum plaster (mix 1:2, 1:3) on metal or wire lath directly attached, furred, or suspended—structural member protected on top with a concrete or gypsum slab not less than 2 inches thick or equivalent; or structural member protected with 1 inch thickness of metal or wire lath and sanded gypsum plaster (mix 1:3).

(k)—Structural member protected with ¾ inch metal or wire lath and sanded gypsum plaster (mix 1:3).

(l)—Structural member protected with 1 inch metal or wire lath and sanded portland cement plaster (mix 1:2½); or ⅝ inch sanded portland cement plaster (mix 1:2, 1:3—with 15 lbs. hydrated lime and 3 lbs. short asbestos fiber per bag of portland cement) on metal or wire lath directly attached, furred, or suspended—structural member protected on top with a concrete or gypsum slab of not less than 2 inches thick or equivalent.

(m)—Structural member protected with 1 inch metal or wire lath and sanded portland cement plaster (mix 1:2½).

(n)—⅞ inch vermiculite-gypsum or perlite-gypsum plaster on self-furring metal lath wrapped around three sides of member.

(o)—Appropriate fire resistance ratings as listed in Design Data—Fire Resistance, 1973-74 Edition, as published by the Gypsum Association, may be accepted as if herein listed.

(p)—3 layers of ⅝" type "X" gypsum wallboard screw attached to to 1⅝" steel studs and No. 20 SWG mesh between the second and third layer.

GENERAL REQUIREMENTS—FIRE-RESISTANCE RATINGS FOR PROTECTED STEEL BEAMS, GIRDERS AND TRUSSES—FOR TABLE 5.

For the foregoing ratings in Table 5, fire-resistance materials shall protect flanges and portions of webs and members not otherwise protected by arches, slabs, or ceilings.

For ratings of Steel Floor and Roof assemblies, see Table 7.

THICKNESS

For block and poured protections, thicknesses in Table 5 refer to thickness of protective material before the application of plaster. Such thickness shall be measured from the extreme outer edge of the member, except that the thickness of protective material required at the extreme edges of lugs, brackets, wind bracing and other connections shall be not less than 1 inch.

REINFORCEMENT

Except as otherwise prescribed, poured protections listed in Table 5 shall be adequately reinforced with 4" x 4" wire mesh weighing not less than 1½ lbs. per square yard, or equivalent reinforcement.

MORTAR

Protective coverings of masonry units shall be solidly bedded and laid in Type M, S, N or O mortar, except that gypsum blocks shall be laid in gypsum mortar and structural clay tile units may likewise be laid in gypsum mortar.

BONDS OR TIES

Block and tile protective coverings shall be securely anchored or bonded by wall ties or metal mesh laid in the horizontal joints, by metal clips connecting one unit to another by outside tie wires not smaller than No. 12 B and S gage (0.08" diameter) with at least one such tie around every course, or by means of specially designed units providing positive anchorage to the member or to other units. Outside tie wires shall in all cases be protected by at least ½" of mortar or plaster.

PLASTER THICKNESS

Except as otherwise prescribed in Table 5, wherever plaster is required to be applied to masonry, gypsum or portland cement plaster not less than ½" thick shall be used except that on gypsum units, gypsum plaster only shall be used. Thickness of plaster applied to metal or wire lath shall be measured from the face of the lath, except as otherwise noted.

METAL OR WIRE LATH

Metal lath or wire lath shall meet the requirements of Section 1001.7 of this Code.

TABLE 6

FIRE RESISTANCE RATINGS FOR REINFORCED CONCRETE COLUMNS, BEAMS, GIRDERS AND TRUSSES

For the fire-resistance ratings of reinforced concrete columns, beams, girders, and trusses indicated, the various thicknesses of fire protection outside of reinforcing steel shall be not less than that prescribed in Table 6, except as provided in Section 1001. Coarse aggregates shall not exceed ¾ inch in diameter for concrete used to meet fire-protection requirements.

Gypsum or portland cement plaster, not less than ¾ inch in thickness, applied on metal lath or wire lath and in contact with the concrete, may be substituted for ½ inch of the required poured protection, provided that such poured protection shall in no case be reduced to less than 1 inch in thickness. Metal or wire lath used for such purposes shall conform to the provisions of Section 1001.7.

TABLE 6 (Continued)

QUALITY OF CONCRETE	Minimum Thickness Outside of Reinforcing Steel (in inches) for Rating Indicated. 4-hr.	3-hr.	2-hr.	1-hr.
Reinforced Concrete Columns (a) Concrete - Coarse aggregate limestone, calcareous gravel (b), trap rock or blast, furnace slag - 12" or larger round or square columns - Unplastered	1½			
Concrete - Coarse aggregate granite, sandstone, or siliceous gravel (c) - 16" or larger round or square columns - Unplastered	2½	1½		
Plastered (e)	1½			
Concrete - Coarse aggregate granite, sandstone, or siliceous gravel (c), Light 2" mesh expanded metal (or equivalent) centrally located in the protective covering. 12" or 14" round or square columns - Unplastered.		1½		
14" or larger round or square columns - Unplastered	1½			
Reinforced Concrete Beams, Girders, and Trusses Concrete coarse aggregate - Grade A Concrete (d) Unplastered	1½	1½	1½	1
Concrete Coarse aggregate - Grade B. Concrete (d) Unplastered	2½	2½	2	1½
Concrete coarse aggregate - Grade B Concrete (d) Unplastered - 3" or smaller metal mesh centrally located in the protective covering.	2	1½	1½	1

Reference Notes to Table 6 — Fire-Resistance Ratings for Reinforced Concrete Columns, Beams, and Trusses

(a)—These ratings apply to columns with standard ties or spirals, and to columns without spirals, if designed on the basis that the protective concrete covering carries no load.

(b)—Gravel containing not more than 10% quartz, chert, or flint.

(c)—Gravel containing 60% or more quartz, chert, or granite.

(d)—Grade A Concrete and Grade B Concrete shall comply with the requirements in Section 1001.2.

(e)—½" coarse aggregate; column covered with 1 inch of 1:2½ (by volume) portland cement and sand or gypsum and sand plaster, with admixture of not over ½ part lime; surface of column hacked or column cast in metal lath serving as a form.

TABLE 7

FIRE RESISTANCE RATINGS FOR FLOOR AND ROOF CONSTRUCTIONS (g) (h)

Where plaster ceiling is required in Table 7 and no thickness is specified, plaster shall be not less than ½ inch thick. On gypsum units, gypsum plaster only shall be used. Thickness of plaster shall be measured from the face of the plaster base.

Wood finish floors may be used with the constructions rated in this Table, unless prohibited elsewhere in this Code.

CONSTRUCTION		Fire Resistance Rating
Reinforced Concrete	4½" slab, expanded slag aggregate, ¾" protection of reinforcement.	4-hr.
	6" slab, air-cooled slag aggregate, 1" protection of reinforcement.	4-hr.
	3" slab, Grade A concrete; Ceiling, 1" vermiculite-gypsum or perlite-gypsum plaster over face of metal lath supported from bottom of slab. (f)	4-hr.
	6" slab, Grade A or Grade B concrete (Section 1001.2), 1" protection of reinforcement.	3-hr.
	6" slab, Grade A or B concrete (Sec. 1001.2) with electrical raceways and junction boxes, 1" protection of reinforcement if marker screws are used.	3-hr. 2-hr.
	2" slab, Grade A concrete; Ceiling, ¾" vermiculite-gypsum or perlite-gypsum plaster over face of metal lath supported from bottom of slab. (f)	3-hr.
	4¾" slab, air-cooled slag aggregate, ¾" protection of reinforcement.	2½-hr.
	4¾" slab, Grade A or Grade B concrete (Section 1001.2), ¾" protection of reinforcement.	2-hr.
	4" slab, Grade A or Grade B concrete (Section 1001.2), ¾" protection of reinforcement.	1-hr.
Monolithic Concrete Slab Joist and Beam Construction	Monolithic reinforced slab, joist and beam with electrical raceways and junction boxes with joists not less than 5" wide ¾" minimum protection for reinforcement and having plastered ceiling over face of metal lath. 3" top slab; ceiling 1" vermiculte-gypsum plaster (mix 1:2, 1:3) 5" slab at underfloor ducts and junction boxes.	4-hr.
	2" top slab; ceiling ¾" vermiculite-gypsum plaster (mix 1:2, 1:3.) 4" slab at underfloor ducts and junction boxes.	3-hr.

TABLE 7 (Continued)

CONSTRUCTION		Fire Resistance Rating
Monolithic Concrete Slab Joist and Beam Construction	Monolithic reinforced slab, joist and beam; joists not less than 5" wide with ¾" minimum protection for reinforcement; and top slab not less than 2½" thick with ⅞" of vermiculite Type MK spray-applied direct to the underside of slab.	2-hr.
	Monolithic reinforced slab, joist and beam; joists not less than 5" wide with ¾" minimum protection for reinforcement and top slab not less than 3" thick.	1-hr.
Concrete Joist Construction	Reinforced concrete top slab on concrete joists not less than 4" wide and not over 30" on centers with ¾" minimum protection for reinforcement, having plastered ceiling over face of metal lath. **2½" top slab;** ceiling ⅞" unsanded gypsum plaster or ⅝" vermiculite-gypsum or perlite-gypsum plaster.	3-hr.
	2¼" top slab; ceiling ⅝" sanded gypsum plaster (mix 1:2, 1:3).	2-hr.
	2" top slab; ceiling ⅝" sanded gypsum plaster (mix 1:2, 1:3) or ⅝" sanded portland cement plaster (mix 1:3 with 15 lbs. of hydrated lime and 3 lbs. of short asbestos fiber per bag of cement.)	1½-hr.
	2½" top slab; ceiling ⅝" Type "X" gypsum wallboard attached to 25 gage galvanized steel channels spaced 24 in. on center. Ceiling located 10 in. below joists. Steel channels suspended by 21 gage galvanized steel straps attached to side of joist by 5/32 in. x 1¼ in. cartridge-driven concrete nails. Gypsum wallboard attached to channels with 1" No. 6 self-tapping screws 8" on center.	2-hr.
Precast Prestressed Concrete Units	**Top Slab**—2" minimum and 6" maximum thickness perlite concrete on precast prestressed double or single stemmed units. **Ceiling** —Not Required U.L. Report R4123-13.	2-hr.
	Top Slab—2" minimum and 6" maximum thickness of vermiculite concrete on precast prestressed double or single stemmed units. **Ceiling** —No ceiling required. U.L. Report R4123-13.	2-hr.
Concrete and Concrete Block	Concrete block and concrete composite construction consisting of concrete block fillers not less than 3⅝" deep with reinforced concrete ribs (¾" protection for reinforcement) and concrete top slab not less than 2½" thick.	3½-hr.

TABLE 7 (Continued)

CONSTRUCTION		Fire Resistance Rating
Concrete and Tile	Tile and concrete composite construction consisting of tile fillers of concrete, gypsum, or structural clay tile not less than 4" deep, with reinforced concrete ribs (¾" protection for reinforcement) and concrete top slab not less than 2" thick.	2½-hr.
	Tile and concrete, composite construction as prescribed above except omitting the top slab and having a gypsum plastered ceiling.	1½-hr.
Gypsum	Poured reinforced gypsum slab at least 3" thick having ¾" minimum protection for reinforcement and with a ceiling of gypsum plaster.	2½-hr.
	2½" Class A (500 PSI) gypsum concrete on ½" gypsum formboard, concrete reinforced with 48-1214 welded wire fabric (self-furring to keep fabric ½" above formboard) sub-purlins (bulb tees) exposed.	2-hr.
	2" Class A (500 PSI) gypsum concrete on ½" gypsum formboard. Concrete reinforced with 48-1214 welded wire fabric, sub-purlins (bulb tees) exposed.	1-hr.
	2" poured reinforced gypsum slab on ½" gypsum formboard spanning between Keydeck truss tee sub-purlins 2' 8¾" on center.	
	Maximum clear span 8'	2-hrs.
	Maximum clear span 9'	1½-hrs.
	Maximum clear span 10'	1-hr.
Brick	Brick arch not less than 4" deep with a level concrete fill 2" above crown.	2½-hr.
Clay Tile	Flat arch of clay tile, 2-cell structural tile not less than 8" deep, with floor fill or incombustible material at least 2" thick and having gypsum plaster ceiling.	2½-hr.
	Segmental arch of clay tile, 2-cell structural tile not less than 6" deep, laid in cement mortar with concrete fill level with crown of arch, and having gypsum plaster ceiling.	2½-hr.
Steel Joist Construction and Formed Steel Members (a) (e) (Using Contact, Furred, or Suspended Ceilings—Constructions With Metal or Wire Lath—Except as Noted)	**Top Slab**—2½" poured concrete top slab or 2" precast gypsum tile, the latter with ½" mortar finish; or 2" reinforced gypsum concrete slab on ½" gypsum wallboard. **Ceiling** —⅞" vermiculite-gypsum or perlite-gypsum plaster over face of metal lath.	4-hr.
	Top Slab—2½" poured concrete top slab, 2½" vermiculite-concrete or perlite-concrete top slab, of 2" precast gypsum tile, the latter with ½" mortar finish. **Ceiling** —⅝" vermiculite-gypsum or perlite-gypsum plaster or ⅞" unsanded gypsum plaster over face of metal lath.	3-hr.

TABLE 7 (Continued)

CONSTRUCTION		Fire Resistance Rating
Steel Joist Construction and Formed Steel Members (a) (e) (Using Contact, Furred, or Suspended Ceilings—Constructions With Metal or Wire Lath—Except as Noted)	**Top Slab**—2" minimum thickness perlite concrete on 28 gauge galvanized steel form units supported by steel joists 4' on center. **Ceiling** —⅞" perlite-gypsum plaster (1:2, 1:2) on expanded metal lath attached to ¾" furring channels wire tied to steel joists.	3-hr.
	Top Slab—2½" poured concrete top slab. **Ceiling** —¾" sanded gypsum plaster (mix 1:2) over face of metal or wire lath.	2½-hr.
	Top Slab—2" poured concrete top slab or 2" precast gypsum tile, the latter with ¼" mortar finish. **Ceiling** —⅝" vermiculite-gypsum or perlite gypsum plaster or ⅞" unsanded gypsum plaster over face of metal or wire lath.	2½-hr.
	Top Slab—2" vermiculite concrete over the top of a metal form. **Ceiling** —¼" vermiculite-gypsum plaster plus ½" of vermiculite acoustical Type Z. Air-duct openings of 31 sq. in. per 100 sq. ft. of ceiling area and recessed light fixtures spaced 25 sq. ft. per 100 sq. ft. of ceiling area.	2-hr.
	Top Slab—2" minimum thickness perlite concrete on 28 gauge galvanized steel form units supported by steel joists 4' on center. **Ceiling** —¾" perlite-gypsum plaster (1:2, 1:2) on expanded metal lath attached to ¾" furring channels wire-tied to lower chord of joists.	2-hr.
	Top Slab—2¼" poured concrete top slab or 2" precast gypsum tile, the latter with ¼" mortar finish. **Ceiling** —⅝" sanded gypsum plaster (mix 1:2, 1:3) over face of metal or wire lath or paper-backed wire fabric.	2-hr.
	Top Slab—2" poured concrete top slab; or 2" precast gypsum tile. **Ceiling** —⅝" sanded gypsum plaster (mix 1:2, 1:3) or ⅝" sanded portland cement plaster (mix 1:2, 1:3—with 15 lbs. hydrated lime and 3 lbs. short asbestos fiber per bag of portland cement) over face of metal lath.	1½-hr.

TABLE 7 (Continued)

CONSTRUCTION		Fire Resistance Rating
Steel Joist Construction and Formed Steel Members (a) (e) (Using Contact, Furred, or Suspended Ceilings—Constructions With Metal or Wire Lath—Except as Noted)	**Top Slab**—1" nominal single wood floor on wood nailers applied to steel deck with insulating paper cemented thereto; or double wood deck on wood nailers attached to steel joists or formed steel members. **Ceiling** —⅝" sanded gypsum plaster (mix 1:2, 1:3) or ⅝" sanded portland cement plaster (mix 1:2, 1:3—with 15 lbs. hydrated lime and 3 lbs. short asbestos fiber per bag of portland cement) over face of metal or wire lath.	1-hr.
	Top Slab—2" poured concrete top slab. **Ceiling** —¾" furring channels, spaced 12", ⅜" perforated gypsum lath attached to furring channels with approved interlocking wire clips; 1" gypsum-perlite or gypsum-vermiculite plaster (100:2, 100:3), reinforced with 1" 20 gage hexagonal mesh attached to furring channels.	4-hr.
Steel Joist Construction and Formed Steel Members (a) (e) (Constructions With Gypsum Lath Ceilings)	**Top Slab**—2" poured concrete top slab. **Ceiling** —¾" furring channels, spaced 12", ⅜" perforated gypsum lath attached to furring channels with approved interlocking wire clips; ⅝" gypsum-perlite or gypsum-vermiculite plaster (100:2, 100:3) reinforced with 14 gage galvanized wire below lath and diagonally between interlocking device of wire clips.	3-hr.
	Top Slab—2" poured concrete top slab. **Ceiling** —¾" furring channels, spaced 16", ⅜" perforated gypsum lath attached to furring channels with approved interlocking wire clips; ⅝" gypsum-perlite or gypsum-vermiculite plaster (100:2, 100:3) reinforced with 14 gage galvanized wire below lath and diagonally between interlocking device of wire clips.	3-hr.
	Top Slab—2" poured concrete top slab. **Ceiling** —¾" furring channels, spaced 16", ⅜" perforated gypsum lath attached to furring channels with approved interlocking wire clips; ½" gypsum-perlite or gypsum-vermiculite plaster (100:2½) reinforced with 1" 20 gage hexagonal mesh attached to furring channels.	3-hr.

TABLE 7 (Continued)

CONSTRUCTION		Fire Resistance Rating
Steel Joist Construction and Formed Steel Members (a) (e) (Constructions With Gypsum Lath Ceilings)	**Top Slab**—2" poured concrete top slab. **Ceiling** —¾" furring channels, spaced 12", ⅜" perforated gypsum lath attached to furring channels with approved interlocking wire clips; ½" gypsum-perlite or vermiculite-plaster (100:2½) reinforced with 14 gage galvanized wire below lath and diagonally between interlocking device of wire clips and ⅝" gypsum acoustical plaster.	2½-hr.
	Top Slab—2" poured concrete top slab. **Ceiling** —¾" furring channels, spaced 16", ⅜" perforated gypsum lath attached to furring channels with approved interlocking wire clips; ½" gypsum-perlite or gypsum-vermiculte plaster (100:2½) reinforced with 14 gage galvanized wire below lath and diagonally between interlocking device of wire clips.	2-hr.
	Top Slab—2" poured concrete top slab. **Ceiling** —¾" furring channels spaced 12" o.c. ⅜" perforated gypsum lath attached to furring channels with approved interlocking wire clips; ⅝" gypsum sand plaster (1:2, 1:3) reinforced with 14 gage galvanized wire below lath and diagonally between interlocking device of wire clips. (NBS Test No. 345).	2-hr.
	Top Slab—2" poured concrete top slab. **Ceiling** —¾" furring channels, spaced 16", ⅜" perforated gypsum lath attached to furring channels with approved interlocking wire clips; 1" of gypsum-perlite or gypsum-vermiculite plaster (100:2, 100:3).	1½-hr.
	Top Slab—2" poured concrete top slab. **Ceiling** —¾" furring channels, spaced 16", ⅜" perforated gypsum lath attached to furring channels with approved interlocking wire clips; ⅝" gypsum-perlite or gypsum-vermiculite plaster (100:2½).	1-hr.
Steel Joist Construction and Formed Steel Members (a) (e) (Construction With Acoustical Tile Ceilings)	**Top Slab**—3" concrete over 10" bar joists spaced 24" o.c. **Ceiling** —"Z" runners spaced 12" o.c. wire clipped to the bottom chord of the bar joists. The suspension system supports ¾" x 12" x 12" acoustical tile listed by U.L. Design No. 77—3 hr., U.L. Test R 4349-19.	3-hr.

TABLE 7 (Continued)

CONSTRUCTION		Fire Resistance Rating
Steel Joist Construction and Formed Steel Members (a) (e) (Construction With Acoustical Tile Ceilings)	**Top Slab**—2½" concrete over 10" bar joists spaced 24" o.c. **Ceiling** —Main tees are spaced 48" or 24" o.c. and supported by hanger wires. Cross tees are spaced 24" o.c. perpendicular to main runners. The suspension system supports ⅝", 24" x 24" or 24" x 48" acoustical lay-in panels listed by U.L. under Design No. 35—2 hr., U.L. Test R 4349-7.	2-hr.
	Top Slab—2½" concrete over 10" bar joists spaced 24" o.c. **Ceiling** —Main tees are spaced 48" or 24" o.c. and supported by hanger wires. Cross tees are spaced 24" o.c. perpendicular to main runners. The suspension system supports ⅝" or 1", 24" x 24" or 24" x 48" acoustical lay-in panels. Protected light fixtures may be installed in ceiling not to exceed 25% of ceiling area. Air duct openings not to exceed 47 square inches per 100 square feet of ceiling area. Listed under U.L. Design 36—2 hr., U.L. Test R4349-9 & 12.	2-hr.
	Top Slab—2" concrete over 10" bar joists spaced 24" o.c. **Ceiling** —"H" runners spaced 24" o.c.: clipped to the joists with wire clips. Concealed cross tees are spaced 12" or 24" o.c. with the ends resting on the lower flange of the "H" runners. The suspension system supports ¾" or ⅝", 12" x 12" to 24" x 24" acoustical tile, listed by U.L. Design No. 23—2 hr. U.L. Test R 4349-4, 8, 11.	2-hr.
	Top Slab—2" poured concrete top slab. **Ceiling** —No. 25 gage steel inverted "J" shaped channels, spaced 12" clip attached to joists and supporting ⅝" or thicker, 12" x 12" noncombustible acoustical tile. Listed by U.L. under Design 16—2 hr. U.L. Test 4337-4.	2-hr.
	Top Slab—2½" poured concrete top slab. **Ceiling** —Exposed steel grid suspension system, listed by U.L., suspended from joists by 12 gage wire hangers spaced 4' o.c., supporting ½" or thicker, 24" x 24" or 48" noncombustible acoustical panels. Listed by U.L. under Design 64— 2 hr. Protected light fixtures may be installed in ceiling not to exceed 8% of ceiling area. Air duct openings not to exceed 28 square inches per 100 square feet of ceiling area. U.L. Test R4337.	2-hr.

TABLE 7 (Continued)

CONSTRUCTION		Fire Resistance Rating
Steel Joist Construction and Formed Steel Members (a) (e) (Construction With Acoustical Tile Ceilings)	**Top Slab**—2" concrete over 10" bar joists spaced 24" o.c. **Ceiling** —"Z" runners 12" o.c. are attached to the bottom chord of the joist with wire clips. The "Z" runners support ¾" x 12" x 12" or ⅝" x 12" x 12" T & G acoustical tile listed by U.L. under Design No. 22—2 hr., U.L. Test R 4349-2.	2-hr.
	Top Slab—2" concrete over 10" bar joists spaced 24" o.c. **Ceiling** —"H" runners spaced 24" o.c. clipped to the joists with wire clips. The concealed cross tees are spaced 12" or 24" o.c. with the ends resting on the lower flange of the "H" runners. The suspension system supports ¾", 12" x 12" to 24" x 24" acoustical tile listed by U.L. under Design No. 200—2 hr., U.L. Test R 4349-18.	2-hr.
	Top Slab—2" concrete over 10" bar joists spaced 24" o.c. **Ceiling** —1½" cold rolled steel channels supported by hanger wires 48" o.c. The "H" runners are spaced 24" o.c. at each intersection with the channel. Cross tee members are spaced 12" o.c., ends resting on the lower flange of the "H" runners. The suspension system supports ⅝" or ¾", 12" x 12" acoustical tile. Protected light fixtures may be installed in ceiling not to exceed 25% of ceiling area. Air duct openings not to exceed 110 square inches per 100 square feet of ceiling area. Listed by U.L. under Design No. 34—2 hr., U.L. Test R4349-6.	2-hr.
	Top Slab—2½" concrete over 8" bar joists spaced 24" o.c. with 10" beam. **Ceiling** —Hanger wires are spaced 48" o.c. and support "H" runners spaced 24" o.c. Concealed cross tees are spaced 24" o.c. with ends resting on lower flange of "H" runners. The suspension system supports ¾" x 24" x 24" acoustical tile. Protected light fixtures may be installed in ceiling not to exceed 25% of ceiling area. Air duct openings not to exceed 47 square inches per 100 square feet of ceiling area. Listed by U.L. under Design No. 80—2 hr., U.L. Test R-4349-11.	2-hr. 2-hr. Beam

TABLE 7 (Continued)

CONSTRUCTION		Fire Resistance Rating
Steel Joist Construction and Formed Steel Members (a) (e) (Construction With Acoustical Tile Ceilings)	**Top Slab**—2" concrete over 10" bar joists spaced 24" o.c. **Ceiling** —Main tees are spaced 48" or 24" o.c. and supported by hanger wires. Cross tees are spaced 24" o.c., perpendicular to main runners. The system supports ½", 24" x 24" or 24" x 48" acoustical lay-in panels. Protected light fixtures may be installed in ceiling not to exceed 8% of ceiling area. Air duct openings not to exceed 25 square inches per 100 square feet of ceiling area. Listed by U.L. under Design No. 43—1 hr., U.L. Test No. R4349.	1-hr.
Steel Joist Construction and Formed Steel Members (a) (e)	**Top Slab**—2" poured concrete top slab. **Ceiling** —¾" acoustical tile (glass fiber) attached to ¾" furring channels, 4' on center, with special clips and splines—2 layers of mineral wool (2" thick each) with aluminum foil center placed on top of unexposed surface of tile. U.L. Design No. 1—2 hr.	2-hr.
	Top Slab—2" poured concrete top slab. **Ceiling** —1 layer ⅝" type "X" gypsum wallboard secured to ⅞" nailing channels at 16" spacing.	1-hr.
	Top Slab—2½" poured concrete top slab. **Ceiling** —⅞" furring channels spaced 12" o.c. ⅝" type "X" gypsum wallboard attached with 1" gypsum wallboard screws.	2-hr.
Gypsum Concrete Deck	**Top Deck**—2½" gypsum concrete on ½" gypsum formboard with welded wire fabric. Subpurlins (bulb tees) exposed.	2-hr.
	Top Deck—2" gypsum concrete on ½" gypsum formboard with welded wire fabric, subpurlins exposed.	1-hr.
Cellular Steel Floor and Roof Units (b) (e)	**Top Deck**—2" cinder concrete (minimum thickness over cells). **Ceiling** —⅞" vermiculite-gypsum or perlite-gypsum plaster (mix 100 lbs. gypsum to 2½ cu. ft. aggregate) over face of metal lath.	4-hr.
	Top Deck—2" concrete (minimum thickness over cells). **Ceiling** —⅞" vermiculite-gypsum or perlite-gypsum plaster over face of metal lath.	4-hr.
	Top Deck—2½" concrete (minimum thickness over cells). **Ceiling** —⅝" vermiculite-gypsum plaster plus ½" vermiculite acoustic plaster over face of metal lath.	4-hr.

TABLE 7 (Continued)

CONSTRUCTION		Fire Resistance Rating
Cellular Steel Floor and Roof Units (b) (e)	**Top Deck**—2" concrete (minimum thickness over cells). **Ceiling** —⅞" unsanded gypsum plaster over face of metal lath.	4-hr.
	Top Deck—2½" concrete (minimum thickness over cells). **Ceiling** —1" perlite-gypsum or vermiculite-gypsum plaster (mix 100 lbs. gypsum to 2½ cu. ft. aggregate) over face of ⅜" ribbed metal lath tied directly to underside of floor units.	4-hr.
	Top Deck—2½" concrete (minimum thickness over cells). **Ceiling** —⅝" perlite-gypsum or vermiculite-gypsum plaster plus ½" gypsum acoustical plaster over face of metal lath.	4-hr.
	Top Deck—2½" concrete (minimum thickness over cells). **Ceiling** —¾" furring channels spaced 12", ⅜" perforated gypsum lath attached to furring channels with approved interlocking clips; ½" vermiculite-gypsum or perlite-gypsum plaster (100:2½) reinforced with 14 gage galvanized wire below lath and diagonally between interlocking device of wire clips.	3-hr.
	Top Deck—3" expanded slag aggregate (minimum thickness over cells).	3-hr.
	Top Slab—4½" concrete (measured from bottom of corrugations). **Ceiling** —1" vermiculite-gypsum or perlite-gypsum plaster over face of furred or suspended metal lath.	4-hr.
	Top Slab—4½" concrete (measured from bottom of corrugations). **Ceiling** —⅜" vermiculite-gypsum or perlite-gypsum plaster over face of metal lath attached directly to bottom of floor units and with plaster pushed through lath to fill the corrugations of the units.	4-hr.
	Top Slab—2½" concrete (minimum thickness over cells). **Ceiling** —⅝" vermiculite Type MK applied direct to underside of floor, following contour.	3-hr.

TABLE 7 (Continued)

CONSTRUCTION		Fire Resistance Rating
Cellular Steel Floor and Roof Units (b) (e)	**Top Deck**—2½" concrete (minimum thickness over cells) with a 10" beam. **Ceiling** —1½" cold-rolled steel channels spaced 48" o.c. supported by hanger wires. "H" runners spaced 24" o.c. clipped at each intersection with the channel. Concealed cross tee members spaced 12" o.c., ends resting on the lower flanges of the "H" runners. The suspension system supports ⅝" or ¾", 12" x 12" T & G acoustical tile. Protected light fixtures may be installed in ceiling not to exceed 25% of ceiling area. Air duct openings not to exceed 110 square inches per 100 square feet of ceiling area. Listed under U.L. Design No. 36—4 hr., U.L. Test R4349-5.	4-hr. 5-hr. Beam
	Top Deck—2½" concrete over deck with 8" beam. **Ceiling** —Main tees spaced 24" o.c. and supported by hanger wires spaced 48" o.c. Cross tees spaced 48" o.c. perpendicular to main tees. The suspension system supports ⅝" x 24" x 48" acoustical lay-in panels. Protected light fixtures may be installed in ceiling not to exceed 25% of ceiling area. Listed under U.L. Design No. 47—3 hr., U.L. Test R4349-10.	3-hr. 4-hr. Beam
	Top Deck—2½" concrete (minimum thickness over cells) with 6" beam. **Ceiling** —1½" cold-rolled steel channels spaced 48" o.c. supported by hanger wires. "H" runners spaced 24" o.c. clipped at each intersection with the channel. Concealed cross tee members spaced 12" o.c., ends resting on the lower flange of the "H" runners. The suspension system supports ¾" x 12" x 12" acoustical tile and is listed by U.L. under Design No. 69—3 hr., U.L. Test No. R 4349.	3-hr. 3-hr. Beam
	Top Deck—2½" concrete (minimum thickness over cells). **Ceiling** —1½" 16 gage carrying channels 4' o.c. suspended by 12 gage wire 4' o.c. with No. 25 gage steel inverted "J" shaped channels spaced 12" o.c. clipped to 1½" channels and supporting ¾" x 12" x 12" noncombustible acoustical tile. Listed by U.L. under Design 25-3 HR. Protected light fixtures may be installed in ceiling not to exceed 25% of ceiling area. Air duct openings not to exceed 150 square inches per 100 square feet of ceiling area. U.L. Test R4337-3.	3-hr.

TABLE 7 (Continued)

CONSTRUCTION		Fire Resistance Rating
Cellular Steel Floor and Roof Units (b) (e)	**Top Deck**—2½" concrete (minimum thickness over cells). **Ceiling** —Exposed steel grid suspension system, listed by U.L., suspended by 12 gage wire 4' o.c., supporting ¾" x 24" x 24" or 48" noncombustible acoustical panels. Listed by U.L. under Design 74-3 HR. Protected light fixtures may be installed in ceiling not to exceed 16% of ceiling area. Air duct openings not to exceed 55 square inches per 100 square feet of ceiling area. UL Test R4337-24.	3-hr.
	Top Deck—2½" concrete (minimum thickness over cells). Trench header duct may be installed in floor to concrete depth. **Ceiling** —1½" 16 gage carrying channels 4' o.c. suspended by 12 gage wire 4' o.c. with No. 25 gage steel inverted "J" shaped channels spaced 12" o.c. clipped to 1½" channels and supporting ¾" x 12" x 12" noncombustible acoustical tile. Listed by U.L. under Design 264-2 HR. Protected light fixtures may be installed in ceiling not to exceed 24% of ceiling area. Air duct openings not to exceed 113 square inches per 100 square feet of ceiling area. UL Test R4337-35.	2-hr.
	Top Deck—2½" concrete (minimum thickness over cells). **Ceiling** —Exposed steel grid suspension system, listed by U.L., suspended by 12 gage wire 4' o.c., supporting ⅝" or thicker 24" x 24" or 48" noncombustible acoustical panels. Listed by U.L. under Design 73-2 HR. Protected light fixtures may be installed in ceiling not to exceed 8% of ceiling area. Air duct openings not to exceed 57 square inches per 100 square feet of ceiling area. UL Test R4337-17.	2-hr.
	Top Deck—2½" concrete (minimum thickness over cells). **Ceiling** —½" vermiculite Type MK applied to the top and sides of fluted sections and ⅝" on the bottom of fluted and flat plate sections.	2-hr.

TABLE 7 (Continued)

CONSTRUCTION		Fire Resistance Rating
Cellular Steel Floor and Roof Units (b) (e)	**Top Deck**—2½" concrete over deck with 8" beam. **Ceiling** —Main tees spaced 24" o.c. and supported by hanger wires spaced 48" o.c. Cross tees spaced 48" or 24" o.c. perpendicular to the main runners. The system supports ⅝", 24" x 24" or 24" x 48" acoustical lay-in panels. Protected light fixtures may be installed in ceiling not to exceed 25% of ceiling area. Air duct openings not to exceed 50 square inches per 100 square feet of ceiling area. Listed by U.L. under Design No. 62—2 hr., U.L. Test R4349-10.	2-hr. 4-hr. Beam
	Top Deck—2½" concrete (minimum thickness over cells) with 6" beam. **Ceiling** —1½" cold rolled steel channels spaced 48" o.c. supported by hanger wires. "H" runners spaced 24" o.c. clipped at each intersection with the channel. Concealed cross tees spaced 12" o.c. with ends resting on lower flange of the "H" runners. The suspension system supports ¾" x 12" x 12" acoustical tile. Protected light fixtures may be installed in ceiling not to exceed 16% of ceiling area. Air duct openings not to exceed 25 square inches per 100 square feet of ceiling area. Listed by U.L. under Design No. 95—2 hr., U.L. Test No. R4349-15.	2-hr. 3-hr Beam
Corrugated Steel Floor Units (e)	**Top Slab**—4½" concrete (measured from bottom of corrugations). **Ceiling** —½" (measured from bottom of corrugations) vermiculite-acoustical plastic or Type MK fireproofing applied directly to underside of steel floor units.	4-hr.
	Top Slab—4½" expanded shale concrete (measured from bottom of corrugations). **Ceiling** —Not required.	3-hr.
	Top Slab—5¼" concrete (Grade A) (measured from bottom of corrugations). **Ceiling** —Not required.	2-hr.
	Top Slab—4½" concrete (Grade A) (measured from bottom of corrugations). **Ceiling** —Not required.	1-hr.

TABLE 7 (Continued)

CONSTRUCTION		Fire Resistance Rating
Combination Flat Plate and Fluted Steel Floor Units	**Top Slab**—2½" concrete (minimum thickness over 1½" deep steel floor). **Ceiling** —⅞" vermiculite Type MK applied direct to underside of floor, following contour.	3-hr.
	Top Slab—2½" concrete (minimum thickness over 3" deep steel floor). **Ceiling** —Vermiculite Type MK applied directly to underside of floor. ⅞" thick following contour of floor (1" across top of fluted sections and 1½" under trench headers.)	3-hr.
	Top Slab—2½" concrete (minimum thickness over 1½" deep steel floor). **Ceiling** —½" vermiculite Type MK applied direct to underside of floor, following contour.	2-hr.
	Top Slab—2½" concrete (minimum thickness over 1½" deep steel floor). **Ceiling** —Vermiculite Type MK applied directly to the underside of floor, 1" on the top and sides of the fluted sections and ¾" on bottom of the fluted and flat plate sections. (1⅛" below flutes filled under trench header). U.L. Report R4374-26.	3-hr.
	Top Slab—2½" concrete (minimum thickness over cells). **Ceiling** —½" Vermiculite Type MK applied to the top and sides of fluted sections and ⅜" on the bottom of fluted and flat plate sections.	2-hr.
	Top Slab—2½" concrete (minimum thickness over 3" deep steel floor). **Ceiling** —Vermiculite Type MK applied direct to underside of floor. 13/16" thick on the top and sides of fluted sections and 11/16" on the bottom of fluted and flat plate sections. (1½" below flutes filled under trench header.)	3-hr.
	Top Slab—2½" concrete (minimum thickness over 1½" deep steel floor). **Ceiling** —Vermiculite Type MK applied direct to underside of floor. ½" thick on the top and sides of fluted sections and ⅜" on the bottom of fluted and flat plate sections.	2-hr.

TABLE 7 (Continued)

CONSTRUCTION		Fire Resistance Rating
Combination Flat Plate and Fluted Steel Floor Units	**Top Slab**—2½" concrete (minimum thickness over 3" deep steel floor). **Ceiling** —Vermiculite Type MK applied directly to underside of floor. 9/16" on the top and sides of fluted sections and 7/16" on the bottom of fluted and flat plate sections. (1" below flutes filled under trench header)	2-hr.
Flat Plate Steel Floor Units	**Top Slab**—2½" concrete (minimum thickness over 3" deep steel floor) **Ceiling** —⅞" vermiculite Type MK applied direct to underside of floor (1½" thick under trench headers, if used).	3-hr.
	Top Slab—2½" concrete (minimum thickness over 3" deep steel floor) **Ceiling** —¾" vermiculite acoustical plastic applied to underside of floor.	2-hr.
	Top Slab—2½" concrete (minimum thickness over 1½" deep steel floor.) **Ceiling** —Vermiculite Type MK applied directly to the underside of floor. ¾" on bottom of flat plate sections. (1⅛" under trench header). U.L. Report R4374-26.	3-hr.
	Top Slab—2½" concrete (minimum thickness over 1½" deep steel floor.) **Ceiling** —Vermiculite Type MK applied directly to the underside of floor. ⅜" on bottom of flat plate sections. (⅞" under trench header). U.L. Report R4374-25.	2-hr.
Fluted Steel Floor Units	**Top Slab**—2½" concrete (minimum thickness over 3" deep steel floor). **Ceiling** —Vermiculite Type MK applied direct to underside of floor. ⅞" thick following contour (1" across top of fluted sections).	3-hr.
Corrugated Steel Roof Units	**Top Slab**—2" vermiculite concrete (minimum thickness over 1½" deep steel deck). **Ceiling** —⅞" vermiculite acoustical plastic or vermiculite Type MK applied direct to underside of deck, following contour.	2-hr.
	Top Slab—3½" perlite concrete (measured from bottom of corrugation, average thickness 2⅝" perlite concrete) reinforced with welded wire mesh. **Ceiling** —Not required.	1½-hr.

TABLE 7 (Continued)

CONSTRUCTION		Fire Resistance Rating
Corrugated Steel Roof Units	**Top Slab**—2½" perlite concrete (measured from top of corrugations) reinforced 19 gage hexagonal wire mesh. **Ceiling** —Not required. U.L. Report R3453-4.	2-hr.
	Top Slab—2¼" vermiculite concrete (measured from top of corrugations). Reinforced with welded wire mesh. **Ceiling** —Not required. U.L. Report R2773-9.	2-hr.
Ribbed Steel Roof Units (e)	**Top Slab**—2" concrete (minimum thickness over 1½" deep steel floor or roof). **Ceiling** —Vermiculite acoustical plastic or vermiculite Type MK fireproofing filling ribs and extending ¾" below.	3-hr.
	Top Slab—3" vermiculite concrete (overall thickness) on ribbed steel roof unit with ribs extending up into concrete. Concrete reinforced with welded wire mesh. **Ceiling** —Not required.	1½-hr.
Steel Roof Deck Construction (c) (e)	**Deck** —Steel deck covered with not less than 2" nominal thickness of vermiculite concrete, or equivalent. **Ceiling** —⅞" vermiculite-gypsum or perlite-gypsum plaster (mix 100 lbs. of gypsum to 2½ cu. ft. of vermiculite or perlite) over face of metal lath.	4-hr.
	Deck —Steel deck covered with not less than 1" nominal thickness of insulation board consisting of shredded wood bonded with portland cement, or equivalent. **Ceiling** —⅞" vermiculite-gypsum or perlite-gypsum plaster (mix 100 lbs. of gypsum to 2½ cu. ft. of vermiculite or perlite) over face of metal lath.	3-hr.
	Deck —Steel deck covered with not less than 1" nominal thickness of insulation board consisting of felted glass fiber or equivalent. **Ceiling** —⅞" vermiculite-gypsum or perlite-gypsum plaster (mix 100 lbs. of gypsum to 2½ cu. ft. of vermiculite or perlite) over face of metal lath	2-hr.
	Deck —Steel deck covered with not less than 1½" nominal thickness of wood fiber-board insulation, or equivalent. **Ceiling** —⅞" sanded gypsum plaster (mix 1:2) over face of metal lath.	2-hr.

TABLE 7 (Continued)

CONSTRUCTION		Fire Resistance Rating
Steel Roof Deck Construction (c) (e)	**Deck** —Steel deck covered with not less than 1½" nominal thickness of insulation consisting of wood fiber and a cement binder, or equivalent. **Ceiling** —¾" sanded gypsum plaster (mix 1:2) over face of metal lath.	2-hr.
	Deck —Steel deck covered with not less than 1" nominal thickness of wood fiber-board insulation, or equivalent. **Ceiling** —⅝" sanded gypsum plaster (mix 1:2, 1:3) over face of metal lath.	1½-hr.
Wood Joist Construction (d)	**Ceiling Protection**—⅝" sanded gypsum plaster (mix 1:2, 1:3) or ⅝" sanded portland cement plaster (mix 1:2, 1:3—with 15 lbs. of hydrated lime and 3 lbs. of short asbestos fiber per bag of portland cement)—plaster shall be applied over face of metal lath which shall be attached to joists as prescribed in Section 1001.7.	1-hr.
	Ceiling Protection—½" sanded gypsum plaster (mix 1:2) applied on ⅜" perforated gypsum lath which shall be nailed to joists in the manner prescribed in Section 1001.6. Joints of gypsum lath shall be covered with 3" strips of metal lath attached with 1¾", No. 12 gage nails having ½" heads, spaced not over 5" apart along joists, and 2 nails per joist for strips running perpendicular to the joists.	1-hr.
	Ceiling Protection—½" vermiculite-gypsum or perlite-gypsum plaster (mix 100 lbs. of gypsum to 2½ cu. ft. of aggregate) applied on ⅜" perforated gypsum lath which shall be nailed to joists in the manner prescribed in Section 1001.6.	1-hr.
	Ceiling Protection—Two layers of ½" gypsum wallboard (see Section 1001.6) separated by 20 gage galvanized wire fabric with 1" hexagonal mesh. First (upper) layer of gypsum wallboard applied to bottom of joists with 5d cement-coated box nails, No. 15 gage, 1⅝" long, with 7/32" diameter heads, at 18" c. to c., or equivalent. Wire fabric nailed over first layer of gypsum wallboard with 8d cement-coated box nails, No. 12½ gage, 2⅜" long, with ¼" diameter heads, at 5" to 7" c. to c., or equivalent. Second layer of gypsum wallboard nailed with 8d cement-coated box nails No. 12½ gage, 2⅜" long, with ¼" diameter heads, at 5" to 7" c. to c., or equivalent.	1-hr.

TABLE 7 (Continued)

CONSTRUCTION		Fire Resistance Rating
Wood Joist Construction (d)	**Ceiling Protection**—⅝" vermiculite-gypsum or perlite-gypsum plaster applied over face of metal lath which shall be attached to joists as prescribed in Section 1001.7.	1-hr.
	Above with nails @ 5" o.c.	1½-hr.
	Ceiling Protection—One layer of ⅝" gypsum wallboard with a specially formulated core which provides greater fire-resistance than regular wallboard of the same thickness. Wallboard nailed with 1⅞ inch nails, 6 inches on centers, or equivalent.	1-hr.
	Ceiling Protection—⅝" gypsum-perlite or vermiculite plaster (mix 1:2) on ⅜" type "X" gypsum lath secured at each bearing with four 1¼" blued lath nails or four 7/16" crown, 16 ga. flattened wire, 1½" long staples with divergent points.	1-hr.
	Ceiling Protection—7/16" sanded gypsum plaster (mix 1:2) on ⅜" type "X" gypsum lath nailed at each bearing with four blued lathing nails. 3" wide strips of 2.5# metal lath applied continuously under each joist, nailed with 11 ga., 1½" long, 7/16" diameter head, galvanized roofing nails at 16" o.c.	1-hr.
	Ceiling Protection—Exposed steel grid suspension system, listed by U.L., suspended by 12 gage wire 4' o.c., supporting ½" or thicker, 24" x 24" or 48" noncombustible acoustical panels. Listed by U.L. under Design 39-1 HR. Protected light fixtures may be installed in ceiling not to exceed 16% of ceiling area. Air duct openings not to exceed 57 square inches per 100 square feet of ceiling area. UL Test R4337-33.	1-hr.
	Ceiling Protection—1½" 16 gage with carrying channels 4' o.c. suspended by 12 gage steel with 25 gage steel inverted "J" shaped channels, spaced 12" o.c. clipped to 1½" channels, supporting ¾" x 12" x 12" noncombustible acoustical tile, listed by U.L. as fire rated. Protected light fixtures may be installed in ceiling not to exceed 16% of ceiling area. Air duct openings not to exceed 55 square inches per 100 square feet of ceiling area. Factory Mutual Test 16738.12.	1-hr.

TABLE 7 (Continued)

CONSTRUCTION		Fire Resistance Rating
Wood Joist Construction (d)	**Flooring** —1" x 4" T & G wood flooring laid over 1" x 6" T & G rough floor applied diagonally to the joists, 2" x 10" joists 16" o.c. **Ceiling** —"H" runners spaced 24" o.c. supported by metal clips 48" o.c.; concealed cross tees are spaced 12" o.c. with the ends resting on the lower flange of the "H" runner. The system supports ¾" x 12" x 12" non-combustible mineral fiber acoustical tile listed by U.L. Design No. 12-1 hr., U.L. Test R4349-3.	1-hr.
	Flooring —1" x 4" T & G wood flooring laid over 1" x 6" T & G rough floor applied diagonally to the joists, 2" x 10" joists 16" o.c. **Ceiling** —Main tees spaced 48" o.c. supported by hanger wires. Cross tees 24" o.c. perpendicular to main tees. Suspension members support ½" or ⅝" 24" x 24" or 24" x 48" acoustical lay-in panels. Protected light fixtures may be installed in ceiling not to exceed 8% of ceiling area. Air duct openings not to exceed 25 square inches per 100 square feet of ceiling area. Listed under U.L. Design No. 22—1 hr. U.L. Test R4349-13.	1-hr.
	Flooring —Plywood Stressed Skin Panels consisting of ⅝" thick interior C-D (exterior glue) Top Stressed Skin on 2" x 6" nominal (minimum) stringers. Adjacent Panel edges joined with 8d common wire nails spaced 6" on center. **Ceiling** —½" thick fiberboard weighing 15 to 18 lbs. per cu. ft. installed with long dimensions parallel to stringers using 5d cooler nails spaced 12" on center. In lieu of fiberboard, ⅜" interior C-D (exterior glue) plywood glued and/or nailed to stringers with the same nail schedule may be used. Second layer of ½" type "X" gypsum-wallboard applied with long dimension perpendicular to joists and attached with 8d cooler nails spaced 6" on center at end joints and 8" on center elsewhere. Wallboard joints staggered with respect to fiberboard joints.	1-hr.
	Flooring —Wood floor consisting of 1⅛" interior grade plywood (exterior glue) with tongue and groove side joints. Butt joints of plywood centered over doubled 2 x 10 joists spaced 4 ft. on center. **Ceiling** —⅝" Type "X" gypsum wallboard attached to ⅞" metal furring channels with 1" wallboard screws spaced 12" on center. Furring channels spaced 24" on center and also located at joints.	1-hr.

TABLE 7 (Continued)

CONSTRUCTION		Fire Resistance Rating
Wood Joist Construction	**Flooring**—Finished floor of 1⅝" perlite-sand concrete (1 part portland cement, 2 parts sand, 3 parts perlite aggregate) reinforced with No. 2160-2-1619 hexagonal shaped wire mesh. Subflooring of 4 x 8 ft. by ⅝" standard plywood with exterior glue (identified in accordance with 1700.3(c)) with polyethylene film between finished floor and subflooring. 2" x 10" joists 16" o.c. laterally braced with 1" x 3" bridging. Glass-fiber insulation secured to joists on both sides up against subflooring.	1-hr.
	Ceiling —⅝" thick gypsum wallboard attached to 2⅝" wide by ½" deep resilient channels. Channels spaced 24" o.c. perpendicular to joists. Wallboard joint sealing system of paper tape embedded in cementitious compound over joints and exposed screw heads.	

Reference Notes to Table 7 — Fire-Resistance Ratings for Floor and Roof Construction

(a)—Steel Joist Construction and Formed Steel Members: Construction:

Top slabs or top decks as prescribed in Table 7, supported on open or solid-web steel joists, formed steel members, or rolled steel beams, having plaster ceilings applied on metal or wire lath attached directly to the under side of the steel members or furred or suspended therefrom, or applied on perforated gypsum lath furred therefrom.

Metal or wire lath of approved weight serving as a form for poured top slabs may be considered as reinforcement. Where a steel deck provides the load carrying properties, reinforcement of the concrete or gypsum fill may be omitted.

Specified slab thicknesses shall be measured from the top of supporting steel members. If precast top decks of concrete or gypsum are used, joints shall be grouted or be so constructed to provide tight end and side joints. Mortar finish shall be applied to top of 2" precast slabs to make up the prescribed thickness of top slab when thicknesses greater than 2" is required.

(b)—Cellular Steel Floor and Roof Units—Construction composed of steel formed into units of two or more cells; depth of units not less than 1⅝ inches and distance between cells not less than 2 inches. The top deck and ceiling protection shall be as prescribed in Table 7. Except as otherwise noted, the ceiling shall be furred or suspended so as to form an airspace between underside of cellular panel and back of lath.

(c)—**Steel Roof Deck Construction**—Steel deck covered with insulation of the kind and thickness not less than that prescribed in Table 7, or equivalent and supported on steel framing, protected with a suspended ceiling of metal and plaster as prescribed for the ratings indicated.

(d)—**Wood Joist Construction** — Double wood floor (sub-floor ¾" wood sheathing, or ½" plywood, finish floor tongue and groove, or ⅝" plywood) with insulating paper between, supported on 2" x 10" wood joists spaced 16" on centers and having fire resistive ceilings as prescribed in Table 7.

(e)—The ratings given in Table 7 for steel joist construction and formed steel members, cellular steel floor and roof units, corrugated steel floor units, corrugated steel roof units, ribbed steel roof units and steel roof deck construction are also applicable to steel beams, girders and trusses protected in the same manner, with non-combustible construction above.

(f)—Where electrical raceways and junction boxes are used in slab an additional thickness of 2" of concrete is required.

(g)—**Fire-Resistance Ratings for Ceilings Protecting Non-Combustible Construction**—The fire-resistance rating of any ceiling described in Table 7 for floor and roof construction shall be accepted as having the same fire resistance rating as a ceiling protecting structural members of steel or other non-combustible construction above the ceiling as is designated for floor and roof construction in Table 7, provided there are no combustible materials above the ceiling.

(h)—Appropriate fire resistance ratings as listed in Design Data—Fire Resistance, 1973-74 Edition, as published by the Gypsum Association, may be accepted as if herein listed.

(i)—Appropriate fire resistance ratings as listed in Performance Data on Architectural Acoustical Materials, Bulletin XXXII-73, as published by the Acoustical and Insulating Materials Association may be accepted as if herein listed.

APPENDIX "C"

WOOD PRESERVATIVES

(a) Standards for wood preservatives and methods of pressure treatment required within the scope of this code shall conform to those contained in the Manual of Recommended Practice of the American Wood Preservers' Association (AWPA).

Accepted preservatives include the following:

1. Creosote......AWPA Standard P1-65
2. Creosote—Coal Tar Solutions......AWPA Standard P2-68
3. Creosote—Petroleum Solution......AWPA Standard P3-67
4. Creosote—Pentachlorophenol Solution......AWPA Standard P11-70
5. Oil-Borne Preservatives (Hydrocarbon Solvents) ......AWPA Standard P8-64
6. Oil-Borne Preservatives......AWPA Standard P9-72
7. Water Borne Preservatives......AWPA Standard P5-72

Standards for methods of treatment, by commodity, include the following:

1. General requirement, all wood products, pressure treatment, AWPA Standard C1-72.
2. Lumber and Timbers, pressure treatment, AWPA Standard C2-72.
3. Piles, pressure treated, AWPA Standard C3-72.
4. Poles, pressure treated, AWPA Standard C4-72.
5. Plywood, pressure treated, AWPA Standard C9-72.
6. Piles and Timber, pressure treated, marine construction, AWPA Standard C18-72.
7. Pole building construction, pressure treatment, AWPA Standard C23-72.
8. Glued laminated wood, pressure treatment, AWPA Standard C28-72.
9. Quality Control Standards published by the American Wood Preservers Bureau for methods of treatment, by commodity include the following:

Softwood Lumber, Timber and Plywood Pressure Treated with Water-Borne Preservatives for Above Ground Use—AWPB Standard LP-2-71.

Softwood Lumber, Timber and Plywood Pressure Treated with Water-Borne Preservatives for Ground Contact AWPB Standard LP-22-71.

Softwood Lumber, Timber and Plywood Pressure Treated with Light Petroleum Solvent-Penta Solution for Above Ground Use—AWPB Standard LP-3-71.

Softwood Lumber, Timber and Plywood Pressure Treated with Light Petroleum Solvent-Penta Solution for Ground Contact—AWPB Standard LP-33-71.

Softwood Lumber, Timber and Plywood Pressure Treated with Volatile Petroleum Solvent (LPG)—Penta Solution for Ground Contact—AWPB Standard LP-4-71.

Softwood Lumber, Timber and Plywood Pressure Treated with Volatile Petroleum Solvent (LPG)—Penta Solution for Ground Contact—AWPB Standard LP-44-71.

Softwood Lumber, Timber and Plywood Pressure Treated with Creosote or Creosote Coal Tar Solutions for Above Ground Use—AWPB Standard LP-5-71.

Softwood Lumber, Timber and Plywood Pressure Treated with Creosote or Creosote Coal Tar Solutions for in Ground Contact Use—AWPB Standard LP-55-71.

Softwood Lumber, Timber and Plywood Pressure Treated with Heavy Petroleum Solvent-Penta Solution for Above Ground Use—AWPB Standard LP-7-71.

Softwood Lumber, Timber and Plywood Pressure Treated with Heavy Petroleum Solvent-Penta Solution for Use in Ground Contact—AWPB Standard LP-77-71.

Softwood Lumber, Timber and Plywood Pressure Treated with Water-Borne Preservatives for Ground Contact Use in Residential and Light Commercial Foundations AWPB Standard FDN-72.

APPENDIX "D"

HURRICANE REQUIREMENTS

Lateral support securely anchored to all walls provides the best, and only sound structural stability against horizontal thrusts, such as winds of exceptional velocity.

This may be obtained by adherence to provisions of Chapters of this Code. They are:

XIII — FOUNDATIONS

XIV — MASONRY CONSTRUCTION

XV — STEEL CONSTRUCTION

XVI — CONCRETE CONSTRUCTION

XVII — WOOD CONSTRUCTION

In the interest of clarity (Sec. 1205.3) the following is recommended:

1. EXTERIOR WALLS AND TIE BEAMS

Exterior walls constructed of hollow concrete masonry units shall be not less than a nominal thickness of 8 inches.

In all one and two family residences a reinforced tie beam shall be placed around the perimeter at each floor and roof level on exterior walls of masonry units. Reinforcing shall be not less than one No. 5 reinforcing bar. Beam size shall not be less than eight inches. This beam may be constructed with approved masonry units poured solid and tamped with concrete.

In all buildings (except one and two family residences) of non-reinforced masonry construction, hollow masonry units, masonry bonded walls and cavity walls shall not exceed 240 square feet without approved vertical and horizontal support where the nominal wall thickness is eight inches and units are laid in Type M, S or N Mortar. Such walls when a nominal thickness of 12 inches and laid in Type M, N, or S Mortar shall not exceed 360 square feet without approved vertical and horizontal support.

Grouted Brick Masonry, when laid in Type M, N or S Mortar, shall be supported vertically and horizontally in areas not to exceed 1.20 times that allowed for hollow unit masonry of comparable wall thickness.

Plain solid Masonry, when laid in Type M, N or S Mortar, shall be supported vertically and horizontally in areas not to exceed 1.10 times that allowed for Hollow unit Masonry of comparable wall thickness.

2. CORNERS—HOLLOW MASONRY CONSTRUCTION

In hollow masonry unit construction each unit cell shall be reinforced with at least one No. 5 bar at all corners; poured solid and tamped with concrete; such reinforcing shall be properly tied into the footing and spandrel beam.

3. WOOD TRUSS RAFTERS—ROOF CONSTRUCTION

All such members shall be securely fastened to the exterior walls with approved hurricane anchors or clips.

APPENDIX "E"

RECOMMENDED GUIDE FOR SOUND ISOLATION IN MULTI-FAMILY DWELLINGS

This Appendix pertains to the minimum allowable sound isolation that is recommended for partitions that separate adjacent units in multi-family dwellings, and similar partitions that separate a dwelling unit from public areas*, service areas**, or commercial facilities.

RECOMMENDATIONS:

(a) All partitions or walls between adjacent dwelling units or between dwelling units and an adjacent public area, hallway, service area, or commercial facility within the structure should have a minimum airborne sound isolation of 45 STC (Sound Transmission Class), as tested in accordance with "Tentative Recommended Practice for Laboratory Measurements of Airborne Sound Transmission, ASTM E90-61T," or latest revision thereof.

(b) All partitions should be constructed substantially as tested, and should not be altered by piercing, flanking, or voiding air space in any way that would lower the Sound Transmission Class.

(c) Partitions, where recommended to provide resistance to airborne sound shall in addition, provide the structural stability, fire resistance and other performance characteristics required elsewhere in this Code.

*Note: Public Areas include hallways, lobbies, storage rooms, stairways and etc.

**Note: Service Areas include boiler rooms, laundries, mechanical equipment rooms, elevator shafts, incinerator shafts, garages and other such uses.

APPENDIX "F"

FALLOUT SHELTER CONSTRUCTION

The following amendments to the Southern Standard Building Code are recommended where provisions are desired to apply to the construction of Fallout Shelters.

F-1. Chapter II, Section 201.2—

Add in alphabetical order: "Fallout Shelter." (See Section 515.2).

F-2. Chapter II, Section 201.2—

Habitable Room: Add after "Cellars" the words: "Fallout Shelters."

F-3. Chapter IV, Section 401.1—

Occupancy Groups: Add after Group H, "Group I—Fallout Shelters."

F-4. Chapter IV, Section 412.4—

Fire-Resistance Rating of Separation:

Under table of minimum requirements, add (in order) "Group I-2 —Single Purpose Fallout Shelters—No requirements for fire-resistive separation."

F-5. Chapter IV

Add new Section 413 as follows:

SECTION 413—GROUP I—FALLOUT SHELTERS

All buildings and structures or parts thereof which are used for fallout shelters and so designated shall be classified in the fallout shelter use group.

413.1—Group I-1—Fallout shelters shall include all dual-use shelters.

413.2—GROUP I-2—Fallout Shelters shall include all single-purpose fallout shelters.

F-6. Chapter V.

Add new Section 517 as follows:

SECTION 517—FALLOUT SHELTERS

This Section shall establish the minimum criteria which must be met before a building space can be constructed, occupied, used or designated as a fallout shelter.

517.1 — SCOPE AND APPLICABILITY

The scope of this section extends to a building or to a building space when it is being used as a fallout shelter in time of national emergency or for reasonable periods of drill and instruction. If the space is being used as a shelter, the provisions of this section shall apply, and if it is not being so used, other applicable provisions of this code shall apply.

517.2 — DEFINITIONS

Fallout Shelter—A fallout shelter is any room, structure or space designated as such and providing its occupants with protection at a minimum protection factor of forty (40) from gamma radiation from fallout from a nuclear explosion as determined by an architect or engineer certified by the Office of Civil Defense as a Qualified Fallout Shelter Analyst.

Dual-use Fallout Shelter—A dual-use fallout shelter is a space having a normal, routine use and occupancy as well as having an emergency use as a fallout shelter.

Single-Purpose Fallout Shelter—A single-purpose fallout shelter is a space having no other use or occupancy than as a fallout shelter.

Protection Factor—A factor used to express the relation between the amount of fallout gamma radiation that would be received by an unprotected person and the amount that would be received by one in a shelter.

Unit of Egress Width—A unit of egress width is 22 inches, the space required for free travel of one file of persons.

517.3 — GENERAL

Nothing in these regulations shall be construed as preventing the dual use or multiple use of normal occupancy space as fallout shelter space, providing the minimum requirements for each such use are met.

517.4 — EXIT FACILITIES

There shall be no fewer than two widely spaced exits from a fallout shelter, leading directly to other spaces of the building or outdoors. In no case shall a single exit be less than 24" wide. In addition, the following requirements must be met:

(a) GROUP I-1—(Dual-use Fallout Shelters)

When requirements for normal occupancy of the space is detailed in Chapter XI exceed the preceding the normal occupancy requirements shall govern.

(b)—GROUP I-2—(Single Purpose Fallout Shelters)

Exits from the fallout shelter shall aggregate at least one unit of egress width for every 200 shelter occupants or fraction thereof. Interior circulation within the fallout shelter shall be governed by requirements of Chapter XI of these regulations.

517.5 — SPACE AND VENTILATION

(a) A minimum of ten (10) square feet of net floor area shall be provided per fallout shelter occupant. Minor partitions, columns, and area for storage of Federal shelter supplies may be included in net area. A minimum of sixty-five (65) cubic feet of volume shall be provided per fallout shelter occupant. A minimum of three (3) cubic feet of fresh air per minute per person shall be provided.

(b) Fallout shelter capacity or occupancy time may be limited by the volume of room and not by its net area. The following table shall be used in determining volume space required per person:

Time for Complete Air Change (Minutes)*	Volume of Space/Person (Cu. Ft.)
1000 or more	500
600	450
400	400
200	300
100	200
60	150
35	100
22	65

*Computed as a ratio: $\frac{\text{Net volume of space (cu. ft.)}}{\text{fresh air supply (cfm.)}}$

517.6 — WINDOWS

No requirements.

517.7 — ILLUMINATION

No special lighting levels are required.

517.8 — SANITATION

Toilets, either flush type operating from the normal water supply system or chemical or other types shall be provided on the basis of one toilet per 50 fallout shelter occupants. Fifty per cent (50%) of the toilets may be provided outside the fallout shelter area. Empty water containers may be considered as fulfilling this requirement.

F-7. Chapter VII, Section 702.3—

Add at end of present section the following:

Group I—Fallout Shelters—Fire-resistive partitions in fallout shelters are not required.

F-8. Chapter XI, Section 1103.1—

Table of maximum distance of travel to an exit—Add:

Group I—Fallout Shelters (single-purpose)....................150

F-9. Chapter XI, Section 1105.1—

Table of Occupancy content floor area per person, add:

Group I—Fallout Shelters..10 sq. ft.

F-10. Chapter XI, Section 1106.(a)

Add at end of first sentence after "fire-resistance":

"; except that such stairways, platforms, landings, and hallways may be enclosed with partitions of not less than one-hour fire-resistance in buildings occupied by more than forty (40) people above or below the story at street level when such occupancy is in connection with use of a designated portion of the building as a fallout shelter in times of emergency."

F-11. Chapter 12.03.1(a)

Table of minimum live loads—Add in alphabetical order:

Fallout Shelters—Group I-1—(dual-use)—To be determined by normal occupancy requirements except that concentrated loads due to water storage and rations shall be considered.

Fallout Shelters—Group I-2 (single-purpose)—40 p.s.f. (concentrated loads due to water storage and rations shall be considered).

F-12. Section 2001.6

Add new paragraph as follows:

(c) For ventilation requirements of fallout shelters, see Section 517.5.

APPENDIX "G"

RECOMMENDED GUIDE FOR THE ESTABLISHMENT OF A FIRE DISTRICT

For the purpose of this Code, the Fire District shall include that territory or area consisting of:

(a) Two or more adjoining blocks, exclusive of intervening streets, where at least 50% of the ground area is built upon and more than 50% of the built-on area is devoted to hotels and motels of Group "A" Occupancy; All Group "B" Occupancies; theatres, night clubs, restaurants of Group "E" Occupancy; garages, express and freight depots, warehouse and storage buildings used for the storage of finished products (not located with and forming a part of a manufactured or industrial plant) or Group "F" Occupancy. Where the average height of a building is two and one-half (2½) stories or more a block should be considered if the ground area built upon is at least 40%.

(b) Where four (4) contiguous blocks or more comprise a Fire District there shall be a buffer zone of two hundred (200) feet around the perimeter of such district. Streets, rights-of-way, and other open spaces not subject to building construction may be included in the two hundred (200) foot buffer zone.

(c) Where blocks adjacent to the Fire District have developed to the extent that at least twenty-five (25) percent of the ground area is built upon, and forty (40) percent or more of the built-on area is devoted to the uses specified in paragraph (a) they may be considered for inclusion in the Fire District, and may form all or a portion of the two hundred (200) foot buffer zone required in paragraph (b).

APPENDIX "H"

CHAPTER IV—CLASSIFICATION OF BUILDINGS BY OCCUPANCY
CHAPTER VI—CLASSIFICATION OF BUILDINGS BY CONSTRUCTION

The purpose of this Appendix "H" is to provide an alternate format of these two Chapters thereby presenting them and their provisions in a condensed and tabular approach for simplification, clarification, reference and use. It is not intended that this Appendix deletes, changes or amends the present Chapters IV and VI which are now in the body of this Code.

Refer to body of Code for all referenced section numbers and chapters, except for Notes related to Tables 400 and 600.

CHAPTER IV

CLASSIFICATION OF BUILDINGS BY OCCUPANCY

SECTION 401 — CLASSIFICATION BY OCCUPANCY OR USE

401.1 — GENERAL

Every new and existing building, structure, or part thereof, shall for the purpose of this Code, be classified according to its use or occupancy as a building or structure of one of the following OCCUPANCY GROUPS:

GROUP A—RESIDENTIAL
GROUP B—BUSINESS
GROUP C—SCHOOL
GROUP D—INSTITUTIONAL
GROUP E—ASSEMBLY
GROUP F—STORAGE
GROUP G—INDUSTRIAL
GROUP H—HAZARDOUS

Each occupancy group is intended to embrace buildings as hereinafter defined and those of similar character or use. Wherever there is any uncertainty as to the classification of a building, the Building Official shall fix the classification within which it falls, according to the relative fire hazard involved.

401.2 — PROTECTIVE REQUIREMENTS—ALL OCCUPANCY GROUPS (Refer to Code).

	SECTION
1. Height and Area	402 to 403.9, inclusive
2. Types of Construction	601 to 610, inclusive

	SECTION
3. Means of Egress	1101 to 1126, inclusive
4. Protection of Vertical Openings	701.1 to 701.4, inclusive
5. Protection of Wall Openings	703.1 to 703.7, inclusive
6. Sprinklers and Standpipes	901.1 to 903.1, inclusive
7. Mixed Occupancy Separations	412 or 413 hereunder
8. Light, Ventilation and Sanitation	2001 to 2002, inclusive
9. Heating Requirements	801 to 825, inclusive
10. Separation of Furnace or Boiler Rooms	812
11. Ratproof Construction	1900 to 1902.3 inclusive

SECTION 402 — HEIGHT AND AREA

402.1 — APPLICATION

(a) The maximum height and area for buildings or structures of the different types of construction, shall be governed by the intended use of the buildings, or occupancy, as provided for in this Chapter and shall not exceed the limits set forth in Table 400 except as provided in Section 403 and 404 hereunder. For the purpose of this Code, each part of a building or structure included within fire walls shall be considered a separate building.

(b) No building hereafter erected shall be extended to exceed the maximum floor area set forth in this chapter, governed by the occupancy and type of construction. However, a building heretofore lawfully erected, which already exceeds such maximum area, may be extended, provided such extension does not exceed the maximum area prescribed and provided such extension is separated from the existing building by a fire wall having a fire-resistive rating of not less than four hours.

(c) No existing building may be increased in height unless the entire building is altered as needed to meet the requirements of this Code for a new building of such increased height and number of stories.

402.2 — DEFINITIONS

For the purpose of this Code, "height" and "area", as applied to a building, has the meaning designated in Chapter II, Definitions.

SECTION 403 — HEIGHT EXCEPTIONS

403.1 — ROOF STRUCTURES

(a) Church spires, chimneys, tanks and supports, aerial supports, parapet walls not over four (4) feet high, bulkheads and penthouses used solely to enclose stairways, tanks, elevator machinery or shafts, or ventilation or air-conditioning apparatus, need not be considered in determining the highest point of the building; provided that the highest point shall be taken to be the highest point of the roof of the highest penthouse when the aggregate area of all penthouses and other roof structures exceeds one-third (1/3) of the area of the roof upon which they stand. (See requirements of Section 713, 714 and 715.)

(b) Where a 1-story automobile parking area (enclosed or open) of Type I or II Construction, with grade entrance is provided under a building of Group A Occupancy, the number of stories to be used in determining the minimum type of construction may be measured from the roof slab of such parking area.

403.2 — MEZZANINES

Mezzanine floors or galleries shall not be regarded as a story unless they exceed thirty-three and one-third (33 1/3) percent of the aggregate ground floor area.

403.3 — BASEMENTS AND CELLARS

(a) The basement or cellar of a building shall not count as a story if the first floor above such basement or cellar is less than seven (7) feet above grade.

(b) Basements or cellars used as classrooms or assembly rooms shall be counted as a story.

403.4 — HEIGHT INCREASE FOR SPRINKLERS

The maximum allowable number of stories may be increased by one story if the building is provided with automatic sprinklers throughout in accordance with Section 901 of Chapter IX, except such height increase shall not be permitted in buildings where the installation of automatic sprinkler equipment is a mandatory requirement of this code, or when the provisions of Section 403.6 are used.

SECTION 404 — AREA EXCEPTIONS

404.1 — GENERAL

The increase of floor areas permitted by this Section may be additive when applicable, except that in buildings where, because of occupancy, type construction or fire district, one-hour fire-resistance construction is a requirement, or automatic sprinkler equipment is required, as specified in Section 901, no increase of area shall be permitted because of such construction or equipment. Except as provided in Section 403.7, the total permissive area shall not exceed 450% of the basic area.

404.2 — AREA INCREASE FOR SEPARATION ON TWO OR MORE SIDES OF A BUILDING

Where streets or public spaces of minimum width not less than 20 feet extend along two or more sides of a building of any occupancy classification, except Group "H" Hazardous, the maximum areas specified in this Chapter and modified as provided in this Section for such buildings may be increased by the percentage specified in Table 404.2 hereunder for each foot by which the minimum width of such streets or public spaces exceeds 20 feet, but such increase shall not exceed the maximum percentage shown in Table 404.2 hereunder.

TABLE 404.2 — RATE OF AREA INCREASE FOR SEPARATION

	Rate for Area Increase For Separation Over 20' Wide	Max. Area Increase
Separation along 2 sides but along not less than 50% of perimeter of building	1% per Ft. over 20 Ft.	50%
Separation along 3 sides but along not less than 75% of perimeter of building	2% per Ft. over 20 Ft.	100%
Separation along all sides or along 100% of perimeter of building	3% per Ft. over 20 Ft.	100%

404.3 — AREA INCREASE FOR SPRINKLERS

The maximum allowable floor and attic area of any occupancy classification, except Group "H"—Hazardous, may be increased by 200% for one story buildings, and by 100% for buildings over one story in height if the building is provided with automatic sprinklers throughout in accordance with Chapter IX, except such area increase shall not be permitted in buildings where the installation of automatic sprinkler equipment is a mandatory requirement of this Code, or when the provisions of Section 402.6 are used.

404.4 — UNLIMITED AREAS

The area of a one-story building of Group "B" Business, Group "F" Storage, or Group "G" Industrial occupancy, located outside the Fire District, shall not be limited provided the building is equipped with an approved automatic sprinkler system throughout or other fire protective systems throughout, in accordance with Section 901 and as approved by the Building Official, and is surrounded on all sides by a permanent open space of not less than sixty (60) feet. Where water may cause or increase a fire, other fire extinguishing systems shall be required in rooms or buildings used for the manufacture or storage of hazardous materials including but not limited to, aluminum powder, calcium carbide, calcium phosphate, metallic sodium and potassium, quicklime, magnesium powder and sodium peroxide.

SECTION 405 — GROUP "A" — RESIDENTIAL

405.1 — SCOPE

Buildings in which families or households live or in which sleeping accommodations are provided, shall be classified as Group A—Residential Occupancy. Group "A"—Residential Occupancy—shall include, among others, the following:

Dwellings	Multiple Dwellings (more than two families)
Dormitories	Lodging Houses
Hotels	Convents
Motels	Monasteries

405.2 — SPECIAL REQUIREMENTS, GROUP "A" OCCUPANCY

Storage and handling of flammable liquids shall be prohibited in every Group "A" Occupancy. Not more than one (1) gallon of flammable liquid, used for cleaning purposes only, may be kept in a residence, provided such flammable liquid is kept in an approved container, used especially for that purpose.

SECTION 406 — GROUP "B" — BUSINESS

406.1 — SCOPE

(a) Buildings which are occupied for business or rendering of professional services shall be classified in Group B-1; buildings which are occupied for the sale or display of merchandise such as department stores, shopping centers, supermarkets, etc., or the supplying of food or drink, shall be classified in Group B-2, except as provided in Section 406.2. Group "B" does not include buildings used for any purpose involving highly combustible, flammable or explosive materials. (See Section 411.)

(b) Group B—Business Occupancy includes, among others, the following:

Group B-1

Greenhouses	Undertaking parlors
Automotive service stations (Filling Stations)	Temporary structures
Banks	Office buildings
Libraries (other than school)	

Group B-2

Stores	Markets
Shops	Restaurants (See exception 406.2)
Bowling Alleys	

406.2 — EXCEPTION

Restaurants or places supplying food or drink that accommodate 75 or more people, or that have a stage, or that provide dancing or entertainment features, shall be classified in Group "E"—Assembly.

406.3 — SPECIAL REQUIREMENTS GROUP "B" OCCUPANCY

	SECTION
1. Temporary Structures	504
2. Automotive Service Stations	505
3. Greenhouses	509
4. Bowling Alleys	,13

5. Storage and handling of flammable liquids shall be prohibited; except in Group B-2 (retail stores) where the quantities of liquids are not in excess of those specified in Section 411.2(b); except automotive service stations which shall conform to Section 501.1 (f).

SECTION 407 – GROUP "C" – SCHOOLS

407.1 — SCOPE

(a) Buildings in which people come together for education or instructional purposes shall be classified in Group "C"—School Occupancy.

(b) Group C—School Occupancy shall include, among others, the following:

Schools	Academies
Colleges	Nursery Schools
Universities	Kindergartens

407.2 — EXCEPTION

(a) Parts of buildings used for the congregating or gathering of 75 or more persons in one room shall be classified as in Group E—Assembly Occupancy—(see Section 409 hereunder), regardless of whether such gathering is of an educational or instructional nature or not.

(b) Schools for business or vocational training shall be classified in the same occupancies and conform to the same requirements as the trade, vocation or business taught.

407.3 — SPECIAL REQUIREMENTS, GROUP "C" OCCUPANCY

	SECTION
1. Classroom Lighting	2001.5

2. Every heating appliance which produces an unprotected open flame shall be prohibited.
3. Storage and handling of flammable liquids shall be prohibited.
4. Where permanent motion picture projectors are used, booths shall be provided, as set forth in Section 512.16.

5. Rooms used for day care nurseries, kindergarten or first grade pupils shall not be located above or below the floor of exit discharge. Rooms used for second grade pupils shall not be located more than one story above the floor of exit discharge.

SECTION 408 — GROUP "D" — INSTITUTIONAL

408.1 — SCOPE

(a) Buildings in which more than six people are detained for penal or correctional purposes; or in which the liberty of the inmates is restricted, or places of involuntary detention, shall be classified in Group D-1.

(b) Buildings in which more than ten people are harbored for medical, charitable or other care or treatment shall be classified in Group D-2.

(c) Group D-1—Institutional Occupancy—shall include, among others, the following:

Insane Asylums
Reformatories
Jails
Prisons

(d) Group D-2—Institutional Occupancy—shall include, among others, the following:

Hospitals
Sanitoriums
Nursing Homes
Orphanages
Homes for the Aged

408.2 — EXCEPTION

Dormitories for doctors, nurses, and able-bodied help (not for patients or inmates) of Institutional buildings shall be classified in Group A—Residential Occupancy.

408.3 — SPECIAL REQUIREMENTS — GROUP "D" OCCUPANCY

	SECTION
1. Special Exit Requirements	1104

2. Storage and Handling of Flammable Liquids shall be prohibited.

SECTION 409 — GROUP "E" — ASSEMBLY

409.1 — SCOPE

(a) Buildings in which provision is made for the congregation or gathering of seventy-five (75) or more persons in one room or space shall be classified in Group E—Assembly Occupancy. Such room or space shall include any occupied connecting room or space in the same story, or in a story or stories above or below, where entrance is common to the rooms or spaces. For general requirements see Section 512.

(b) Group E—Assembly Occupancy shall include, among others, the following:

Passenger Depots	Motion Picture Houses
Stadiums and Grandstands	Public Assembly Halls
Restaurants (large)	Churches
Amusement Park Buildings	Museums
Tents (Assembly)	Auditoriums
Theaters	Dance Halls
Gymnasiums	Recreation Halls

409.2 — SUB-CLASSIFICATIONS

Group E shall be divided into two sub-classifications as set forth in this Section, both of which shall comply with the requirements for Group E Occupancy unless otherwise specified:

(a) **Group E-1—Large Assembly**

Group E-1 shall include places of public assembly having a working stage (see definition) and having a capacity of seven hundred (700) or more people; also, Group E-1 shall include places of assembly having a non-working stage but having a capacity of 1,000 or more people.

(b) **Group E-2—Small Assembly**

Group E-2 shall include places of assembly having a capacity of 75 or more persons but having a capacity less than designated for Group E-1.

409.3 — METHOD OF DETERMINING CAPACITY

The capacity of occupant content of places of assembly shall be as set forth in Chapter XI.

409.4 — SPECIAL REQUIREMENTS—GROUP "E" OCCUPANCY

	SECTION
1. Tents	504.2
2. Stadiums and Grandstands	510
3. Amusement Park Buildings	511
4. Churches	514

5. Storage and handling of flammable liquids shall be prohibited.

SECTION 410 — GROUP "F" — STORAGE

410.1 — SCOPE

(a) Buildings which are used for the storage of goods, wares or merchandise, excepting limited storage incidental to the display, sale or manufacture of such goods, wares or merchandise, shall be classified in Group "F"—Storage Occupancy. Group "F" does not include buildings used to store highly combustible, flammable or explosive products or materials (see Section 411).

(b) Group "F"—Storage Occupancy—shall include, among others, the following:

Airplane Hangars	Warehouses
Coal Pockets	Storage Buildings
Garages	Freight Depots

410.2 — SPECIAL REQUIREMENTS—GROUP "F"—STORAGE

	SECTION
1. Airplane Hangars	502
2. Coal Pockets	503
3. Garages	
Private	506
Public Parking Decks	507
Public	508
4. Storage and Handling of Flammable Liquids shall conform to	501.1 (f)

SECTION 411 — GROUP "G" — INDUSTRIAL

411.1 — SCOPE

Buildings in which work or labor is performed in connection with the fabrication, assembly, processing, etc., of products or materials shall be classified in Group "G"—Industrial Occupancy. Group "G" does not include buildings used for any purpose involving highly combustible, flammable, or explosive products or materials (See Section 411). Group "G" Occupancy shall include, among others, the following:

Manufacturing Plant	Processing Plant
Factory	Mill
Assembly Plant	Power Plant

411.2 — SPECIAL REQUIREMENTS GROUP "G"—INDUSTRIAL

	SECTION
1. Storage and Handling of Flammable Liquids shall conform to	501.1 (f)
2. Posting Floor Loads	110

SECTION 412 — GROUP "H" — SPECIAL HAZARDOUS

412.1 — SCOPE

(a) Buildings or structures used for purposes that involve highly combustible, flammable or explosive products or materials or that constitute exceptional fire hazards, because of the form, character or volume stored, processed or manufactured, shall be classified in Group "H"—Special Hazardous Occupancy. For special requirements, see Section 501.

(b) Group "H"—Special Hazardous Occupancy—shall include, among others, the following:

Dry Cleaning Establishments
Grain Elevators
Storage or use of Highly Combustible Materials
Storage of Nitrate Film

412.2 — HAZARDOUS MATERIALS AND PRODUCTS

The processing, manufacturing or storing of the following materials, among others, shall be classified as a special fire hazard, (Group "H"—Special Hazardous Occupancy), because of the highly combustible and explosive quality of the materials involved. They shall be classified in two groups as set forth below:

(a) Buildings or structures in which the following materials or products are stored or handled shall be classified as Group "H"—Special Hazardous Occupancy, regardless of the volume or quantity stored or handled:

Acids: Sulphuric, Nitric or Hydroflouric—Calcium Carbide—Asphaltum—Celluloid. Chemicals: Poisonous, combustible, explosive—Cereal, flour or feed mills—Cork—Excelsior—Petroleum Products (Not retail gasoline service stations) Pyroxylin and Pyroxylin Plastic Products—Shoddy Mills—Loose Spices and Vegetable Stocks (not retail stores)—Explosives—Feather Renovating—Films (see Section 501.3) Fireworks. Gas: Poisonous irritating and flammable—Jute—Kapok—Munitions—Naval Stores—Nitrate of Soda. Paper: Baled Waste—Oakum or Hemp Processing—Paints: Manufacturing—Starch Mills—Waste Paper—Rubber Manufacturing Plants.

(b) Buildings or structures in which the following materials or products are stored or handled in quantities in excess of the cubic foot areas set forth below shall be classified as Group "H"—Special Hazardous Occupancy:

	In Excess Of
Acetylene gas in pressure release containers	2000 cu. ft.
Artificial flowers	1200 cu. ft.
Artificial leather	600 cu. ft.
Bags—Burlap, paper or cotton	1000 cu. ft.
Barrels—Second hand	1800 cu. ft.
Brooms and Broom corn	1200 cu. ft.
Cotton (Loose) Wadding or Waste	400 cu. ft.
Drugs	6500 cu. ft.
Enameling	2400 cu. ft.
Fertilizer (Bags only)	2700 cu. ft.
Lacquers (Separate containers only)	3600 cu. ft.
Matches	300 cu. ft.
Mattresses	4800 cu. ft.
Paints (Separate metal containers only)	6000 cu. ft.
Spray painting shops	3200 cu. ft.
Tires (Recapping)	Section 517
Tires—Storage	12,000 cu. ft.
Varnish—Turpentine (Separate metal containers only)	2400 cu. ft.

SECTION 413 — MIXED OCCUPANCY AND SEPARATION REQUIREMENTS

413.1 — DEFINITIONS

When a building is used for two or more occupancies, classified within different occupancy groups, it shall be considered a mixed occupancy.

413.2 — LIMITATIONS

A mixed occupancy building shall be governed by the height and area limitations applying to the principal use therein. Accessory occupancies shall not exceed the area limitations or be located at a story height greater than that permitted for such accessory occupancy and the type of construction being used.

413.3 — FIRE RESISTIVE RATING OF SEPARATION

(a) The minimum fire resistance of constructions separating any two occupancies horizontally and/or vertically in a building of mixed occupancy, shall be the higher rating required for the occupancies being separated, as specified below:

Minimum Requirements*

Group A—Residential	1 Hr. fire-resistive separation
Group B—Business	1 Hr. fire-resistive separation
Group C—School	2 Hr. fire-resistive separation
Group D—Institutional	2 Hr. fire-resistive separation
Group E-1—Large Assembly	4 Hr. fire-resistive separation
Group E-2—Small Assembly	2 Hr. fire-resistive separation
Group F—Storage	4 Hr. fire-resistive separation
Group G—Industrial	2 Hr. fire-resistive separation
Group H—Special Hazardous	4 Hr. fire-resistive separation

(b) A separation between a private garage and any occupancy shall be the minimum fire-resistance specified above for the occupancy except in the case of a one or two-family dwelling no fire-resistive separation shall be required.

(c) No separation shall be required between a Sunday School room or rooms and a Church Auditorium of Group E-2 Small Assembly.

(d) Portions of buildings used as accessory offices or for customary non-hazardous uses necessary for transacting the principal business of Group F Storage and Group G Industrial occupancies may be separated by partitions of non-combustible construction without fire-protection or by partitions constructed of materials as permitted in the type of construction used.

(e) When in a building, or portion of a building of a single occupancy classification, enclosed spaces are provided for separate tenants, such spaces shall be separated by not less than one-hour fire resistance.

(f) A separation between a public garage used exclusively for the storage of passenger vehicles that will accommodate not more than nine (9) passengers and any other occupancy other than Group E-1, Group F and Group H shall be two (2) hours.

* For materials and assemblies to provide the required fire-resistive construction, see Chapter X.

413.4 — PROTECTION OF HORIZONTAL OPENINGS

For requirements governing the protection of door openings in walls and partitions separating mixed occupancies, see Section 703.4.

413.5 — PROTECTION OF VERTICAL OPENINGS

For requirements governing the protection of vertical openings, see Section 701.

TABLE 400 — ALLOWABLE HEIGHT AND AREA

Figures Within Parentheses are the Heights in Stories
Figures Opposite Parentheses is Allowable Area for that Story or Stories
Letters in Table refer to Table 400 Notes
For Exceptions to Allowable Height and Area See Sections 403 and 404 herein.

USE GROUP	TYPE OF CONSTRUCTION								
	Type I	Type II	Type III	Type IV		Type V		Type VI	
				1-Hour Protected	Unprotected	1-Hour Protected	Unprotected	1-Hour Protected	Unprotected
A RESIDENTIAL	No Limits	No Area Limit 80 Feet Height Limit	(3) 9,750 (2) 15,000 (1) 18,000	a. (3) to (5) 9,750 (2) 15,000 (1) 18,000	a. h. (3) to (5) 6,500 (2) 10,000 (1) 12,000	a. (3) to (5) 9,600 (2) 15,000 (1) 18,000	a. h. (3) to (5) 6,400 (2) 10,000 (1) 12,000	(3) 6,400 (2) 10,500 (1) 10,500	(2) 7,000 (1) 7,000
B-1 BUSINESS l.	r. No Limits	c. No Area Limit 80 Feet Height Limit	(3) to (5) 13,500 (2) 18,000 (1) 25,500	a. (3) to (5) 9,000 (2) 18,000 (1) 25,500	a. h. (3) to (5) 6,000 (2) 12,000 (1) 17,000	a. (3) to (5) 9,000 (2) 16,500 (1) 21,000	a. h. (3) to (5) 6,000 (2) 11,000 (1) 14,000	(2) 10,500 (1) 13,500	(2) 7,000 (1) 9,000
B-2 MERCANTILE l.	No Limits	c. No Area Limit 80 Feet Height Limit	(3) to (5) 8,000 (2) 12,000 (1) 13,500	a. (3) to (5) 6,000 (2) 12,000 (1) 13,500	a. h. (3) to (5) 4,000 (2) 8,000 (1) 9,000	a. (3) to (5) 6,000 (2) 12,000 (1) 13,500	a. h. (3) to (5) 4,000 (2) 8,000 (1) 9,000	(2) 6,000 (1) 9,000	(2) 4,000 (1) 6,000

TABLE 400 (Continued)

USE GROUP	TYPE OF CONSTRUCTION								
	Type I	Type II	Type III	Type IV		Type V		Type VI	
				1-Hour Protected	Unprotected	1-Hour Protected	Unprotected	1-Hour Protected	Unprotected
C SCHOOL	r. No Limits	No Area Limit 80 Feet Height Limit	d. k. (2) 12,000 (1) 18,000	k. (2) 12,000 (1) 18,000	d. k. p. (1) 12,000	k. (2) 12,000 (1) 18,000	d. k. p. (1) 12,000	(2) 8,000 (1) 12,000	d. (1) 8,000
D-1 INSTITUTIONAL	No Limits	No Area Limit 80 Feet Height Limit	Not Permitted	Not Permitted	Not Permitted	Not Permitted	Not Permitted	Not Permitted	Not Permitted
D-2 INSTITUTIONAL	No Limits	No Area Limit 80 Feet Height Limit	(2) 8,000 (1) 12,000	(2) 4,000 (1) 7,000	Not Permitted	(2) 4,000 (1) 7,000	Not Permitted	(1) 5,000	Not Permitted
E-1 LARGE ASSEMBLY (Working Stage)	r. No Limits	No Area Limit 80 Feet Height Limit	Not Permitted	Not Permitted	Not Permitted	Not Permitted	Not Permitted	Not Permitted	Not Permitted
E-1 LARGE ASSEMBLY (Non-Working Stage)	r. No Limits	No Area Limit 80 Feet Height Limit	e. (1) 12,000	(1) 12,000	f. (1) 8,000	(1) 12,000	f. (1) 8,000	Not Permitted	Not Permitted

TABLE 400 (Continued)

USE GROUP	TYPE OF CONSTRUCTION								
	Type I	Type II	Type III	Type IV		Type V		Type VI	
				1-Hour Protected	Unprotected	1-Hour Protected	Unprotected	1-Hour Protected	Unprotected
E-2 SMALL ASSEMBLY (Working Stage)	r. No Limits	No Area Limit 80 Feet Height Limit	m. (1) 10,000	m. (1) 9,000	f. m. (1) 6,000	m. (1) 9,000	f. m. (1) 6,000	(1) 4,500	(1) 3,000
E-2 SMALL ASSEMBLY (Non-Working Stage)	r. No Limits	No Area Limit 80 Feet Height Limit	e. m. (2) 12,000 (1) 12,000	m. (2) 12,000 (1) 12,000	f. m. (2) 8,000 (1) 8,000	m. (2) 12,000 (1) 12,000	f. m. (2) 8,000 (1) 8,000	g. (1) 7,500	g. (1) 5,000
F STORAGE l. s.	No Limits	(3) to (6) 20,000 (2) 30,000 (1) 30,000	h. (3) to (6) 15,000 (2) 24,000 (1) 24,000	h. (4) 15,000 (3) 15,000 (2) 24,000 (1) 24,000	h. (4) 10,000 (3) 10,000 (2) 16,000 (1) 16,000	h. (4) 15,000 (3) 15,000 (2) 24,000 (1) 24,000	h. (4) 10,000 (3) 10,000 (2) 16,000 (1) 16,000	(1) 9,000	(1) 6,000
G INDUSTRIAL l.	r. No Limits	80 Feet Height Limit (3) to (80') 20,000 (2) 30,000 (1) No Limit	i. (3) to (6) 15,000 (2) 22,500 (1) 31,500	h. (4) 15,000 (3) 15,000 (2) 22,500 (1) 31,500	h. (4) 10,000 (3) 10,000 (2) 15,000 (1) 21,000	h. (4) 13,500 (3) 13,500 (2) 18,000 (1) 22,500	h. (4) 9,000 (3) 9,000 (2) 12,000 (1) 15,000	(1) 15,000	(1) 10,000
H SPECIAL HAZARDOUS j. n.	(4) 5,000 (3) 5,000 (2) 7,500 (1) 11,500	(3) 4,000 (2) 6,000 (1) 8,300	(2) 6,000 (1) 7,500	(1) 5,000	(1) 5,000	(1) 5,000	(1) 5,000	Not Permitted	Not Permitted

TABLE 400 — NOTES

a. When five or more stories in height 2-hour fire-resistive floors shall be required over the basement or cellar.

c. The height of Type II construction for buildings of Group "B"—Business Occupancies shall not be limited provided the fire-resistance of all columns shall be not less than 3 hours and of the other structural members including floors shall be not less than shown in Chapter VI, but in no case less than 2 hours except that roofs shall be of not less than 1½ hours fire-resistive construction.

d. Floors located immediately above useable space in basement or cellars shall have a fire-resistive rating of not less than 1-hour except where an approved automatic sprinkler system is provided.

e. Buildings which are surrounded on all sides by a permanent open space of not less than sixty (60) feet, and are provided with an approved automatic sprinkler system shall not be limited in area.

f. 1-hour fire-resistive floors shall be provided.

g. May be increased 33⅓ percent for places of worship.

h. When three stories or more in height an approved automatic sprinkler system shall be installed throughout the building. (Does not apply to Public Parking Decks.)

i. When four or more stories in height an approved automatic sprinkler system shall be installed throughout the building.

j. No modification in area shall be permitted in Group "H" special hazardous occupancies.

k. The area of a one-story Type III, IV or V building may be increased (100%) one hundred percent if the building is surrounded on all sides by a permanent open space of not less than (60) sixty feet, and there are not less than two exits provided from each classroom, one of which opens directly to the exterior of the building. For other allowable increases, see Section 403 of Code.

l. See Section 404.4 herein for unlimited area provisions.

m. See Section 512.2 of Code for exception to area limitations.

n. See Chapter V for special detail requirements of Group "H" Special Hazardous Occupancies.

p. At least one hour interior fire resistive construction shall be used throughout all Group "C"—Schools two or more stories in height.

r. In Type I Construction, partitions, columns, trusses, girders, beams and floors may be reduced by one hour if the building is equipped with an approved automatic sprinker system throughout, but no component or assembly may be less than one hour.

s. See Section 507.2 for Public Parking Decks.

CHAPTER VI

CLASSIFICATION OF BUILDINGS BY CONSTRUCTION

SECTION 601 — CLASSIFICATION BY TYPE OF CONSTRUCTION

601.1 — TYPES

All buildings shall be classified into six general types according to the character of materials employed and their method of assembly, as follows:

TYPE I

TYPE II

TYPE III

TYPE IV
- ONE-HOUR PROTECTED
- UNPROTECTED

TYPE V
- ONE-HOUR PROTECTED
- UNPROTECTED

TYPE VI
- ONE-HOUR PROTECTED
- UNPROTECTED

601.2 — FIRE RESISTIVE REQUIREMENTS

All fire-resistive requirements are expressed in terms of the number of hours of satisfactory performance in accordance with the "Standard Methods of Fire Tests of Building Construction and Materials of the American Society for Testing and Materials, ASTM E119-69".

601.3 — MATERIALS AND CONSTRUCTION APPROVED FOR FIRE PROTECTION

(a) The degree of fire resistance and the materials, assemblies, and constructions providing such resistance shall be defined in Chapter X of this Code, except that other materials, assemblies, and constructions shall be approved, provided test data of a recognized engineering or testing laboratory are submitted, establishing that they develop the required fire-resistance ratings under tests made in accordance with the "Standard Methods of Fire Tests of Building Construction and Materials, ASTM E119-69".

(b) Where structural requirements necessitate assemblies providing greater fire resistance than specified in this Chapter, such structural requirements shall govern.

601.4 — FIRE DISTRICT — SECTION 301

601.5 — HEIGHT AND AREA — SECTIONS 402 TO 404 INCLUSIVE and TABLE 402

601.6 — REGULATIONS GOVERNING EXTERIOR USE OF COMBUSTIBLE MATERIALS (Refer to Code).

a. Skylights .... Section 707
b. Dormer Windows .... Section 709
c. Gutters and Leaders........ Section 712
d. Towers, Spires and Cupolas........ Section 713
e. Tanks .... Section 714
f. Cooling Towers .... Section 715
g. Roof Coverings........ Sections 301 and 706

601.7 — REGULATIONS GOVERNING INTERIOR USE OF COMBUSTIBLE MATERIALS (Refer to Code).

a. Group E, Assembly Occupancies........ Section 512
b. Floor Finish .... Section 704.2
c. Ceilings and Interior Wall Finish........ Section 704
d. In Group H, Special Hazardous Occupancy, only non-combustible finishes shall be used.
e. Vertical Openings .... Section 701
f. Partitions .... Section 702

601.8 — STRUCTURAL AND ENGINEERING REQUIREMENTS

a. Minimum Design Loads........ Chapter XII
b. Foundations .... Chapter XIII
c. Masonry and Veneered Walls........ Chapter XIV
d. Steel .... Chapter XV
e. Concrete .... Chapter XVI
f. Wood .... Chapter XVII
g. Lathing and Plastering........ Chapter XVIII
h. Safeguards During Construction........ Chapter XXI
i. Elevators and Escalators........ Chapter XXIV
j. Plastics .... Chapter XXVI
k. Glass .... Chapter XXVII
l. Aluminum .... Chapter XXVIII

601.9 — FIRE PROTECTIVE REQUIREMENTS (Refer to Code).

a. Roof Coverings........ Sections 301 and 706
b. Protection of Wall Openings........ Section 703
c. Firestopping........ Sections 705 and 1703.1
d. Special Occupancies........ Chapter V
e. Means of Egress........ Chapter XI
f. Plastics .... Chapter XXVI
g. Sprinklers and Standpipes........ Chapter IX
h. Separation of Furnace or Boiler Rooms........ Section 812

SECTION 602 – TYPE I CONSTRUCTION

Type I Construction, is that in which all exterior walls are masonry or reinforced concrete, or of other approved materials or combination of materials, and in which all the structural members are of noncombustible materials, and provide fire resistance not less than specified in Table 600. Fire retardant treated wood may be used as specified in Section 602.5.

SECTION 603 – TYPE II CONSTRUCTION

Type II Construction, is that in which all exterior walls are of masonry or reinforced concrete, or of other approved materials or combination of materials and in which all the structural members are of noncombustible materials, and provide fire resistance not less than specified in Table 600. Fire retardant treated wood may be used as specified in Table 603.5.

SECTION 604 – TYPE III CONSTRUCTION

604.1 — GENERAL

Type III Construction, is that type in which fire resistance is attained by the sizes of heavy timber members (sawn or glued-laminated) being not less than indicated in this Section and Table 600, or by providing fire resistance not less than one-hour where materials other than wood are used; by the avoidance of concealed spaces under floors and roofs; by the use of approved fastenings, construction details, and adhesives for structural members; and by providing the required degree of fire resistance in exterior and interior walls.

604.2 — COLUMNS

Columns shall be continuous or superimposed by means of properly designed reinforced concrete or metal caps, or by timber splice plates affixed to the columns by means of timber connectors or by other approved methods.

604.3 — WALLS

(a) Exterior walls shall extend not less than eighteen (18) inches above the roof, except that parapet walls need not be constructed on buildings where the roof slopes more than four (4) inches vertical to twelve (12) inches horizontal from the back of the exterior wall of such building, or where the exterior of such buildings is located thirty (30) feet or more distance from the property line or other building on the same property, or faces on an alley or public way thirty (30) feet or more in width.

(b) Party and fire walls shall extend not less than three feet above the roof.

604.4 — FLOOR DECKS

Heavy timber floors shall be of sawn or glued-laminated plank, splined, or tongued and grooved of not less than 3 inches, nominal, in thickness or of planks not less than 4 inches, nominal, in width set on edge and well spiked together. The planks shall be laid so that no continuous line of joints will occur except at points of support.

Planks shall be covered with 1 inch, nominal, tongued and grooved flooring laid crosswise or diagonally. Planks and flooring shall not extend closer than ½ inch to walls to provide an expansion joint, and the joint shall be covered at top and bottom.

604.5 — ROOF DECKS

(a) Heavy timber roof decks shall be sawn or glued-laminated, splined or tongue and grooved plank, not less than 2 inches, nominal, in thickness, one and one-eighth (1⅛) inches thick interior plywood (exterior glue), or of planks not less than 3 inches, nominal, in width, set on edge and spiked together, as required for floors.

(b) Other types of roof decking may be used that provide equivalent fire resistance, if not more combustible than two (2) inch nominal wood sheathing, and is used within spans which have been proved structurally safe by approved tests.

604.6 — FRAMING

(a) For minimum sizes of columns, trusses, girders and beams see Table 600.

(b) Framed or glued-laminated arches which spring from grade or the floor line and support floor loads shall be not less than 8 inches, nominal, in any dimension, or be of one-hour protected.

(c) Framed or glued-laminated arches for roof construction which spring from grade or the floor line and do not support floor loads shall have members not less than 6 inches, nominal, in width and not less than 8 inches, nominal, in depth for the lower half of the height and not less than 6 inches, nominal, in depth for the upper half.

(d) Framed or glued-laminated arches for roof construction which spring from the top of walls or wall abutments, framed timber trusses, and other roof framing which do not support floor loads, shall have members not less than 4 inches, nominal, in width and not less than 6 inches, nominal, in depth. Spaced members may be composed of two or more pieces not less than 3 inches, nominal, in thickness when blocked solidly throughout their intervening spaces or when such spaces are tightly closed by a continuous wood cover plate of not less than 2 inches, nominal, in thickness, secured to the underside of the members. Splice plates shall be no less than 3 inches, nominal, in thickness. When protected by approved automatic sprinklers under the roof deck, framing members shall be not less than 3 inches, nominal, in width.

SECTION 605 — TYPE IV CONSTRUCTION

Type IV Construction, is that in which all structural members, including wall framing, floors, roofs and their supports, shall be of steel, iron or other metal, or of other noncombustible materials, and in which the exterior surface of the building is of steel, iron or other metal, or of asbestos, masonry, reinforced concrete, or other noncombustible materials, and are fire protected as required in Table 600.

SECTION 606 — TYPE V CONSTRUCTION

Type V Construction, is that construction not meeting the requirements of Type III, but in which the exterior walls are of masonry or reinforced concrete or of approved materials or assembly of materials that provide fire resistance as required in Table 600, and in which the interior framing is partially or wholly of unprotected wood, or of unprotected iron or steel, and are fire protected as required in Table 600.

SECTION 607 — TYPE VI CONSTRUCTION

Type VI Construction, is that in which the enclosing walls are of wood or other combustible materials, including construction having exterior masonry veneer, stucco, or metal, which is dependent upon wood for support, stability or rigidity, and in which interior framing is of wood or other combustible materials and are fire protected as required in Table 600.

SECTION 608 — EXCEPTIONS TO FIRE PROTECTION

608.1 — ELEVATOR FRAMES

Structural members of frames for elevators will not be required to have the fire protection required for structural steel, provided such members are erected within an enclosure of the prescribed fire resistance rating. Section 701—Enclosure of Vertical Openings.

608.2 — LINTELS

Lintels over openings in walls shall be protected to provide a fire resistance rating at least equal to that required for beams, except that when such lintels are used over openings less than four (4) feet wide, such protection may be omitted. The outer member of an assembled steel lintel, which supports face masonry that is securely bonded to backing need not be protected, provided that the load carrying member of such lintel is protected as herein required.

608.3 — UNPROTECTED EXTERIOR WALLS OR PANELS

Unprotected walls or panels may be permitted in exterior nonbearing walls under the following conditions:

(1) Provided such walls are of non-combustible material or of exterior grade Fire Retardant Treated Wood.

(2) Provided such walls face a street or permanent open space of 30 feet or more in width.

(3) Provided that in buildings three stories or more in height, exterior openings located in a story above a Group D, Institutional; Group F, Storage; Group G, Industrial; Group H, Hazardous Occupancy are separated from such an occupancy by a 2-hour fire-resistive wall construction not less than 3 feet in height.

608.4 — WOOD VENEERS ON EXTERIOR WALL PANELS

(a) Wood veneers of not less than one (1) inch nominal thickness or three-eighths (⅜) inch exterior type plywood or particleboard may be used on exterior walls under the following conditions:

(1) The wall to which the veneer is attached faces a street or permanent open space of thirty (30) feet or more in width.

(2) The veneer does not exceed two stories in height, measured from grade.

(3) The veneer is attached to or furred from a non-combustible backing of the fire resistance required by other provisions of this Chapter.

(4) Where open or spaced wood veneers (without concealed spaces) are used, they shall not project more than twenty-four (24) inches from the building wall.

(b) Where the wood veneer is furred from the wall and forms a solid surface, the distance between the back of the veneer and the wall shall not exceed one and five-eighths (1⅝) inches and the space thereby created shall be firestopped in accordance with Section 1703 and arranged so that there will be no open space exceeding one hundred (100) sq. ft. Where wood furring strips are used, they shall be of approved wood or natural decay-resistance of pressure treated wood.

608.5 — UNUSABLE SPACE

In one hour fire resistant construction the ceiling may be omitted over unusable crawl space and flooring may be omitted when unusuable attic space occurs above.

SECTION 609 – MIXED TYPES OF CONSTRUCTION

609.1

When two or more types of construction not separated by fire walls occur in the same building not classified as a mixed occupancy, the entire building shall then be subject to the occupancy restrictions of the least fire resistive type of construction used in the building.

609.2

Where a building is constructed of more than one type of construction, the following limitations shall be observed:

TYPE I construction shall not be supported by any other type.

TYPE II construction shall not be supported by construction other than Type I or Type II.

TYPE III construction shall not be supported by construction other than Type I, Type II, or Type III.

TYPE IV construction shall not be supported by Type V, or Type VI.

TYPE V construction shall not be supported by Type VI.

SECTION 610 — BUILDINGS LOCATED ON THE SAME LOT

Where the exterior walls of two or more buildings located on the same lot face one another, and one of the walls is not constructed as required for a fire wall, a common-property line shall be assumed between them. The fire resistance requirements for such facing walls and for the protection of openings therein shall be the same as required by this code for walls and openings facing common-property lines, except as provided in Table 600.

TABLE 600 — FIRE PROTECTIVE REQUIREMENTS
REQUIRED FIRE RESISTANCE IN HOURS

STRUCTURAL MEMBER	Type I	Type II	Type III	Type IV		Type V		Type VI	
				Protected 1-Hour	Unprotected	1-Hour Protected	Unprotected	1-Hour Protected	Unprotected
	Sec. 602	Sec. 603	Sec. 604	Sec. 605	Sec. 605	Sec. 606	Sec. 606	Sec. 607	Sec. 607
Party and Fire Walls	4 hr. a.	4 hr. a.	4 hr. a.	4 hr. a.	4 hr. a.	4 hr. a.	4 hr. a.	4 hr. a.	4 hr. a.
Exterior Bearing Walls (See Section 608)	4 hr.	3 hr. b.	3 hr. Sec. 604.3	1 hr. c.	NC c. d.	3 hr. e.	3 hr. e.	1 hr.	0 hr. f.
Exceptions: Walls facing on street or public place				Sec. 507.2	Sec. 507.2				
30 ft. or more in width 50 ft. or more in width	3 hr. 2 hr.	2 hr. b.				2 hr. e.	2 hr. e.		
Where horizontal separation of more than 3 ft. is provided			2 hr.						
Exterior Non-Bearing Walls (See Section 608)	2 hr.	2 hr.	3 hr.	1 hr. c.	NC c. d.	1 hr. e. g.	NC e.g.	1 hr.	0 hr. f.
Exceptions:	Sec. 507.2	Sec. 507.2	Sec. 604.3						
Where protection of wall openings is not required by Sec. 703	1 hr.	1 hr.							

TABLE 600 — (Continued)

STRUCTURAL MEMBER	Type I	Type II	Type III	Type IV		Type V		Type VI	
				1-Hour Protected	Unprotected	1-Hour Protected	Unprotected	1-Hour Protected	Unprotected
	Sec. 602	Sec. 603	Sec. 604	Sec. 605	Sec. 605	Sec. 606	Sec. 606	Sec. 607	Sec. 607
Where horizontal separation of more than:									
3 ft. but less than 20 ft.			2 hr.						
20 ft. but less than 30 ft.			1 hr.						
30 ft. or more is provided	NC Sec. 608.3	NC Sec. 608.3	0 hr. Sec. 608.3	NC Sec. 608.3	NC Sec. 608.3	NC Sec. 608.3	NC Sec. 608.3		
Inner Court Walls	3 hr.	2 hr.	3 hr. Sec. 604.3	Same as for exterior walls					
Penthouse Walls	2 hr. h.	2 hr. h.	2 hr. h.	1 hr.	NC				
Partitions									
Interior Bearing	4 hr.	3 hr.	3 hr.	1 hr.	NC	1 hr. i	0 hr.	1 hr.	0 hr.
Interior Non-Bearing	← See Sections 412, 701 and 702 of Code →								
Columns			Sec. 604.2						
Supporting Masonry or Bearing Walls	4 hr.	3 hr.	2 hr.	2 hr. k.	2 hr. k.	m.	m.	1k.	1k.
Supporting Roofs Only	3 hr.	2 hr.	6x8 or 1 hr. j	1 hr.	NC	1 hr.	0 hr.	1 hr.	0 hr.
Other Columns	4 hr.	2 hr.	8x8 or 1 hr. j	1 hr.	NC	1 hr.	0 hr.	1 hr.	0 hr.

TABLE 600 — (Continued)

STRUCTURAL MEMBER	Type I	Type II	Type III	Type IV		Type V		Type VI	
				1-Hour Protected	Unprotected	1-Hour Protected	Unprotected	1-Hour Protected	Unprotected
	Sec. 602	Sec. 603	Sec. 604	Sec. 605	Sec. 605	Sec. 606	Sec. 606	Sec. 607	Sec. 607
Trusses, Girders and Beams			Sec. 604.6						
Supporting Masonry or Bearing Walls, Columns, Girders or Trusses	4 hr.	3 hr.	2 hr.	2 hr. k.	2 hr. k.	2 hr. k.	2 hr. k.	1 hr. k.	1 hr. k.
Supporting Roofs: Trusses and Girders	2 hr. n.	1 hr. p.	4x6 or 1 hr. j.	1 hr.	NC	1 hr.	0 hr.	1 hr.	0 hr.
Beams	1½ hr. n.	1 hr. p.	4x6 or 1 hr. j.	1 hr.	NC	1 hr.	0 hr.	1 hr.	0 hr.
Other Trusses	2½ hr.	1½ hr.	8x8 or 1 hr. j.	1 hr.	NC	1 hr.	0 hr.	1 hr.	0 hr.
Other Girders and Beams	2½ hr.	1½ hr.	6x10 or 1 hr. j.	1 hr.	NC	1 hr.	0 hr. q.	1 hr.	0 hr.
Arches			j. Sec. 604.6						
Floors	2½ hr.	1½ hr.	Sec. 604.4	1 hr.	NC	1 hr.	0 hr. q.	1 hr.	0 hr.
Roofs			j.						
Deck Construction	1½ hr. n.	1 hr. p.	Sec. 604.5	1 hr.	NC	1 hr.	0 hr.	1 hr.	0 hr.
High Above Floor			Sec. 604.5	1 hr.	NC	1 hr.	0 hr.	1 hr.	0 hr.

NC designates that no fire resistance rating is required, but construction shall be noncombustible

TABLE 600 — REFERENCE NOTES

a. Party and Fire Walls shall extend not less than three (3) feet above the roof, except that fire walls need not extend above the roof where the roof is of noncombustible construction for the area within forty (40) feet of each side of the wall.

b. Exterior walls of Type II buildings not over three (3) stories in height of Group A, B, C, D, or E occupancy may be constructed of framed wall assemblies, that have fire resistance against outside exposure, as specified in Table 600, and not less than 1½ hours fire resistance inside the building.

c. For Type IV buildings located within the fire district, exterior walls shall provide the following fire resistance against outside fire exposure:

	Required Fire Resistance (Hours)	
Exterior Walls		Fire District
(See Section 608.3)	2 c	All walls except:
	N.C.	Walls of one story buildings not exceeding 2000 sq. ft. in area facing and more than 15 feet from a common property or interior lot line.

d. Walls facing and within 8 ft. of common property lines shall provide 1-hour fire resistance against outside exposure.

e. Exterior walls shall extend not less than eighteen (18) inches above the roof, except that parapet walls need not be constructed on buildings where the roof slopes more than four (4) inches vertical to twelve (12) inches horizontal from the back of the exterior wall of such buildings or where the exterior wall of such building is located fifteen (15) feet or more distant from the property line or is located on an alley or public way of fifteen feet or more in width.

f. One hour fire resistive exterior walls shall be provided in one of the exposed walls where two or more buildings of Type VI construction on the same lot are located within six (6) feet of each other and all exterior walls located less than three (3) feet from property lines.

g. For Type V buildings located within the fire district, exterior walls shall provide 2 hour fire resistance.

h. Where penthouse walls are set back less than five (5) feet from exterior walls, they shall conform to the fire resistance requirements for exterior walls.

i. The use of combustible construction for interior bearing partitions shall be limited to the support of not more than 2 floors and a roof.

j. Where horizontal separation of twenty (20) feet or more is provided, wood columns, arches, beams, and roof deck conforming to heavy timber sizes may be used externally.

k. This requirement applies only to structural members supporting masonry walls, except that this does not apply in one (1) story buildings or where the only masonry supported is a masonry veneer.

m. Same rating as required for wall it supports.

n. In two (2) story buildings approved Fire Retardant Treated Wood may be used.

In buildings of Group C and E (School and Assembly) occupancies where structural members supporting a roof only are not less than 18' clear above any floor or balcony, one hour fire resistance shall be provided; where the clear distance is 25' or more, fire protection of structcral members supporting roof construction may be omitted.

p. In two (2) story buildings approved Fire Retardant Treated Wood may be used.

In buildings of Group C and E (School and Assembly) occupancies, fireproofing may be omitted where structural members supporting a roof only and are twenty (20) feet or more clear above any floor or balcony.

In one (1) story buildings structural members of heavy timber sizes may be used as an alternate to unprotected structural roof members.

q. When over one story in height, except one and two family dwellings, floors located immediately above useable spaces in basements or cellars and above furnaces shall have one hour fire protection except where basement or cellar is equipped with an approved automatic sprinkler system.

APPENDIX "I"

MOBILE HOMES STANDARDS

I-1 — PURPOSE. APPLICATION AND SCOPE

(a) The requirements set forth in this appendix shall apply specifically to all new mobile homes, and shall provide minimum standards for construction of mobile homes.

I-2 — GENERAL

Mobile homes shall comply with the Standard for mobile homes, ANSI Standard A119.1, 1969 and the amendments herein.

I-3 — STRUCTURAL DESIGN

All mobile homes shall be designed to meet loading conditions as specified in Part 1 of ANSI Standard A119.1. Non-designed units must meet the Southern Building Code requirements for small structures.

I-4 — DEFINITIONS

Mobile Home—A vehicular, portable structure built on a chassis and designed to be used without a permanent foundation as a dwelling when connected to indicated utilities. (Per ANSI Standard A119.1)

I-5 — IDENTIFICATION

A mobile home unit may bear the label or seal of compliance with ANSI Standard for Mobile Homes A119.1, of a recognized independent, engineering testing laboratory and/or agency having follow-up inspection services. Such label or seal shall be deemed to be in full compliance with the standards for mobile homes prescribed by the Southern Building Code Congress within the governmental jurisdiction which has approved such independent recognized testing laboratory for this service.

All mobile home units bearing such label or seal shall be acceptable within such city, county or state jurisdiction, which has approved such labeling service.

Any mobile home unit not bearing such label or seal of independent testing laboratory approved by the governmental jurisdiction wherein the unit is to be erected, is subject to inspection in the same manner as other structures.

I-6 — TESTS

All tests shall be conducted in accordance with Section 2504 (b).

APPENDIX "J"

STANDARDS FOR DEMOLITION

J-1 — PURPOSE AND SCOPE

The standards set forth in this appendix shall apply to the demolition of buildings or structures for which a permit is required under Section 105.1 of this Code.

J-2 — DEFINITION

Demolition. The act of demolishing or razing of a building or structure, or portion thereof to the ground level.

J-3 — TIME LIMIT

Notwithstanding the provision of Section 106.3 the Building Official may impose a time limit as an additional condition of a permit for completion of demolition work once such work shall have commenced, provided that for cause one or more extensions of time, for periods not exceeding thirty days each, may be allowed in writing by the Building Official.

J-4 — STANDARDS

(a) Demolition work having commenced shall be pursued diligently and without unreasonable interruption with due regard to safety. It is the intent of this section to limit the existence of an unsafe condition or nuisance on the premises during the period of demolition operations.

(b) Any surface holes or irregularities, wells, septic tanks, basements, cellars, sidewalk vaults, or coal chutes remaining after demolition of any building or structure shall be filled with material as approved by the Building Official, and shall be graded in such manner that will provide effective surface drainage.

(c) All debris and accumulation of material resulting from demolition of any building or structure shall be removed from the premises.

(d) All building sewers shall be effectively plugged with concrete at the property line, or as may be required by the Building Official.

J-5 — INSPECTIONS

In lieu of the inspections required by Section 108.2 (c) of this Code, the Building Official shall make the following inspections upon notification from the permit holder or his agent:

Initial Inspection: To be made after all utility connections have been disconnected and secured in such manner that no unsafe or unsanitary conditions shall exist during or remain after demolition operations.

Final Inspection: To be made after all demolition work is completed.

CROSS INDEX

A

Section

Section

Section

E

Section

Section

H

Section

Section

Section

R

Section

Section

U

V

Section

W

Section

AMENDMENTS

To

Standard Building Code

(FORMERLY SOUTHERN STANDARD BUILDING CODE)

1975 REVISION

TO

1973 EDITION

SOUTHERN BUILDING CODE CONGRESS INTERNATIONAL, INC.

3617 - 8TH AVENUE, SOUTH

BIRMINGHAM, ALABAMA 35222

PHONE (205) 252-8930

PREFACE

This pamphlet contains those amendments officially approved since the publication of the STANDARD BUILDING CODE 1973 EDITION.

These amendments represent those resulting from public hearings held at the Annual Research Conferences of the Southern Building Code Congress in Jacksonville, Florida, and Birmingham, Alabama.

The Standard Building Code is kept modernized and up-to-date through annual study and review. Amendments to the Building Code become official changes after public hearings before those active members in attendance at the Annual Research Conferences and ratification by written ballot of the active membership of the Southern Building Code Congress International.

It is suggested that cities, counties, states and other governmental entities maintain their Building Code up-to-date by reference to the identifying cover sheet and contents of this pamphlet.

PRICE $2.00 EACH

REFERENCE INDEX

Change addresses of following associations to read:

American Iron and Steel Institute
1000 16th Street, N.W., Washington, D.C. 20036

Gypsum Association
1603 Orrington Avenue, Evanston, Illinois 60201

National Fire Protection Association
470 Atlantic Avenue, Boston Massachusetts 02210

National Mineral Wool Insulation Association, Inc.
382 Springfield Avenue — Suite 312 Basset Building
Summit, New Jersey 07901

National Particleboard Association
2306 Perkins Place, Silver Springs, Maryland

105.3 — DRAWINGS AND SPECIFICATIONS — Page 1 - 7

Section 105.3(c) (1974 Meeting)

In paragraph (c) the second sentence, change the phrase "exceeding two stories" to read "three (3) stories or more".

Section 105.3—Add new subsection (d) to read as follows: (1973 Meeting)

(d) Plans for all buildings shall indicate how required structural and fire-resistive integrity will be maintained where a penetration of a required fire-resistive wall, floor or partition will be made for electrical, mechanical, plumbing and communication conduits, pipes and systems and also indicate in sufficient detail how the fire integrity will be maintained where required fire-resistive floors intersect the exterior walls.

107.4 — SCHEDULE OF PERMIT FEES — Page 1 - 10

Delete present Section 107.4 in its entirety and substitute new Section 107.4 as follows: (1973 Meeting)

107.4 — SCHEDULE OF PERMIT FEES

On all buildings, structures or alterations requiring a building permit, as set forth in Section 105, a fee for each building permit shall be paid as required at the time of filing application, in accordance with the schedule as established by the authority having jurisdiction. (See Appendix "M" for recommended fee schedules.)

CHAPTER II — DEFINITIONS

SECTION 201 — DEFINITIONS

Revise the following definitions: (1974 Meeting)

Revise the definition of "Public Parking Decks" to read as follows and relocate alphabetically:

AUTOMOBILE PARKING STRUCTURE — means a structure used for the parking or storage of automobiles.

Revise definition of Condominium as follows:

CONDOMINIUM DWELLING UNIT for the purpose of this code is an apartment as defined in this code.

Revise the definition of "Fire Retardant (Pressure Treated) Wood" so that it reads:

FIRE RETARDANT TREATED WOOD shall be defined as any wood product which, when impregnated with chemicals by an approved pressure process, or by other approved means during manufacture shall, when tested in accordance with ASTM test designation E-84, have a flamespread equivalent to 25 or less and show no evidence of progressive combustion when the test is continued for a period of 30 minutes.

All fire retardant treated wood products shall bear identification showing the fire hazard classification thereof issued by an approved agency having a re-examination service.

Where fire retardant treated wood products are to be subjected to sustained high humidity or exposed to weather it shall be further identified as Exterior type and to indicate that there is no increase in the listed fire hazard classification when subjected to the ASTM D2898-72, Methods of Test for Durability of Fire Retardant treatment of Wood.

Subsequent to treatment, fire retardant treated lumber and plywood shall be dried to a moisture content of 19% or less for lumber and 15% or less for plywood.

SMOKE DETECTOR—A smoke detector is an approved listed detector sensing either visible or invisible particles of combustion.

TOWNHOUSE is a single-family dwelling unit constructed in a series or group of attached units with property lines separating such units.

CHAPTER III — FIRE DISTRICT

SECTION 301 — GENERAL BUILDING RESTRICTIONS— WITHIN THE FIRE DISTRICT

301.3 — OTHER SPECIFIC REQUIREMENTS

Section 301.3(e) In the first line change "over one (1) story in height" to read "two (2) stories or more in height". (1974 Meeting)

Add new paragraph (f) Section 301.3 as follows: (1974 Meeting)

(f) TYPE IV BUILDINGS

Within the Fire District the exterior walls of Type IV Buildings shall provide the following fire resistance:

		FIRE DISTRICT
Exterior Walls (Except as provided in Section 608.3)	2	A.) Building More Than One Story OR More Than 2000 Sq. Ft. in Area.
		B.) One Story Buildings Not Exceeding 2000 Sq. Ft. in Area.
	2	1.) Walls facing and within 15 feet of common property or interior lot line.
	N.C.	2). Other walls.

CHAPTER IV — CLASSIFICATION OF BUILDINGS BY OCCUPANCY

403.7 — UNLIMITED AREAS — Page 4 - 4

Amend Section 403.7 by adding a third paragraph as follows: (1973 Meeting)

In Group "F" Storage and Group "G" Industrial occupancies where non-combustible products are manufactured or stored (such as metal processing and manufacturing plants, and metal products are not stored in combustible wrappings, containers or palletized) the sprinkler system may be omittted upon approval of the Building Official.

404.3 — SPECIAL REQUIREMENTS, GROUP "A" OCCUPANCY — Page 4 - 5

Add paragraph 3 to read as follows: (1973 Meeting)

3. Fire Detection Systems required for single, multiple family dwellings and townhouses. Section 1127.

412.8 — SEPARATION BETWEEN TOWNHOUSES — Page 4 - 25

Add the following new section: (1973 Meeting)

412.8 — SEPARATION BETWEEN TOWNHOUSES

Each townhouse unit shall be considered a separate dwelling unit and may be separated from adjoining units by the use of exterior walls meeting the requirements for zero clearance from property lines as required by the type of construction and fire protection requirements, or when not more than three (3) stories in height, may be separated by a single wall meeting the following requirements.

1. Such wall shall provide not less than two (2) hours fire resistance and shall not contain plumbing, piping, ducts or electrical hardware.

2. Such wall shall extend from the foundation to the underside of the roof sheathing, and the underside of the roof shall have at least one hour fire resistance for a width not less than four (4) feet on each side of such wall.

3. Each dwelling unit sharing such wall shall be designed and constructed to maintain its structural integrity independent of the unit on the opposite side of the wall.

CHAPTER V — SPECIAL OCCUPANCY REQUIREMENTS

SECTION 508 — PUBLIC GARAGES

Combine paragraphs (h) and (i) as follows: (1974 Meeting)

Heating equipment, other than unit heaters suspended at least 8' above the garage floor, shall be placed in a separated room cut off by four-hour fire rated construction and 4" reinforced concrete floor and ceiling construction.

Entrance shall be from the outside or by means of a doorway with sill raised at least 8" above the garage floor level or through a vestibule providing two doorway separations.

Doors for the protection of the interior openings shall have a minimum fire protection rating of three hours.

Nothing in this section shall prohibit the installation of other than unit heaters in garages when the heating equipment is listed for such use and is installed in accordance with its listing and the provisions of this code and other applicable ordinances.

Delete Section 504.3 and Section 510 and substitute the following: (1974 Meeting)

SECTION 510 – REVIEWING STANDS, GRANDSTANDS AND BLEACHERS

a) Scope. Reviewing stands, grandstands and bleachers shall conform to the provisions of this Section.

b) Definitions:

BLEACHERS. Bleachers are seating facilities without backrests in which less than 3 square feet is assigned per person for computing the occupant load.

FOOTBOARDS. Footboards are that part of a raised seating facility other than an aisle or cross aisle upon which the occupant of the stands walks to reach a seat or upon which he may rest his feet.

GRANDSTANDS. Grandstands are seating facilities wherein an area of 3 square feet or more is provided for each person and where rows of seats are on an increasing height level by a terraced arrangement of walking platforms and seats.

OPEN AIR GRANDSTANDS AND BLEACHERS. Open air grandstands and bleachers shall refer to seating facilities which are located so that the side toward which the audience faces is unroofed and without an enclosing wall.

PERMANENT. Permanent stands are those seating facilities which remain at a location for more than 90 days.

REVIEWING STANDS. Reviewing stands are elevated platforms accommodating not more than 50 persons. Seating facilities, if provided, are normally in the nature of loose chairs. Reviewing stands accommodating more than 50 persons shall be regulated as grandstands.

SAFE DISPERSAL AREA. Safe dispersal area shall mean an area which will accommodate a number of persons equal to the total capacity of the stand and building which it serves in such a manner that no person within the area need be closer than 50 feet from the stand or building. Dispersal areas are based upon an area of not less than 3 square feet per person.

TEMPORARY. Temporary seating facilities are those which are intended for use at a location for not more than 90 days.

c. Height of Grandstands and Bleachers:

Grandstands and bleachers, other than those of open skeleton frame type, when more than one story in height or 400 square feet in area, shall be of not less than one-hour fire resistive construction. When the space under such structures is used for any purpose, it shall be separated from all parts of the grandstand or bleacher, including exits, by walls, floors and ceilings of not less than one-hour fire resistive construction.

EXCEPTION: 1. Exits under temporary grandstands need not be separated.

2. The underside of continuous steel deck grandstands, when erected outdoors, need not be fire protected when occupied for public toilets.

Grandstands or bleachers employing combustible members in the structural frame shall be limited to 11 rows or 9 feet in height. Seatboards, toeboards, bearing or base pads and footboards may be of combustible materials.

d) Design Requirements. See Chapter XII.

e) General Requirements. 1. Row spacing. There shall be a clear space of not less than 12 inches measured horizontally between the back or backrest of each seat and the front of the seat immediately behind it. The minimum spacing of rows of seats measured from back to back shall be:

A. Twenty-two inches for seats without backrests.

B. Thirty inches for seats with backrests.

C. Thirty-three inches for chair seating.

2. Rise between rows. The maximum rise from one row of seats to the next shall not exceed 16 inches unless the seat spacing from back to back measured horizontally is 40 inches or more.

3. Seating capacity determination. Where bench type seating is used, the number of seats shall be based on one person for each 18 inches of length of the bench.

4. Aisles. A. Aisles required. Aisles shall be provided in all seating facilities except that aisles may be omitted when all of the following conditions exist:

(i) Seats are without backrests.

(ii) The rise from row to row does not exceed 12 inches per row.

(iii) The number of rows does not exceed 11 in height.

(iv) The top seating board is not over 10 feet above grade.

(v) The first seating board is not more than 20 inches above grade.

B. Obstructions. No obstruction shall be placed in the required width of any aisle or exitway.

C. Width. Aisles serving seats on both sides shall have a minimum width of 44 inches. When serving seats on only one side, the aisle shall have a minimum width of 36 inches.

5. Cross aisles and vomitories. Cross aisles and vomitories shall be not less than 54 inches in clear width and shall extend to an exit, enclosed stairway or exterior perimeter ramp.

6. Stairs and ramps. All stairs have a maximum rise of every step in a stairway of 7½ inches and a run of not less than 10 inches with a ¼ inch tolerance.

Ramps shall not exceed a slope of one vertical to ten horizontal and shall have landings of at least 5 feet length measured in the direction of the ramp run at the top and bottom and at least one intermediate landing shall be provided for each 5 feet of rise.

7. Guardrails. Perimeter guardrails or enclosing walls shall be provided for all portions of elevated seating facilities more than 30 inches above grade. Guardrails shall be not less than 42 inches high and shall comply with Section 1204.2.

EXCEPTION: Guardrails at the front and not at the end of an aisle may have a height of 36 inches with a midrail.

8. Toeboards. A 4-inch high vertical barrier shall be installed along the edge of walking surfaces wherever guardrails are required.

9. Footboards. Footboards shall be provided for all rows of seats above the third row or beginning at such a point where the seating plank is more than 2 feet above grade. Where the same platform is used for both seating and footrests, footrests will not be required provided each level or platform is not less than 24 inches wide. Footboards in bleachers at a level below the seat board it serves are not to be considered as walking platforms but shall be not less than a structural grade of 2-inch by 8-inch lumber or equivalent. When bleachers exceed 11 rows in height, a walking platform not less than 18 inches in width shall be provided.

f). Special Requirements. 1. Grandstands and bleachers within buildings. Grandstands within a building shall comply with the other applicable sections of Chapter XI.

EXCEPTION: When seats are without backrests there may be nine seats between any seat and an aisle.

2. Open air grandstands and bleachers. A. Number of seats between aisles. The number of seats between any seat and an aisle shall not be greater than 20 when the seats are without backrests and nine if the seats have backrests.

B. Dead ends. Dead ends in vertical aisles shall not exceed a depth of 16 rows for permanent grandstands and 26 rows for temporary grandstands.

C. Distance to exit. The line of travel from any seat to a safe dispersal area exit ramp, enclosed stairway or vomitory shall not be more than 200 feet. When the seats have no backrests, the distance may be measured by direct line.

D. Safe dispersal area. Each safe dispersal area shall have a minimum of two exits. If more than 6000 persons are to be accommodated within a dispersal area, there shall be a minimum of of three exits and for more than 9000 persons there shall be at least four exits. The aggregate clear width of exits from a safe dispersal area shall be determined on the basis of not less than one exit unit of 22 inches for each 500 persons to be accommodated and no exit shall be less than 44 inches in width.

E. Two exits required. Two exits shall be provided from every stand which accommodates more than 300 persons.

F. Three exits required. Three exits shall be required where a grandstand or section thereof accommodates more than 1000 persons.

G. Four exits required. Four exits shall be provided where a grandstand or section thereof accommodates more than 3000 persons.

H. Determination of exit width. The total width of exits in feet shall be not less than the total occupant load served divided by 150 when exiting by stairs and divided by 200 when exiting by ramps, corridors, tunnels or vomitories.

I. Minimum exit width. No exit shall be less than 44 inches in width.

SECTION 512.2 — EXCEPTION TO AREA LIMITATIONS

Add a new second paragraph as follows: (1974 Meeting)

One story buildings used for participation sports activities such as swimming, tennis, skating and similar activities, limited in occupant content to those participating in the sports activity, and with no spectator seating permitted, may be unlimited in area when of Types III, IV or V Construction and are surrounded on all sides by not less than thirty (30) feet of permanent public space.

Revise Section 518 as follows: (1974 Meeting)

SECTION 518 – SPECIAL PROVISIONS FOR HIGH RISE GROUP "A" OCCUPANCY AND GROUP "B-1" OFFICE BUILDINGS

a.) **Scope**

These requirements shall apply to all apartment houses, hotels and office buildings having floors used for human occupancy located more than six stories or 75 feet above the average grade. Such buildings will be required to provide either sprinkler protection or safe areas of refuge (compartmentation).

Such buildings more than 12 stories or 150 feet in height shall be provided with a complete automatic sprinkler system conforming to the requirements of this Section.

All mechanical and electrical equipment shall be approved and installed in accordance with approved plans and specifications pursuant to this Section and shall be tested and proved to be in proper working condition to the satisfaction of the Building Official before issuance of the Certificate of Occupancy.

b.) **Automatic Fire Sprinklers**

An approved complete automatic sprinkler system shall be provided throughout the building. The sprinkler system shall be designed using the parameters set forth in the NFPA 13-1974 and the following:

1. Shut off valves and water flow devices shall be provided at the riser connection on each floor. In addition to actuating a local alarm on the floor upon which the water flow is detected, such valves shall be supervised by a continuously manned control station or by a central station. The sprinkler riser may be combined with the Fire Department standpipe riser.

2. Each floor shall be supplied by two or more risers. In Seismic Zones 2 and 3, an approved check valve shall be provided at each point of connection of the sprinkler system to the standpipe in such a manner that one riser can remain operational if a break occurs in the other riser.

3. Piping may be of copper tube or steel pipe with no copper tube less than ¾" diameter and no steel pipe less than 1" diameter.

4. Pitching of lines is not required.

5. In Seismic Zones 2 and 3, an on-site supply of water equal to a 20 minute demand or 15,000 gallons on a combined sprinkler and standpipe, whichever is smaller, shall be provided. This supply shall be automatically available if the principal supply fails.

6. A fire pump shall be required when the off-site water supply and pressure is not adequate for the demands of the sprinkler system and the Fire Department hose outlets combined unless the system is provided with an alternate source of water such as a reservoir, gravity tank or pressure tank meeting the requirements of the cited standard.

The required capacity of the fire pumps shall be maintained when there is a 15 PSI drop from the normal operating pressure of the water source.

c.) **Smoke Detection Systems**

At least one approved listed smoke detector shall be installed:

1. In every room containing mechanical equipment, boiler, electrical, telephone, elevator or similar equipment.

2. At each opening where a plenum connects to a vertical shaft.

3. In the return air portion of every air conditioning and mechanical ventilation system that serves other than the floor on which the equipment is located.

4. In every elevator lobby.

The actuation of any detector shall operate the voice alarm system and shall cause such other operations as are necessary to prevent the recirculation of smoke and any other functions required by this Code.

d). **Alarm and Communication Systems**

These systems shall be so designed and installed that damage to any station will not affect the operation of the remainder.

The voice alarm and public address system may be a combined system. When approved, the Fire Department communications system may be combined with the voice alarm system and the public address system.

Three communication systems shall be provided as follows:
1. Voice Alarm System

The operation of any smoke detector, sprinkler, waterflow device or manual fire alarm system, shall automatically activate a voice alarm system. The voice alarm system shall provide a predetermined message on a selective basis to the area where the alarm originated.

The voice alarm shall provide information and give direction to the occupants.

The Central Control Station shall contain controls for the voice alarm system so that a selective or general voice alarm may be manually initiated.

The system shall be continuously electrically supervised against component failure of the audiopath including amplifiers, speaker wiring, switches and electrical contacts and shall detect opens and shorts which might impair the function of the system.

Activation of the system shall automatically sound an alert signal to the desired areas followed by voice instructions giving appropriate information.

The alarm shall be designed to be heard clearly by all occupants within the building or designated portions thereof as is required for the Public Address System.

2. Public Address System

A public address communication system designed to be clearly heard by all occupants of the building shall operate from the Central Control Station. It shall be established on a selective or general basis to the following terminal areas:

a. Elevators

b. Elevator lobbies

c. Corridors

d. Exit stairways

e. Rooms and tenant spaces exceeding 1,000 sq. ft. in area

f. Dwelling units in apartment houses

g. Hotel guest rooms or suites

3. Fire Department Communication System

A two-way fire department communication system shall be provided for Fire Department use. It shall operate between the Central Control Station and every elevator, elevator lobby, entry to every enclosed exit stairway and in corridors.

e.) **Central Control Station**

A Central Control Station for Fire Department operations, separated from the remainder of the building by a two hour fire resistive occupancy separation, shall be provided in an approved location. It shall contain:

1. The voice alarm and public address system panels.
2. The Fire Department communications panel.
3. Fire detection and alarm system annunciator panels.
4. Status indicator and controls for elevators.
5. Status indicators and control for air handling systems.
6. Controls for unlocking all stairway doors simultaneously.
7. Sprinkler valve and waterflow detector display panels.
8. Standby power controls.
9. A public service telephone.

f.) Smoke Control

Natural or mechanical ventilation for the removal of products of combustion shall be provided in every story and shall consist of one of the following:

1. Panels or windows in the exterior walls which can be opened from an approved location other than the fire floor. Such venting facilities shall be provided at the rate of 20 square feet per 50 lineal feet of exterior wall in each story and distributed around the perimeter at not more than 50 feet intervals. Such panels and their controls shall be clearly identified.

EXCEPTION: When a complete automatic fire sprinkler system is installed, windows or panels manually openable from within the fire floor or approved fixed tempered glass may be used in lieu of the remotely operated openable panels and windows.

2. When a complete and approved automatic sprinkler system is installed, the mechanical air handling equipment may be designed to accomplish smoke removal. Under fire conditions, the return and exhaust air shall be moved directly to the outside without recirculation to other sections of the building. The supply fans shall provide 100% outside air. The area involved shall have a minimum of an air change every 10 minutes to meet this requirement.

3. Any other approved design which will produce equivalent results and which is acceptable to the Building Official.

g.) Elevators

General

Elevators and elevator lobbies shall comply with the following provisions:

NOTE: A bank of elevators is a group of elevators or a single elevator controlled by a common operating system, that is, all those elevators which respond to a single call button constitute a bank of elevators. There is no limit on the number of cars which may be in a bank or group but there may not be more than four cars within a common hoistway.

1. Openings in the elevator lobby shall be limited to those required for access to the elevators and for egress from the building. The lobby and openings therein shall be designed as required for a corridor.

EXCEPTION: When the building is provided with a service elevator serving every story and isolated from the passenger elevators, and that service elevator is separated from the remainder of the building by one hour fire-resistive construction with protected openings, the passenger elevator lobby doors may be eliminated.

The elevator lobby shall have no openings except those serving the elevator cars. Exit stairways, chutes, janitor closets, guest rooms, service rooms, etc., shall not open into the elevator lobby.

2. Each elevator lobby or entrance area shall be provided with an approved and listed smoke detector located on the ceiling. When activated, elevator doors are not to open and all cars serving that lobby are to return to the ground floor. The smoke detector is to operate by the time that the optical density exceeds 0.03 per foot. The detector may serve to close the lobby doors.

3. Each elevator call station shall have an illuminated sign which flashes on and off to show the words "EMERGENCY—USE THE EXIT STAIRS" when an elevator lobby smoke detector is activated. The signs on all floors served by that group shall flash. The words shall be in minimum ½ inch block letters.

4. Elevator hoistways shall not be vented through an elevator machine room. Cable slots entering the machine room shall be sleeved beneath the machine room floor to inhibit the passage of smoke into the machine room.

5. At least one elevator car in each building serving all floors shall have a minimum inside car platform of 4' 3" deep by 6' 8" wide with a minimum clear opening width of 42" unless otherwise designed to provide equivalent utility, to accommodate an ambulance stretcher (minimum size 22" x 78") in its horizontal position. This elevator shall be identified.

h.) Standby Power and Light

1. On-site Generated Power

A permanently installed standby power generation system conforming to NFPA 70-1971 shall be provided. The system shall be equipped with suitable means for automatically starting the generator set upon failure of the normal electrical service and for automatic transfer and operation of electrical functions. System supervisions devices with manual start and transfer features shall be provided at the Central Control Station.

An on-premise fuel supply sufficient for not less than two hours full demand operation of the system shall be provided. All power lighting, signal and communication facilities provided under the requirements of this section shall be transferable to the standby power system.

2. Standby Power Loads

The following loads are classified as standby power loads. The transition time from the instant of failure of the normal power source to the generator source shall not exceed 60 seconds. The standby power load shall be sized to supply the following:

a. The elevator required under g. 5.

b. Mechanical air handling systems required to be operating during an emergency.

c. Fire pumps.

3. Emergency Power Loads

The transition time from the instant of failure of the normal power source to the generating source shall not exceed 10 seconds unless otherwise specifically noted. The following loads are classified as emergency Power Loads:

a. Voice alarm system

b. Voice communication system

c. Fire alarm systems

d. Fire detection systems

e. Elevator car lighting

f. Escape route lighting and exit sign illumination which shall both operate within one second.

4. Wiring shall be run in fire resistive enclosures containing no other combustibles.

i.) Exits

1. All stairwav doors which are to be locked from the stairwav side shall have the capabilitv of being unlocked simultaneously without unlatching upon a signal from the Central Control Station.

2. Emergency telephones available to the public shall be provided at not less than every fifth floor in each required stairway.

3. Required stairways shall be pressurized to a minimum of 0.15 and a maximum of 0.50 inches of water column when the shaft is provided with a minimum of 2500 cfm mechanical exhaust at the roof when a smokeproof enclosure is not provided.

j.) Seismic Considerations

In Seismic Zones 2 and 3 the anchorage of the following mechanical and electrical equipment required by the section shall be designed in accordance with Section 1206 for a lateral force based on a "Cp" value of 0.5 unless data substantiating a lesser value is furnished:

1. Elevator drive and suspension systems.
2. Standby power and lighting facilities.
3. Fire pumps and other fire protection equipment.

k.) Areas of Refuge (Compartmentation) Alternate

1. In lieu of complete automatic protection, areas of refuge may be provided in buildings which do not exceed 12 stories or 150 feet in height. To provide such areas of refuge, a story exceeding 30,000 square feet in area shall be divided into two or more areas of approximately the same size but not exceeding 30,000 square feet each by means of a horizontal exit wall.

2. When a story or compartmented portion of a story is less than 30,000 square feet in area, adjacent stories or compartmented portions thereof may be combined to form one area of refuge.

The area of refuge provided by combining floors shall not exceed 30,000 square feet or five stories. To isolate one combination of stories from another, the separation may be obtained by one of the following:

a. Interrupting the stairshaft with smoke barriers.

b. Discontinuous stairshafts with flights of stairs interrupted by a corridor connecting flights of stairs.

c. All stairways constructed as smokeproof enclosures. Any combination of horizontal and vertical separations may be employed which will limit the size of any area of refuge to 30,000 square feet.

3. Openings in exterior walls, where such openings are within five feet of each other horizontally on adjacent floors or located vertically above one another, shall be protected by approved flame barriers either extending 30 inches beyond the exterior wall in the plane of the floor or by approved vertical panels not less than three feet in height above the floor.

4. Horizontal exit walls used for compartmenting a building shall have a fire-resistance rating of not less than two hours. Duct penetrations of this wall shall not be permitted. Piping and conduit may penetrate or pass through the wall only if the openings are caulked with impervious noncombustible materials sufficiently tight to prevent the transfer of smoke or combustion gases from one side of the wall to the other and are so maintained. The fire door serving as the horizontal exit between compartments shall be so installed, fitted and gasketed that it will provide a substantial barrier to the passage of smoke.

5. The fire resistance of the floor or the floor-ceiling assembly shall extend to and be tight against the exterior wall so that the fire resistive integrity is maintained. No penetrations or other installations which will impair the fire-resistive integrity of the floor or floor-ceiling assembly are permitted.

6. A manual fire alarm system (pull boxes) shall be provided.

l.) Alternates Permitted

1. When a complete approved automatic sprinkler system, complying with this Section, is installed in a building, the following modifications of Code requirements are acceptable:

a. Fixed tempered glass may be used in lieu of openable panels for smoke control purposes.

b. The 1½" wet standpipe, hoze and nozzles is not required, however, the Fire Department risers and hose connections in required stairways are to be provided.

c. The manually operated fire alarm system required in the compartmented building is not required.

d. Spandrel walls, eyebrows and compartmentation are not required, however, the fire resistance of the floors and juncture of exterior walls with each floor must be maintained.

e. Fire dampers, other than those needed to protect floor-ceiling assemblies (to maintain the fire resistance of the assembly) are not required except for those which may be necessary to by-pass smoke to the outside; to convert from recirculated air to 100% outside air; and that which may be required to protect the fresh air supply intake against smoke which may be outside the building.

m.) **Special Provisions for Apartment Houses and Hotels**

In apartment houses and hotels, where each dwelling unit or guest room is provided with an approved and listed smoke detector to arouse the occupants (local alarm) and an approved heat detector is connected to both a central station supervisory service and to the central control station annunciator panel, sprinklers may be omitted within the guest room, suite or dwelling unit. Such limited sprinkler protection does not permit any of the alternates listed under (1) other than the omission of the manual fire alarm system.

CHAPTER VI — CLASSIFICATION OF BUILDINGS BY CONSTRUCTION

Substitute the following for Chapter VI in its entirety: (1974 Meeting)

CHAPTER VI — CLASSIFICATION OF BUILDINGS BY CONSTRUCTION

SECTION 601 — CLASSIFICATION BY TYPE OF CONSTRUCTION

601.1 — TYPES

Every building shall be classified by the Building Official into one of the types of construction as set forth in this section:

TYPE I

TYPE II

TYPE III

TYPE IV
- ONE-HOUR PROTECTED
- UNPROTECTED

TYPE V
- ONE-HOUR PROTECTED
- UNPROTECTED

TYPE VI

ONE-HOUR PROTECTED
UNPROTECTED

601.2 — FIRE RESISTIVE REQUIREMENTS

All fire-resistive requirements are expressed in terms of the number of hours of satisfactory performance in accordance with the "Standard Methods of Fire Tests of Building Construction and Materials of the American Society for Testing and Materials, ASTM E119-71".

601.3 — MATERIALS AND CONSTRUCTION APPROVED FOR FIRE PROTECTION

(a) The degree of fire resistance and the materials, assemblies, and constructions providing such resistance shall be defined in Chapter X of this Code, except that other materials, assemblies, and constructions shall be approved, provided test data of a recognized engineering or testing laboratory are submitted, establishing that they develop the required fire-resistance ratings under tests made in accordance with the "Standard Method of Fire Tests of Building Construction and Materials, ASTM E119-71".

(b) Where structural requirements necessitate assemblies providing greater fire resistance than specified in this Chapter, such structural requirements shall govern.

601.4 — FIRE DISTRICT — SECTION 301

601.5 — HEIGHT AND AREA — CHAPTER IV

601.6 — REGULATIONS GOVERNING EXTERIOR USE OF COMBUSTIBLE MATERIALS

a. Skylights .... Section 707

b. Dormer Windows .... Section 709

c. Gutters and Leaders .... Section 712

d. Towers, Spires and Cupolas .... Section 713

e. Tanks .... Section 714

f. Cooling Towers .... Section 715

g. Roof Coverings .... Sections 301 and 706

601.7 — REGULATIONS GOVERNING INTERIOR USE OF COMBUSTIBLE MATERIALS

a. Group E, Assembly Occupancies .... Section 512

b. Floor Finish .... Section 704.2

c. Ceilings and Interior Wall Finish .... Section 704

d. In Group H, Special Hazardous Occupancy, only non-combustible finishes shall be used.

e. Vertical Openings .... Section 701

f. Partitions .... Section 702

601.8 — STRUCTURAL AND ENGINEERING REQUIREMENTS

a. Minimum Design Loads .... Chapter XII

b. Foundations .... Chapter XIII

c. Masonry and Veneered Walls .... Chapter XIV

d. Steel .... Chapter XV

e. Concrete .... Chapter XVI

f. Wood .... Chapter XVII

g. Lathing and Plastering .... Chapter XVIII

h. Safeguards During Construction .... Chapter XXI

i. Elevators and Escalators .... Chapter XXIV

j. Plastics .... Chapter XXVI

k. Glass .... Chapter XXVII

l. Aluminum .... Chapter XXVIII

601.9 — FIRE PROTECTIVE REQUIREMENTS

a. Roof Coverings .... Sections 301 and 706

b. Protection of Wall Openings .... Section 703

c. Firestopping .... Sections 705 and 1703.1

d. Special Occupancies .... Chapter V

e. Means of Egress .... Chapter XI

f. Plastics .... Chapter XXVI

g. Sprinklers and Standpipes .... Chapter IX

h. Separation of Furnace or Boiler Rooms .... Section 312

SECTION 602 — TYPE I CONSTRUCTION

Type I Construction is that type of construction in which the structural members including exterior walls, interior bearing walls, columns, floors and roofs are of noncombustible materials and are protected so as to have fire resistance not less than that specified for the structural elements as specified in Table 600. Reference Note: Materials may be used as specified in Table 600, or as permitted in this chapter.

SECTION 603 — TYPE II CONSTRUCTION

Type II Construction is that type of construction in which the structural members including exterior walls, interior bearing walls, columns, floors and roofs are of noncombustible materials and are protected so as to have fire resistance not less than that specified for the structural elements as specified in Table 600. Reference Note: Materials may be used as specified in Table 600, or as permitted in this chapter.

SECTION 604 — TYPE III CONSTRUCTION

604.1 — GENERAL

Type III Construction, is that type in which fire resistance is attained by the sizes of heavy timber members (sawn or glued-laminated) being not less than indicated in this Section, or by providing fire resistance not less than one-hour where materials other than wood of heavy timber sizes are used; by the avoidance of concealed spaces under floors and roofs; by the use of approved fastenings, construction details and adhesives for structural members; and by providing the required degree of fire resistance in exterior and interior walls. Reference Note: Materials may be used as specified in Table 600, or as permitted in this chapter.

604.2 — COLUMNS

(a) Wood columns may be sawn or glued-laminated and shall be not less than 8 inches, nominal, in any dimension when supporting floor loads, and not less than 6 inches, nominal, in width and 8 inches, nominal, in depth when supporting roof and ceiling loads only.

(b) Columns shall be continuous or superimposed throughout all stories by means of reinforced concrete or metal caps with brackets, or shall be connected by properly designed steel or iron caps, with pintles and base plates, or by timber splice plates affixed to the columns by means of metal connectors housed within the contact faces, or by other approved methods.

604.3 — FLOOR FRAMING

(a) Beams and girders of wood may be sawn or glued-laminated and shall be not less than 6 inches, nominal, in width and not less than 10 inches, nominal, in depth.

(b) Framed or glued-laminated arches which spring from the floor line and support floor loads shall be not less than 8 inches, nominal, in any dimension.

(c) Framed timber trusses supporting floor loads shall have members of not less than 8 inches, nominal, in any dimension.

604.4 — ROOF FRAMING

(a) Framed or glued-laminated arches for roof construction which spring from the floor line and do not support floor loads shall have members not less than 6 inches, nominal, in width and 8 inches, nominal, in depth for the lower half of the height and not less than 6 inches, nominal, in any dimension for the upper half of the height.

(b) Framed or glued-laminated arches for roof construction which spring from the top of walls or wall abutments, framed timber trusses and other roof framing which do not support floor loads, shall have members not less than 4 inches, nominal, in width and not less than 6 inches, nominal, in depth. Spaced members may be composed of two or more pieces not less than 3 inches, nominal, in thickness when blocked solidly throughout their intervening spaces or when such spaces are tightly closed by a continuous wood cover plate of not less than 2 inches, nominal, in thickness, secured to the underside of the members. Splice plates shall be no less than 3 inches, nominal, in thickness. When protected by approved automatic sprinklers under the roof deck, such framing members shall be not less than 3 inches, nominal, in width.

604.5 — CONSTRUCTION DETAILS

(a) Wall plate boxes of self-releasing type, or approved hangers, shall be provided where beams and girders enter masonry. An air space of ½ inch shall be provided at the top, ends and sides of the member unless approved durable or treated wood is used.

(b) Girders and beams shall be closely fitted around columns and adjoining ends shall be cross-tied to each other, or inter-tied by caps or ties, to transfer horizontal loads across the joint. Wood bolsters may be placed on tops of columns which support roof loads only.

(c) Where intermediate beams are used to support floors, they shall rest on top of the girders, or shall be supported by ledgers or blocks securely fastened to the sides of the girders, or they may be supported by approved metal hangers into which the ends of the beams shall be fitted closely.

(d) Columns, beams, girders, arches and trusses of material other than wood shall have a fire resistance rating of not less than one hour.

(e) Wood beams and girders supported by walls required to have a fire resistance rating of two hours or more shall have not less than 4 inches of solid masonry between their ends and the outside face of the wall, and between adjacent beams.

(f) Adequate roof anchorage shall be provided.

604.6 — FLOOR DECKS

Floors shall be without concealed spaces. They shall be of sawn or glued-laminated plank, splined, or tongue and grooved, of not less than 3 inches, nominal, in thickness, or of planks not less than 4 inches, nominal, in width set on edge and well spiked together. The planks shall be laid so that no continuous line of joints will occur except at points of support and they shall not be spiked to supporting girders. Planks shall be covered with 1-inch, nominal, tongue and grooved flooring laid crosswise or diagonally or with ½ inch plywood. Planks and flooring shall not extend closer than ½ inch to walls to provide an expansion joint, and the joint shall be covered at top or bottom.

604.7 — ROOF DECKS

Roofs shall be without concealed spaces and roof decks shall be sawn or glued-laminated, splined or tongue and grooved plank, not less than 2 inches, nominal, in thickness, or of planks not less than 3 inches, nominal, in width set on edge and spiked together as required for floors of 1⅛ inch tongue and grooved plywood bonded with exterior glue. Other types of decking may be used when approved by the Building Official.

SECTION 605 — TYPE IV CONSTRUCTION

Type IV Construction is construction in which the structural members including exterior walls, interior bearing walls, columns, floors and roofs are of noncombustible materials. Type IV Construction may be protected or unprotected. Fire resistance requirements for structural elements of Type IV Construction shall be as specified in Table 600. Reference Note: Materials may be used as specified in Table 600, or as permitted in this Chapter.

SECTION 606 – TYPE V CONSTRUCTION

Type V Construction is construction in which the exterior bearing and nonbearing walls and bearing portions of interior walls are of noncombustible material and have fire resistance not less than that specified in Table 600; and floors, roofs and interior framing are wholly or partly of wood or other approved materials. Type V Construction may be either protected or unprotected. Fire resistance requirements for structural elements of Type V Construction shall be as specified in Table 600. Reference Note: Materials may be used as specified in Table 600, or as permitted in this Chapter.

SECTION 607 – TYPE VI CONSTRUCTION

Type VI Construction is construction in which the exterior bearing and nonbearing walls and partitions, floors and roofs and their supports are wholly or partly of wood or other approved materials. Type VI Construction may be either protected or unprotected. Fire resistance requirements for structural elements of Type VI Construction shall be as specified in Table 600. Reference Note: Materials may be used as specified in Table 600, or as permitted in this Chapter.

SECTION 608 – EXCEPTIONS TO FIRE PROTECTION

608.1 — ELEVATOR FRAMES

Structural members of frames for elevators will not be required to have the fire protection required for structural steel, provided such members are erected within an enclosure of the prescribed fire resistance rating. (Section 701—Enclosure of Vertical Openings.)

608.2 — LINTELS

Lintels over openings in walls shall be protected to provide a fire resistance rating at least equal to that required for beams, except that when such lintels are used over openings less than four (4) feet wide, such protection may be omitted. The outer member of an assembled steel lintel, which supports face masonry that is securely bonded to backing need not be protected, provided that the load carrying member of such lintel is protected as herein required.

608.3 — UNPROTECTED EXTERIOR WALLS OR PANELS

Unprotected walls or panels may be permitted in exterior non-bearing walls when all the following conditions are met:

(1) Provided such walls are of noncombustible material or of exterior grade Fire Retardant Treated Wood, and

(2) Provided such walls face a street or permanent open space of 30 feet or more in width, and

(3) Provided that in buildings three stories or more in height, exterior openings located in a story above a Group D, Institutional; Group F, Storage; Group G, Industrial; Group H, Hazardous Occupancy are separated from such an occupancy by a 2 hour fire-resistive wall construction not less than 3 feet in height.

608.4 — WOOD VENEERS ON EXTERIOR WALL PANELS

(a) Wood veneers of not less than one (1) inch nominal thickness or three-eighths (⅜) inch exterior type plywood or particleboard may be used on exterior walls when all the following conditions are met:

(1) The wall to which the veneer is attached faces a street or permanent open space of thirty (30) feet or more in width, and

(2) The veneer does not exceed two stories in height, measured from grade, and

(3) The veneer is attached to or furred from a noncombustible backing of the fire resistance required by other provisions of this Chapter, and

(4) Where open or spaced wood veneers (without concealed spaces) are used, they shall not project more than twenty-four (24) inches from the building wall.

(b) Where the wood veneer is furred from the wall and forms a solid surface, the distance between the back of the veneer and the wall shall not exceed one and five-eighths (1⅝) inches and the space thereby created shall be firestopped in accordance with Section 1703 and arranged so that there will be no open space exceeding one hundred (100) sq. ft. Where wood furring strips are used, they shall be of approved wood of natural decay-resistance or pressure treated wood.

608.5 — UNUSABLE SPACE

In one hour fire resistance construction the ceiling may be omitted over unusable crawl space and flooring may be omitted when unusable attic space occurs above.

SECTION 609 – MIXED TYPES OF CONSTRUCTION

609.1 — HORIZONTAL SEPARATION

When two or more types of construction not separated by fire walls occur in the same building, the entire building shall then be subject to the restrictions, based on occupancy of the least fire resistive type of construction used in the building.

609.2 — VERTICAL SEPARATION

(a) Where a building is constructed of more than one type of construction, the following limitations shall apply:

TYPE I construction shall not be supported by any other type.

TYPE II construction shall not be supported by construction other than Type I or Type II.

TYPE III construction shall not be supported by construction other than Type I, Type II or Type III.

TYPE IV construction shall not be supported by construction other than Type I, II, III, or IV.

TYPE V construction shall not be supported by construction other than Type I, II, III, IV or V.

(b) When types of construction of lower classification are erected above higher classifications, the entire building shall then be subject to the restrictions of the lowest fire resistive type of construction used in the building.

SECTION 610 – BUILDINGS LOCATED ON THE SAME LOT

Where the exterior walls of two or more buildings located on the same lot face one another, and one of the walls is not constructed as required for a fire wall, a common-property line shall be assumed between them. The fire resistance requirements for such facing walls and for the protection of openings therein shall be the same as required by this code for walls and openings facing common-property lines, as provided in Table 600.

TABLE 600 — FIRE PROTECTIVE REQUIREMENTS REQUIRED FIRE RESISTANCE IN HOURS

STRUCTURAL ELEMENT	Type I	Type II	Type III	Type IV		Type V		Type VI	
				1-Hour Protected	Unprotected	1-Hour Protected	Unprotected	1-Hour	Unprotected Protected
PARTY AND FIRE WALLS (a)	4	4	4	4	4	4	4	4	4
INTERIOR BEARING WALLS						(i)	(i)		
Supporting more than one floor, columns or other bearing walls	4	3	2	1	0	1	0	1	0
Supporting one floor only	3	2	1	1	0	1	0	1	0
Supporting a roof only	3	2	1	1	0	1	0	1	0
INTERIOR NONBEARING PARTITIONS			See	Sections	412, 701	and 702			
COLUMNS (e)			See Sec. 604						
Supporting more than one floor or other columns	4	3	H(d)	1	0	1	0	1	0
Supporting one floor only	3	2	H(d)	1	0	1	0	1	0
Supporting a roof only	3	2	H(d)	1	0	1	0	1	0
BEAMS, GIRDERS, TRUSSES & ARCHES (e)			See Sec. 604						
Supporting more than one floor or columns	4	3	H(d)	1	0	1	0	1	0
Supporting one floor only	3	2	H(d)	1	0	1	0	1	0
Supporting a roof only	2(f)	1½(f,g)	H(d)	1	0	1	0	1	0

*—See Section 608.3
NC—Non Combustible
(NL)—No Limits
H—Heavy Timber Sizes
See Section 604

TABLE 600 — FIRE PROTECTIVE REQUIREMENTS — Continued

STRUCTURAL ELEMENT	Type I	Type II	Type III	Type IV		Type V		Type VI	
				1-Hour Protected	Unprotected	1-Hour Protected	Unprotected	1-Hour Protected	Unprotected
FLOOR CONSTRUCTION	3	2	See Sec. 604 H	1	0	1	0	1	0
ROOF CONSTRUCTION (h)	2(f)	1½(f,g)	See Sec. 604 H(d)	1	0	1	0	1	0
EXTERIOR BEARING WALLS (h)	(%	indicates	percent	of wall	opening	permitted)			
Horizontal separation—(distance from common property line or assumed property line). 0 ft. to 3 ft. (c) 3 ft. to 10 ft. (c) 10 ft. to 20 ft. (c) 20 ft. to 30 ft. over 30 ft.	4(0%) 4(10%) 4(20%) 4(40%) 4(NL)	3(0%) 3(10%) 3(20%) 3(40%) 3(NL)	3(0%)(b) 2(10%)(b) 2(20%)(b) 1(40%) 1(NL)	2(0%) 1(10%) 1(20%) 1(40%) 1(NL)	1(0%) 1(10%) NC(20%) NC(40%) NC(NL)	3(0%)(b) 2(10%)(b) 2(20%)(b) 1(40%) 1(NL)	3(0%)(b) 2(10%)(b) 2(20%)(b) 1(40%) 1(NL)	1(0%) 1(20%) 1(40%) 1(60%) 1(NL)	1(0%) 0(20%) 0(40%) 0(60%) 0(NL)
EXTERIOR NONBEARING WALLS (h)	(%	indicates	percent	of wall	opening	permitted)			
Horizontal separation—(distance from common property line or assumed property line). 0 ft. to 3 ft. (c) over 3 ft. to 10 ft. (c) over 10 ft. to 20 ft. (c) over 20 ft. to 30 ft. over 30 ft.	3-(0%) 2-(10%) 2-(20%) 1-(40%) NC* (NL)	3-(0%) 2-(10%) 2-(20%) 1-(40%) NC* (NL)	3-(0%) 2-(10%) 2-(20%) 1-(40%) NC* (NL)	2-(0%) 1-(10%) 1-(20%) NC-(40%) NC* (NL)	1-(0%) 1-(10%) NC-(20%) NC-(40%) NC* (NL)	3-(0%) 2-(10%) 2-(20%) 1-(40%) NC* (NL)	3-(0%) 2-(10%) 2-(20%) 1-(40%) NC* (NL)	1-(0%) 1-(20%) 1-(40%) 0-(60%) 0-(NL)	1-(0%) 0-(20%) 0-(40%) 0-(60%) O-(NL)

*—See Section 608.3
NC—Non Combustible
(NL)—No Limits
H—Heavy Timber Sizes
See Section 604

TABLE 600 — REFERENCE NOTES

a. Party and Fire Walls shall extend not less than three (3) feet above the roof, except that fire walls need not extend above the roof where the roof is of noncombustible construction for the area within forty (40) feet of each side of the wall.

b. Exterior walls shall extend not less than eighteen (18) inches above the roof, except that parapet walls need not be constructed on buildings where the roof slopes more than four (4) inches vertical to twelve (12) inches horizontal from the back of the exterior wall of such buildings or where the exterior wall of such building is located fifteen (15) feet or more distant from the common property line or the center line of a public way.

c. See Section 703 for protection of wall openings.

d. Where horizontal separation of twenty (20) feet or more is provided, wood columns, arches, beams, and roof deck conforming to heavy timber sizes may be used externally.

e. Columns, beams, trusses and girders supporting masonry or concrete walls 8 feet or more in height shall have the fire rating required for the wall they support, but in no case less than one (1) hour.

f. In two (2) story buildings approved Fire Retardant Treated Wood may be used.

In buildings of Group C and E (School and Assembly occupancies, fireproofing may be omitted where structural members support a roof only and are twenty (20) feet or more clear above any floor or balcony.

g. In one (1) story buildings structural members of heavy timber sizes may be used as an alternate to unprotected structural roof members.

h. See Section 717 for Penthouses and Roof Structures.

i. The use of combustible construction for interior bearing partitions shall be limited to the support of not more than 2 floors and a roof.

CHAPTER VII — FIRE PROTECTION REQUIREMENTS

Revise Section 701.1(b) to read as follows: (1974 Meeting)

(b) Except as prescribed by Sections 701.2, 701.3, and 701.4 in buildings of four (4) or more stories in height, other than one and two family dwellings, all vertical shafts extending through more than one story shall be enclosed throughout their length with constructions of not less than 2-hour fire resistance. In buildings not over three stories in height, other than one and two family dwellings, such enclosure shall provide not less than 1-hour fire resistance. A shaft that does not extend through the roof shall have its top enclosed with construction having fire resistance at least equal to that of the enclosing walls.

SECTION 701.3 — ELEVATOR ENCLOSURES

Amend Section 701.3 as follows: (1974 Meeting)

(a) Not more than four (4) elevators shall be located in any one hoistway. Such hoistway enclosure shall have at least two-hour fire resistance, except in residential buildings, Group A or business building, Group B, not over three (3) stories in height, enclosure may be of construction having one-hour fire resistance.

(b) Where an elevator is installed in a blind hoistway or on the outside of a building, there shall be installed in the blind portion of the hoistway or blank face of the building, an emergency door at every third floor but not more than 36 feet apart at least 30 inches wide and 78 inches high conforming to the requirements of Section 703.6.

(c) Elevators shall not be in a common enclosing shaft with a stairway, and the path of travel from one flight of stairs to the next shall not pass directly in front of elevator doors.

(d) No change.

(e) First paragraph—no change. Second paragraph—Where the machine room is placed over the hoistway at the top, the floor over the hoistway shall have the same fire resistance as the hoistway walls. Openings essential for ropes, cables, sheaves and similar elevator equipment shall be kept to the minimum size practical.

(f) Hoistways shall be vented as required by the American Standard Safety Code for Elevators, A17.1-1971 Section 100.4. Such venting shall not be through the machine room.

SECTION 702 — PARTITIONS

Section 702.1. Add a new paragraph as follows: (1974 Meeting)

(h) View panels in one (1) hour fire resistive partitions shall be limited to 1296 square inches of ¼" labeled wire glass in 18 gage steel frames, but shall not exceed 25% of the wall area separating a room from a corridor.

702.3 — PARTITION REQUIREMENTS BY OCCUPANCY

Revise Section 702.3—Partition Requirements By Occupancy as follows: (1974 Meeting)

702.3 Group B—Change "more than one story" to read "two (2) stories or more".

Group C—Change "over one story" to read "two stories or more".

Revise Section Group "C"—Schools to add a paragraph as follows:

Partitions along hallways, used for exit access, in all buildings housing Group "C" School Occupancies shall be of not less than one (1) hour fire resistant construction.

Add the following note at the end of this section:

Note: See Section 1104.5 Exit Access Corridors for the use of public corridors for return and/or exhaust air from adjoining spaces.

703.2(b)

703.2(b) Class B Openings—At the end of this paragraph add the following: (1974 Meeting)

(For exceptions see paragraph 1106(f).)

703.4 — PROTECTION OF DOOR OPENINGS IN WALLS AND PARTITIONS — Page 7 - 7

703.4(b)

Amend Subsection (b) to read as follows: (1974 Meeting)

In 4 hour and 3 hour fire resistive walls or partitions, doors meeting a fire protection rating of 3 hours shall be used. No opening shall exceed one hundred twenty (120) square feet in area with no dimension greater than (12) feet, and the aggregate width of all openings at any level shall not exceed twenty-five percent of the length of such wall or partition.

Remainder of paragraph (b) to remain as written.

Amend sub-section (d) to read as follows:

(d) In one hour fire resistive corridors and smoke barriers unless otherwise specified, all door openings shall be protected with a tight fitting smoke and draft assembly (including door frame and hardware) having a fire protection rating of 20 minutes when tested in accordonce with ASTM E-152 without the hose stream. Said doors shall be equipped with an approved self-closing device.

703.8 — FIRE DAMPERS — Page 7 - 8

Delete Section 703.8 in its entirety and substitute new Section 703.8 as follows: (1973 Meeting)

703.8 — FIRE DAMPERS

Except when proper fire tests have shown that fire dampers are not necessary to maintain the required fire restistance of construction, fire dampers complying with the requirements of UL 555 shall be installed in the following locations:

(a) Duct penetrations of fire walls. When a fire wall is of 3 hour or greater fire resistance, a fire door meeting the requirements of UL 10 (b) 1970 shall be used.

(b) Ducts passing through Occupancy separations.

(c) Ducts penetrating fire-rated corridor walls.

(d) Duct penetrations of fire-rated shafts. When return air subducts extend 22 inches vertically in a vented shaft, fire dampers may be omitted.

Change paragraph (e) to read as follows: (1974 Meeting)

(e) Ducts penetrating the ceiling membrane fire protection, which is a required part of the floor/ceiling or roof/ceiling assembly.

Add new paragraph (f) as follows: (1973 Meeting)

(f) Grills or Louvres. Grills, louvres or other openings occurring in fire walls or partitions shall have the same damper requirements as for ducts when occurring in smoke barriers, vertical enclosures, corridor walls and required tenant separations.

SECTION 704 — RESTRICTIONS ON INTERIOR USE OF COMBUSTIBLE MATERIALS

Amend Section 704.3 by adding a new paragraph to read: (1974 Meeting)

(4) Foam plastics shall not be used as interior finish except as provided in Section 718.

Revise Section 704 by adding a new subsection to read as follows: (1973 Meeting)

704.4 — PLUMBING, ELECTRICAL AND AIR HANDLING SYSTEMS IN RATED ASSEMBLIES

In Type I and Type II Construction, materials used for piping, conduit, raceways or duct systems which do not qualify as noncombustible in accordance with the requirements of part one (1) of the definition of Noncombustible Material contained in Section 201.2, shall not penetrate or be concealed within any fire-resistive assembly unless enclosed by or totally embedded within materials which comply with the requirements of part one (1) of the definition of Noncombustible Material of Section 201.2, unless such materials and methods of penetration have been tested in accordance with Section 1001 (a) (c).

SECTION 717

Add new Section 717 as follows: (1974 Meeting)

SECTION 717 — PENTHOUSES AND ROOF STRUCTURES

717.1 — GENERAL REQUIREMENTS

(a) Height. No penthouse or other projection above the roof in structures of other than Type I construction shall exceed 28 feet in height above the roof when used as an enclosure for tanks or for elevators which run to the roof and in all other cases shall not extend more than 12 feet in height above the roof.

(b) Area. The aggregate area of all penthouses and other roof structures shall not exceed 33⅓ percent of the area of the supporting roof.

(c) Prohibited Uses. No penthouse, bulkhead, or any other similar projection above the roof shall be used for purposes other than shelter of mechanical equipment or shelter of vertical shaft openings in the roof. Penthouses or bulkheads used for purposes other than permitted by this Section shall conform to the requirements of this Code for an additional story.

717.2 — TYPES OF CONSTRUCTION REQUIRED

Roof structures shall be constructed with walls, floors and roof as required for the main portion of the building.
EXCEPTIONS:

1. On Buildings of Type I and Type II construction, the exterior walls and roofs of penthouses which are more than 5 feet and less than 20 feet from a common property line shall be of at least 1-hour non-combustible construction. Walls and roofs which are over 20 feet from a common property line may be of non-combustible construction. All interior framing and walls shall be non-combustible construction.

2. On buildings of Type III, IV and V construction, the exterior walls and roofs of penthouses which are more than 5 feet and less than 20 feet from a common property line shall be at least 1-hour construction. Walls which are over 20 feet from a common property line may be of heavy timber construction or non-combustible construction. Roofs may be of wood frame construction. All interior framing and walls shall be heavy timber construction or non-combustible construction.

3. Enclosures housing only mechanical equipment and located at least 20 feet from adjacent property lines may be of unprotected non-combustible construction.

On one-story buildings, unroofed mechanical equipment screens, fences or similar enclosures may be of combustible construction when located at least 20 feet from adjacent property lines and when not exceeding 4 feet in height above the roof surface.

The restrictions of this subsection shall not prohibit the placing of wood flagpoles or similar structures on the roof of any building.

717.3 — TOWERS AND SPIRES

Towers and Spires when enclosed shall have exterior walls as required for the building to which they are attached. (See Section 713.) The roof covering of spires shall be as required for the main roof of the rest of the structure.

SECTION 718

Add a new Section 718—Foam Plastics Insulation, to read: (1974 Meeting)

SECTION 718 — FOAM PLASTICS INSULATION

a. General. Approved foam plastics shall be those which have a flame-spread rating of 200 or less and a smoke density rating no greater than 450 when tested in accordance with ASTM E-84-1970.

b. Specific Requirements. The following requirements shall apply to all uses of foam plastic in or on walls, ceilings or both unless specifically approved. Such approval shall be based on acceptable diversified tests such as, but not limited to, tunnel tests conducted in accordance with ASTM E-84-1970, Standard Method of Test for Surface Burning Characteristics of Building Materials, full scale corner tests and an ignition temperature test.

1. Foam plastic having a flame-spread of 75 or less may be used in the following locations:

a) Within the cavity of a masonry or concrete wall.

b) On the room side surface of conforming walls or ceilings provided the foam plastic is fully protected from the interior of the building by a thermal barrier of ½ inch gypsum wallboard or other approved material having a finish rating of not less than 15 minutes installed so as to remain in place for the required period of time.

c) Within the wall cavity of combustible non-fire resistive wall construction provided the protection is applied as described in item (b) above.

d) Within the cavity of walls classified as combustible fire resistive construction provided fire tests are conducted in accordance with ASTM E-119-1971, Fire Tests of Building Construction and Materials, and the protection from the interior of the building is at least equivalent to that required in item (b) above.

2. Foam plastic insulation having a flame-spread of 75 or less when tested in a thickness of 4 inches, may be used in thickness up to 10 inches for use in cold storage rooms, food processing rooms, ice plants and similar rooms when the room is protected with automatic sprinklers and the insulation is covered with ½ inch Portland cement plaster or other approved material having a finish rating of not less than 15 minutes. Thermal barriers shall be installed in a manner that will assure that they remain in place for the required period of time.

3. Foam plastic insulation having a flame-spread of 25 or less may be used in a thickness of not more than 4 inches in or on walls when the foam plastic is covered by a metal facing of not less than 0.032 inches of aluminum or 26 gauge galvanized steel sheet and the insulated area is protected with automatic sprinklers. Such walls shall not be used where non-combustible or fire-resistive construction is required.

c. Class A, B or C roof coverings may include foam plastic if complying with the provisions of Section 706.2, 706.3 or 706.4.

CHAPTER VIII — CHIMNEYS, FIREPLACES AND HEATING EQUIPMENT

824.2 — FIRE PROTECTION

Delete paragraph (c) as it now stands and substitute the following: (1974 Meeting)

(c) Ducts passing through walls, floors and/or partitions shall be sealed in accordance with the requirements of NFPA 90B, 1973 "Warm Air Heating and Air Conditioning Systems, Residence Type" or NFPA 90A, 1974 "Air Conditioning and Ventilating Systems".

CHAPTER IX — SPRINKLERS AND STANDPIPES

901.8 — OTHER OCCUPANCY SPRINKLER REQUIREMENTS

Revise this Section as follows: (1974 Meeting)

(a)—GROUP "B-2"—BUSINESS—MERCANTILE

An approved automatic sprinkler system shall be provided in stores and similar occupancies where stocks of combustible materials are on display for public sale and where the story floor area exceeds 15,000 square feet.

(b)—GROUP "D-2"—INSTITUTIONAL

Approved automatic sprinkler systems shall be provided in all hospitals, sanitoriums, convalescent, nursing homes and homes for the aged having more than 10 patients. In hospitals only, of Type I or Type II Construction the automatic sprinkler system may be omitted from operating, delivery, x-ray, cardiac and intensive care rooms and patient sleeping rooms not exceeding 600 square feet in area when each room is provided with an approved listed smoke detector connected to the alarm system.

(c)—GROUP "E-1"—LARGE ASSEMBLY OCCUPANCY

An approved automatic sprinkler system shall be provided in Group E-1 Large Assembly Occupancies over areas which could be used for the display, sale or storage of combustible materials when such display, sale or storage floor area exceeds 15,000 square feet.

CHAPTER XI — MEANS OF EGRESS REQUIREMENTS

1103.2(a) — MINIMUM NUMBER OF EXITS

Amend Section 1103.2 (a) as follows: (1974 Meeting)

(a) Every room or floor space of a building, occupied by seventy five (75) persons or more, or occupied by a Group H, Hazardous occupancy, shall have not less than two (2) independent exits accessible to each tenant area.

1103.2 (c) — 1 and 2

Amend this section to read: (1974 Meeting)

Where one exit is permitted:

1. In Group A, Residential occupancies having no more than four dwelling units per floor, with the total floor area not exceeding 3500 square feet, may be served by one common exit. Such buildings shall not exceed two stories in height.

The maximum distance of travel to reach the exit from the entrance door to any living unit shall not exceed thirty (30) feet.

2. In Group B-1 occupancies having no floor area over 3500 sq. ft. served by that exit and not over two stories in height provided the occupant content shall not exceed forty (40) persons above the street floor. Maximum distance of travel to the exit shall not exceed seventy-five (75) feet.

1104.1 (b)

Delete paragraph (b) as it now stands and substitute the following: (1974 Meeting)

1104.1 — INSTITUTIONAL OCCUPANCIES

(b) All Group D-2, Institutional Occupancies, shall have smoke partitions to divide into at least two compartments every story used by inpatients for sleeping and/or treatment, and any story having an occupant load of 50 or more persons, and to limit on any story the maximum area of each smoke compartment to not more than 22,500 square feet, of which both the length and width shall not be more than 150 feet. At least 30 net square feet per occupant for the total of bed and/or litter patients shall be provided, on each side of the smoke partition, and on stories not housing bed and/or litter patients at least six (6) square feet per occupant in adjoining compartments.

(c) Smoke partitions shall be constructed of non-combustible construction of at least one hour fire resistance rating, except that in Type V and Type VI construction the partitions may be of combustible construction of one hour fire resistance. The partitions shall form an effective membrane continuous from outside wall to outside wall and from floor slab to floor/roof slab/deck thereby including continuity through all concealed spaces, such as those found, above suspended ceilings, and including intertitial structural and mechanical spaces. Transfer grilles, whether equipped with fusible link-operated dampers or not, shall not be used in these partitions.

(d) Smoke partitions shall have openings in the corridors only. Corridor doors shall be a pair of swinging type doors, each swinging in opposite direction from each other, and the minimum width of each door shall be 44" for hospitals and 32" for residential custodial care institutions. Doors shall have a fire resistance rating of 20 minutes, shall have vision panels of ¼" labeled wire glass mounted in steel frames, and the glass area shall be limited to 720 square inches in each door. The doors shall close the opening with only the clearance necessary for proper operation under self-closing, and shall be without undercuts, louvres and/or grilles, babbets or bevels. Astragals are required at the meeting edges, and stops are required on the head and sides of the door frame. Positive latching hardware is not required, and center mullions are prohibited.

(f) Corridor doors in smoke partitions shall be provided with door holding devices of the fail-safe type, which shall release the doors causing it to close upon the actuation of smoke detectors sensing either visible or invisible particles of combustion, as well as upon the application of a manual pull of fifty (50) pounds against the hold-open device.

1104.5 — EXIT ACCESS CORRIDORS

In paragraph (a) change the word "public" to read "exit access". (1974 Meeting)

1104.6 — EMERGENCY EGRESS OPENINGS

Add a new Section 1104.6 as follows: (1974 Meeting)

Every sleeping room in one and two-family residences shall have at least one operable window or exterior door approval for emergency egress or rescue. The units must be operable from the inside to a full clear opening without the use of separate tools. Where windows are provided as a means of egress or rescue, they shall have a sill height of not more than 44 inches above the floor.

All egress or rescue windows from sleeping rooms must have a minimum net clear opening of 5.7 square feet. The minimum net clear opening height dimension shall be 24 inches. The minimum net clear opening width dimension shall be 20 inches.

1104.7

Amend Section 1104—Special Exit Requirements by adding a new Section 1104.7 — Smokeproof Enclosures as follows: (1974 Meeting)

1104.7 — SMOKEPROOF ENCLOSURES

(a) General. A smokeproof enclosure shall consist of a vestibule and continuous stairway enclosed from the highest point to the lowest point by walls of two-hour fire-resistive construction. The supporting frame shall be protected as set forth in Chapter VI.

(b) Where Required. Where a floor of any story is located more than 75 feet above the highest grade, one of the required exits shall be a smokeproof enclosure. When a smokeproof enclosure is required it shall be used to meet the requirements of Section 1120.

(c) Construction. Stairs in smokeproof enclosures shall be of noncombustible construction.

(d) Outlet. A smokeproof enclosure shall exit into a public way or into an exit passageway leading to a public way. The exit passageway shall be without other openings and shall have walls, floors, and ceiling of two-hour fire resistance.

(e) Barrier. A stairway in a smokeproof enclosure shall not continue below the grade level unless an approved barrier is provided at the ground level to prevent persons from accidentally continuing into the basement.

(f) Access. Access to the stairway shall be by way of a vestibule or by way of an open exterior balcony of noncombustible materials.

(g) Smokeproof Enclosures by Natural Ventilation. 1. Doors. Doors to both the vestibule and to the stairway shall have a one-hour fire-resistive rating and have closing devices as specified in Section (h) 6 below.

2. Open air vestibule. The vestibule shall have a minimum of 16 square feet of opening, in a wall facing an exterior court, yard or public way at least 20 feet in width.

(h) Smokeproof Enclosures by Mechanical Ventilation. 1. Doors. The door from the building into the vestibule shall have a one and one-half hour fire-resistive rating and have closing devices as specified in Section 1117.1 (f).

The door from the vestibule to the stairway shall be as required by Section 703.4. Wired glass, if provided, shall not exceed 100 square inches in area and shall be set in a steel frame. The door shall be provided with a drop sill or other provision to minimize air leakage.

2. Vestibule size. The vestibule shall have a minimum dimension of 44 inches in width and 72 inches in direction of exit travel.

3. Vestibule ventilation. The vestibule shall be provided with not less than one air change per minute and the exhaust shall be 150 per-

cent of the supply. Supply air shall enter and exhaust air shall discharge from the vestibule through separate, tightly constructed ducts used only for that purpose. Supply air shall enter the vestibule within 6 inches of the floor level. The top of the exhaust register shall be located at the top of the smoke trap but no more than 6 inches down from the top of the trap and shall be entirely within the smoke trap area. Doors, when in the open position, shall not obstruct duct openings. Duct openings may be provided with controlling dampers if needed, to meet the design requirements but are not otherwise required.

NOTE: For buildings where such air changes would result in excessively large duct and blower requirements, a specially engineered system may be used. Such an engineered system shall provide 2500 cfm exhaust from a vestibule when in emergency operation and shall be sized to handle three vestibules simultaneously and the smoke detector located outside each vestibule shall release to open the supply and exhaust duct dampers in that affected vestibule.

4. Smoke trap. The vestibule ceiling shall be at least 20 inches higher than the door opening into the vestibule to serve as a smoke and heat trap and to provide an upward moving air column.

5. Stair shaft air movement system. The stair shaft shall be provided with mechanical supply and exhaust air. There shall be a minimum of 2500 cfm discharge at the top of the shaft. The supply shall be sufficient to provide a minimum of .05 inch of water column with respect to atmospheric pressure with all doors closed and a minimum of .10-inch water column difference between the stair shaft and the vestibule.

6. Exit doors. The exit doors into the vestibule and into the stair shaft shall close automatically when released by activation of a detector meeting the requirements of Section 1117.1 (f). The door holding devices shall be of an approved type which will release the doors so that they will close in the event of a power failure.

7. Operation of ventilating equipment. Vestibule and stair shaft mechanical ventilation may be inactive or may operate at reduced levels for normal operations as approved by the Building Official; but when the detectors referred to in paragraph 6 either fail or are activated, the mechanical equipment shall operate at the levels specified in paragraphs 3 and 5.

8. Standby power. Mechanical ventilation equipment shall be provided by an approved self-contained generator set to operate whenever there is a loss of power in the normal house current. The generator shall be in a separate room having a minimum one-hour fire-

resistive occupancy separation and shall have a minimum fuel supply adequate to operate the equipment for two hours.

9. Acceptance and testing. Before the mechanical equipment is accepted by the Building Official, it shall be tested in his presence to confirm that the mechanical equipment is operating in compliance with these requirements.

10. Emergency lighting. The stair shaft and the vestibule shall be provided with emergency lighting. The standby generator which is installed for the smokeproof enclosure mechanical ventilation equipment may be used for standby emergency lighting power supply.

11. Air-conditioned buildings. In buildings with air conditioning systems or pressure air supply, serving more than one story, a smoke detector shall be placed in the return air prior to exhausting from the building or being diluted by outside air. Upon activation the detector shall cause the return air to exhaust completely from the building without any recirculation through the building. Such devices may be installed in each room or space served by a return air duct.

1105.1 — OCCUPANT CONTENT

Change Group A—Residential—Minimum Occupant Content Floor Area Per Person from 125 sq. ft. to read 200 sq. ft. (1974 Meeting)

SECTION 1106 – EXIT ENCLOSURES

Amend Section 1106(f) to read as follows: (1974 Meeting)

(f) In stair enclosure walls or partitions protecting the stair from the interior of the building, no openings except the necessary doorways shall be permitted. Fire windows of the fixed or automatic closing type may be installed in stair enclosures provided they open to the exterior of the building and are located at least ten (10) feet from any other wall opening.

Doorways to the interior of the building shall be equipped with approved fire assemblies having a 1 hour (B) classification where 1 hour walls are involved and 1½ hour (B) assemblies where 2 hour construction is provided. Such doors are to be self-closing and are to be so maintained or shall be automatic closing by actuation of an approved smoke detector.

1115.3 — TREADS AND RISERS — Page 11 - 12

Amend Section 1115.3 (c) and add new Note 1: (1973 Meeting)

(c) The use of winders or circular stairways is prohibited in stairways serving as required exits. Within individual dwelling units, winders may be used if the required width of tread is provided at a point not more than twelve inches (12") from the side of the stairway where the treads are narrower, but in no case shall any width of tread be less than six inches (6") at any point.

(1) Circular stairs may be used in an individual dwelling unit as an exit providing the minimum width of tread is not less than ten inches (10") and the smaller radius is not less than twice the width of the stairway. All treads in any one flight between landings shall have identical dimensions within a one-quarter inch (¼") tolerance.

1117.1 — DOORWAYS, GENERAL — Page 11 - 14

Amend paragraph (f) to read as follows: (1973 Meeting)

(f) All doors in smoke barriers, horizontal exits, stairway enclosures and other doors opening between rooms and fire-rated corridors shall be self-closing and so maintained or shall be provided with approved door holding devices of the fail safe type which will release the door(s) causing it to close when activated by approved listed smoke detectors sensing visible or invisible particles of combustion.

1117.3 — POWER OPERATED DOORS

Add new second paragraph to this section to read as follows: (1974 Meeting)

Power operating sliding doors may be used provided the sliding leaf is equipped with an emergency swing (panic release) feature.

SECTION 1126 – FIRE ALARM

Revise Section 1126 by deleting a portion of the first sentence and adding a note at the end of the existing requirements as follows: (1974 Meeting)

(a) A manual fire alarm system in accordance with provisions of "Local Protective Signalling Systems", NFPA 72-A, 1974, shall be installed in all the following buildings:

The requirements as contained within the Codes for Group "A" through Group "H" remain the same.

Add the following note after Group "H"—Hazardous Occupancies:

NOTE: The manual fire alarm system may be omitted, where required above, in buildings equipped with an automatic fire alarm or automatic sprinkler system covering all areas. Actuation of the automatic fire alarm or automatic sprinkler system shall sound an internal evacuation alarm.

1126 (a) Group G:—Delete the word "over" in the first sentence, which will make the requirement read "two stories or more".

SECTION 1127 — FIRE DETECTION SYSTEMS

Amend Section 1127 as follows: (1974 Meeting)

Every dwelling and every dwelling unit within an apartment house, condominium and townhouse and every basement or cellar within such dwellings shall be provided with an approved listed smoke detector, installed in accordance with the manufacturer's recommendation and listing. When actuated the detector shall provide an audible alarm.

CHAPTER XII — MINIMUM DESIGN LOADS

1203.5 — LOAD TESTS AND SECTION 2504 (b) — TESTS — Page 12 - 4, Page 25 - 1

Add a sentence at the end of Sections 1203.5 and 2504 (b) to read: (1973 Meeting)

Such load tests of constructions are not intended to be used as a method for determination of acceptable allowable working stresses of a material as an alternate to the established standard method of determining such stresses.

SECTION 1205 — WIND LOADS — Page 12 - 5

Delete Section 1205 in its entirety and substitute a new Section 1205 as follows: (1973 Meeting)

SECTION 1205 — WIND LOADS

1205.1 — DESIGN FOR WIND LOADS REQUIRED

(a) Buildings and structures and every part thereof shall be designed to withstand the forces of wind pressure assumed in any direction. No allowance shall be made for the effect of shielding by other structures.

(b) The floor, roof or other horizontal bracing system shall be designed and constructed to transfer horizontal forces to the parts of the structural frame designed to carry the forces to the ground. Where horizontal or vertical shear-resisting elements are designed to transfer forces through diaphragm action, the analysis shall include the design of chord members at or near the extremities of the diaphragm and the method by which the forces are transferred to the resisting elements. The total shear in any horizontal plane shall be distributed to the various elements of the lateral force-resisting system in proprotion to their rigidities, taking into consideration the rigidity of the horizontal bracing system or diaphragm.

1205.2 — VELOCITY PRESSURES

(a) The basic velocity pressures to be used to determine the minimum design unit pressures shall be as set forth in Table 1205.1. The Building Official may accept a design based on lower pressures, the validity of which is shown by nationally recognized data. The Building Official may require evidence to support the design pressures used in the design of structures not included in this Section.

Section 1205.2—Velocity Pressures

After paragraph (a) Alternate — delete reference to ANSI A 58-1—1972 and substitute therefor the following: (1974 Meeting)

ALTERNATE: NAVFAC DM-2, October, 1970, may be used in lieu of this section for the design of one and two story structures provided that Table 1205.1 shall be used to determine the basic wind load. When used, the plans shall show that structural design is based on NAVFAC DM-2, October, 1970.

(b) Structural members and systems providing the stability for the building or structure shall be designed for the velocity pressures set forth in Table 1205.1, multiplied by the appropriate shape factor set forth in Table 1205.2, 1205.3, and 1205.4, except as required in Paragraphs (e) and (f) of this Subsection.

(c) Components such as purlins, girts, and similar secondary members that transfer wind loads to the primary structural frame or system shall be designed for the velocity pressures set forth in Table 1205.1, multiplied by the appropriate shape factor set forth in Table 1205.5, except as specified in Paragraphs (e) and (f) of this Subsection.

(d) Curved roofs shall be designed for wind forces computed on the basis that the curved section is divided into not less than five equal segments. The force on each segment shall be determined by use of the shape factors set forth in Table 1205.4 for inclined surfaces, appropriate to the slope of the chords of the segment.

(e) Where rigid awnings, canopies, and canopy shutters are intended to be positioned to close an opening during periods of high wind, they shall be designed as required in Paragraph (b) of this Subsection. Where rigid awnings, canopies, and canopy shutters are intended to be removed or repositioned during periods of high wind, they shall be designed in their open position to withstand a pressure of 60 percent of that set forth in Table 1205.1, with applicable shape factors, but not less than 15 psf. Shape factors for these structures shall be as set forth in Tables 1205.2, 1205.3, 1205.4, and 1205.5.

(f) Screened enclosures and supports for screening shall be designed to withstand the loads set forth in Table 1205.6, multiplied by a shape factor of 1.3 inward or outward to the walls and 0.7 aplide upward or downward to the roof. These factors apply only to enclosures at or near grade level.

(g) For determining stresses, all vertical design loads, except the roof live load and crane loads, shall be considered to be acting simultaneously with the wind pressure.

Section 1205.2 Velocity Pressures—Add the following sentence to 1205.2 (g): (1974 Meeting)

Where simultaneous combination of vertical design loads and wind pressure would produce a design stress less than either when acting alone, then the single condition resulting in the maximum design stress shall be used.

TABLE 1205.1

BASIC WIND LOAD PRESSURES IN POUNDS PER SQUARE FOOT[b,c]

(See Figure 1205.1)

Height Above Ground[a], Feet	100 Yr. Recurrence of Fastest Mile of Wind, MPH						
	70	80	90	100	110	120	130
0 - 30	10	13	16	20	24	29	34
31 - 50	14	18	22	28	34	40	47
51 - 100	16	21	27	33	40	48	56
101 - 200	20	26	33	40	49	58	68
201 - 300	23	30	38	47	57	67	79
301 - 400	25	33	42	52	62	74	87
401 - 500	27	36	45	55	67	80	94
501 - 800	30	39	50	62	74	89	104
801 - 1,000	33	43	55	68	82	97	114
Over 1,000	34	45	56	70	84	100	117

a—Measured above the average level of the ground adjacent to the structure.

b—To be modified by shape factors.

c—Velocity pressures are based on the formula

$P = 0.00256 \times V^2 \times \left\{\frac{H}{30}\right\}^{2/7}$ where:

V = Wind Speed in MPH; and

H = the height above grade (in feet) of the pressure being computed.

FIGURE 1205.1

FIGURE 1205.1 — BASIC WIND SPEEDS IN MILES PER HOUR

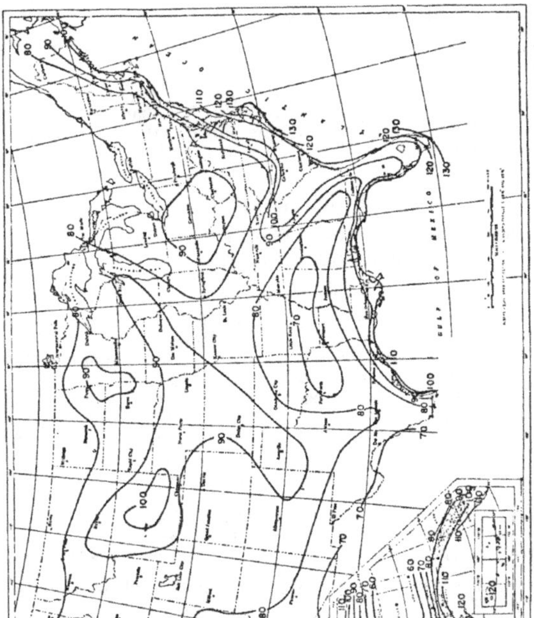

Annual Extreme Fastest-Mile Speed 30 Feet Above Ground, 100-Year Mean Recurrence Interval

TABLE 1205.2

SHAPE FACTORS FOR PRIMARY FRAMES AND SYSTEMS— VERTICAL SURFACES

Vertical Surface	Factor
Rectangular prismatic structures*[a]	+1.3[b]*
Cylinders	+ 0.7
Flat surfaces with no appreciable depth, including signs and fences	1.4
Partially open surfaces[c]* [percent solid]	
10%	0.35
20%	0.55
40%	0.80
60%	1.00
80%	1.20
100%	1.30

*See Table 1205.4 for footnotes.

TABLE 1205.3

SHAPE FACTORS FOR PRIMARY FRAMES AND SYSTEMS— HORIZONTAL SURFACES

	Factor	
Horizontal Surface[d]*	Normal to Windward 1/3 of Surface	Normal to Leeward 2/3 of Surface
Enclosed buildings	— 1.0[b]	— 0.75
Buildings with one or more sides open	— 1.5	— 1.25
Overhangs and eaves	— 1.5	— 1.50

*See Table 1205.4 for footnotes.

TABLE 1205.4

SHAPE FACTORS FOR PRIMARY FRAMES AND SYSTEMS— INCLINED SURFACES

Inclined Surface[e] Degree from Horizontal	Factor Normal to Windward Surface	Factor Normal to Leeward Surface
70-90	+ 0.80	— 0.70
60-70	+ 0.65	— 0.70
50-60	+ 0.65	— 0.70
40-50	+ 0.25	— 0.70
30-40	— 0.25	— 0.70
20-30	— 0.75	— 0.70
10-20	— 0.93	— 0.70
Overhangs and Eaves	— 1.50	— 1.50

[a]Includes + 0.8 on windward and — 0.5 on leeward sides.

[b]+ indicates forces inward, — indicates forces outward.

[c]Shape factor to be applied to gross area of surface.

[d]Includes surfaces with less than 10° inclination to horizontal.

[e]For buildings with one or more sides open, add — 1.0 to all negative factors.

TABLE 1205.5

SHAPE FACTORS FOR COMPONENTS TRANSFERRING WIND LOADS TO THE STRUCTURAL FRAME OR SYSTEM

Vertical Surface	Pressure Inward	Pressure Outward
Exterior walls of closed buildings, including fixed glass, glazing and supporting members	+ 1.1[c]	— 1.1
Operative windows and sliding glass doors, including parts	+ 1.1	— 0.55
Exterior walls of buildings with one or more sides open	+ 1.1	— 1.5
Horizontal surfaces	Table 1205.3	
Inclined surfaces	Table 1205.4	

[c]+ signifies forces inward.
— signifies forces outward.

TABLE 1205.6

WIND PRESSURES FOR SCREENED ENCLOSURES[a]

Percent Open[b]	Wind Pressure [psf]
Less than 40	30
50 to 60	20
Over 60	10

[a]See Subsection 1205.2 (f) for shape factors; load to be applied to gross screened area.

[b]Percentage of gross area.

1205.3 — STABILITY

(a) Calculations to determine overturning and uplift forces shall be made with the shape factors set forth in Tables 1205.2, 1205.3 and 1205.4.

(b) The overturning moment calculated from the wind pressure shall not exceed 2/3 of the dead load resisting moment.

(c) The uplift forces calculated from the wind pressure shall not exceed 2/3 of the resisting dead loads.

(d) Anchorage of the roof to walls and columns and of walls and columns to the foundation to resist overturning, uplift, and sliding forces shall be provided.

SECTION 1206 — EARTHQUAKE LOADS

Delete reference (see Figure 1) at the end of the section and delete Figure 1 on page 12-8. (1974 Meeting)

CHAPTER XIII

1302.1 — GENERAL — Page 13 - 2

Amend Section 1302.1 so as to read as follows: (1973 Meeting)

1302.1 — GENERAL

Except in the case of temporary structures or secondary buildings not over one story in height and not exceeding 400 square feet in area, footings and foundations, unless specifically provided shall be constructed of grillages of steel, of masonry or of reinforced concrete or other approved materials (one and two family dwellings may not be required to have reinforced concrete footings or grillage of steel) in no case less than 12 inches below grade. In geographical areas subject to severe freezes, the bottom of foundations shall be placed below the frost line establishd by local records. Masonry units in foundation walls and footings shall be laid up in Type M, S or N mortar. The base area of all footings and foundations shall be proportioned as specified in Section 1302.3.

1303.2(b) — STEEL PIPES

Revise date of material specification for steel pipe in first sentence of Section 1303.2(b) as follows: (1974 Meeting)

(b) Steel Pipe Piles—Steel pipe piles shall consist of steel pipe conforming to the "Specifications for Welded and Seamless Steel Pipe Piles, ASTM A22-73". They may be driven either open-ended or with ends closed. Steel pipe piles diven open-ended shall have a nominal outside diameter of not less than 10 inches and a minimum wall thickness of not less than 0.25 inch for diameters less than 14 inches and a minimum wall thickness of not less than .375 inch for diameters 14 inches and over. Pipe of less wall thickness may be driven open-ended if a suitable cutting shoe is provided. If steel pipe piles are to be dirven with closed ends, a forged, cast steel, flat plate or other end closure of approved design shall be used. Steel pipe piles driven with ends closed may be of smaller sizes and wall thickness than specified above but no such pile of uniform section shall have a nominal outside diameter of less than 8 inches.

1303.2(c)

Revise dates of material specifications for steel piling in first sentence of Section 1303.2 (c) to read as follows: (1974 Meeting)

(c) Rolled Structural Steel Piles—Steel used in rolled structural steel piles shall conform to one of the following standard specifications:

"Specifications for Structural Steel, ASTM A36-70a"

"Specification for High-Strength Low Alloy Columbian Vanadium Steels of Structural Quality, ASTM A572-73."

Sections of such piles shall be of H form, with flange projection not exceeding fourteen times the minimum thickness of metal in either web or flange and with total flange width at least 85% of the depth of the section. No section shall have a nominal thickness of metal less than ⅜ inch, nor a nominal depth in the direction of the web less than 8 inches. Other structural sections or combinations of sections having flange widths and depths of not less than 10 inches and thickness of metal not less than ⅜ inch may also be used.

CHAPTER XIV — MASONRY CONSTRUCTION

Page 14 - 2

Revise Section 1402.6 to read: (1973 Meeting)

1402.6 — PLAIN CONCRETE

Concrete that is either unreinforced or contains less reinforcement than the minimum amount specified for reinforced concrete (Chapter XVI) shall be classed as plain concrete. Plain concrete used for structural purposes shall conform to the requirements of Section 1602 and have a minimum compressive strength, f'_c, of 2500 psi.

Page 14 - 7

Revise Section 1403.5 to read: (1973 Meeting)

1403.5 — ALLOWABLE STRESSES IN PLAIN CONCRETE

Unless designed in accordance with the provisions of Chapter XVI, structural members of plain concrete shall be proportioned for allowable stresses not to exceed 25 percent for compression and 3 percent for tension in extreme fiber in flexure of the compressive strength of the concrete. When the ratio of height to thickness exceeds 10, the precentage for compression stress shall be reduced proportionately to 18 percent for a ratio of height to thickness of 20.

CHAPTER XV — STEEL CONSTRUCTION

SECTION 1504

Change the title and date of referenced specification 1504 to read as follows: (1974 Meeting)

SECTION 1504 — COLD-FORMED STAINLESS STEEL CONSTRUCTION

The design of cold-formed stainless steel structural members shall conform to the "Specification for the Design of Cold-Formed Stainless Steel Structural Members," AISI-1974 Edition.

SECTION 1506 — WELDING

Revise the title and date of the referenced specification in Section 1506 to read as follows:

Details of design workmanship and technique for welding, inspection of welding and qualification of welding operators shall conform to the recommendations of the "Structural Welding Code," AWS D1.1-72 including AWS D1.1-Rev. 1-73 and AWS D1.1-Rev. 2-74.

SECTION 1507 — STRUCTURAL APPLICATIONS OF STEEL CABLES FOR BUILDINGS — Page 15 - 1

Add new Section 1507 to read as follows: (1973 Meeting)

1507 — STRUCTURAL APPLICATION OF STEEL CABLES FOR BUILDINGS

The design, deflection, connections, protective coatings, fabrication, erection and inspection of building construction utilizing steel cables shall conform to Part I of the "Criteria for Structural Applications of Steel Cables for Buildings," AISI—1973 edition.

CHAPTER XVI — CONCRETE CONSTRUCTION

SECTION 1601 — GENERAL

Revise the first sentence of Section 1601 as follows: (1974 Meeting)

All structures of reinforced concrete, including prestressed concrete, shall be designed and constructed in accordance with the provisions of "Building Code Requirements for Reinforced Concrete (ACI 318-71)" as amended by the "1973 Supplement to Building Code Requirements for Reinforced Concrete (ACI 318-71)."

Add a second paragraph to Section 1601 as follows: (1973 Meeting)

Structural members of plain concrete may be designed and constructed in accordance with the provisions of "Building Code Requirements for Structural Plain Concrete," ACI 322-72.

CHAPTER XVII

1700.1(f)

Change standards referred to under American Plywood Association as follows: (1974 Meeting)

Plywood Construction Systems for Commercial and Industrial Buildings (P310) 1973. Residential Construction Guide (Q450) 1973.

1700.3(c)

Change the fourth line to read as follows: (1974 Meeting)

. . . its type in Product Standard PS 1-74 for Construction and Industrial Plywood. . . .

Also, editorially change PS 1-66 to PS 1-74 where it occurs in Tables 2 and 3 of Supplement to Chapter XVII (pages 17-27, 28).

1700.3(e)

Revise paragraph (e) to read as follows: (1974 Meeting)

(e) Fiberboard for its various uses shall conform to "Voluntary Product Standard, Cellulosic Fiber Insulating Board, PS 57-73" and the applicable "Standards of IB Specification No. 1-1970, No. 2-1972, or No. 3-1972."

1702.8.1 — ACCESS TO CRAWL SPACE — Page 17 - 6

Add new Section 1702.8.1 as follows: (1973 Meeting)

1702.8.1 — ACCESS TO CRAWL SPACE

Usable crawl spaces under buildings without basements shall be provided with a minimum of one (1) access opening not less than eighteen (18") inches by twenty-four (24") inches. Access opening shall be readily accessible and provided with a door or device that may be easily removed or operated.

TABLE 1705.6A — ALLOWABLE SPANS FOR PLYWOOD FLOOR AND ROOF SHEATHING, etc. — Page 17 - 21

Change the second sentence footnote 4: (1973 Meeting)

. . . Allowable uniform load based on deflection of 1/360 of span is 165 psf.

TABLE 1705.6 A

Change the fourth line of the Footnote 5 to read as follows: (1974 Meeting)

. . ., and the live loads does not exceed 25 pounds per square foot. For other grades, a minimum thickness of ½" 5-ply is required.

TABLE 1705.6A — ALLOWABLE SPANS FOR PLYWOOD FLOOR AND ROOF SHEATHING, etc. — Page 17 - 21

Insert new footnote (8) to Table 1705.6A: (1973 Meeting)

(8) For joists spaced 24" o.c. plywood sheathing with Identification Index numbers 42/20 or greater can be used for subfloors when supporting 1½" lightweight concrete.

TABLE 1705.6B — ALLOWABLE SPANS FOR PLYWOOD COMBINATION SUBFLOOR-UNDERLAYMENT — Page 17 - 22

Change the second sentence of footnote 1 as follows: (1973 Meeting)

. . . Allowable uniform load based on deflection of 1/360 of span is 125 psf . . .

TABLE 1705.6B — ALLOWABLE SPANS FOR PLYWOOD COMBINATION SUBFLOOR-UNDERLAYMENT — Page 17 - 22

Add new sentence to footnote 1 as follows: (1973 Meeting)

Except for 1/2 inch, Underlayment Grade and C-C (plugged) panels may be of nominal thickness 1/32 inch less than the nominal thicknesses shown when marked with the reduced thickness.

1706.7 (a) — EXTERIOR WALL COVERINGS — Page 17 - 13

Revise 1706.7 (a) to read: (1973 Meeting)

(a) Weatherboarding. Wood siding patterns known as rustic drop siding or shiplap shall have an average thickness in place of not less than 19/32" and shall have a minimum thickness of not less than 3/8". Bevel siding shall have a minimum thickness measured at the butt section of not less than 7/16" and a tip thickness of not less than 3/16". Siding of lesser dimensions may be used provided such wall covering is placed over sheathing which conforms to the provisions of Section 1706.2.

CHAPTER XX — LIGHT, VENTILATION AND SANITATION

2001.1.1

Add new Section 2001.1.1 as follows: (1974 Meeting)

2001.1.1 — ROOM DIMENSIONS

(a) Ceiling Heights. Habitable (space) rooms, storage rooms and laundry rooms shall have a ceiling height of not less than 7 feet 6 inches. Hallways, corridors, bathrooms and water closet rooms shall have a ceiling height of not less than 7 feet measured to the lowest projection from the ceiling.

If any room in a building has a sloping ceiling, the prescribed ceiling height for the room is required in only one-half the area thereof. No portion of the room measuring less than 5 feet from the finished floor to the finished ceiling shall be included in any computation of the minimum area thereof.

If any room has a furred ceiling, the prescribed ceiling height is required in two-thirds the area thereof, but in no case shall the height of the furred ceiling be less than 7 feet.

(b) Floor Area. Every dwelling unit shall have at least one room which shall have not less than 150 square feet of floor area. Other habitable rooms except kitchens shall have an area of not less than 70 square feet.

(c) Width. No habitable room other than a kitchen shall be less than 7 feet in any dimension.

CHAPTER XXI — SAFEGUARDS DURING CONSTRUCTION

2101.9(c)

2101.9 (c) change "over three (3) stories" to read four (4) stories or more." (1974 Meeting)

CHAPTER XXV — PREFABRICATED CONSTRUCTION AND MOBILE HOMES OVER EIGHT (8) FEET WIDE

SECTION 2505 — PLYWOOD

Change the paragraph to read as follows: (1974 Meeting)

Plywood shall conform to U. S. Product Standard PS 1-74, Construction and Industrial Plywood. Plywood of species not covered in PS 1-74, when used structurally, shall be identified as to veneer grade and glue type by an approved agency and shall meet the performance standards in U. S. Product Standard PS 1-74 for its type.

CHAPTER XXVII — GLASS

2703.1

Delete present Section 2703.1 in its entirety and substitute a new Section 2703.1 as follows: (1974 Meeting)

2703.1 — IMPACT LOADS

(a) Glazing in hazardous locations such as glass doors, including fixed glazed side light panels immediately adjacent to such doors, fixed glass panels adjacent to any floor area normally used as a walking surface, sliding glass door units, including fixed glass panels which are a part of such units, shower doors, tub enclosures and storm doors shall be safety glazing materials and meet the requirements of Table 2706.

EXCEPTIONS: 1. Glass lights located not less than 18 inches above the adjacent finished floor walking surface.

2. Glass lights when the least dimension is no greater than 18 inches.

3. Glass lights when by comparative tests are proven to produce equivalent performance to those requirements in Table 2706.

(b) Where the risk of accidental breakage may involve hazard to human safety such as in explosive process areas, in skylights over public use areas, underwater view areas, etc., detailed drawings, specifications and test data shall be submitted by a licensed engineer experienced in this work to the Building Official for approval.

APPENDIX "B"

TABLE 1 — FIRE RESISTANCE RATINGS FOR NON-LOAD BEARING WALLS AND PARTITIONS — Page B-1

Add the following fire ratings to Appendix B, Table 1—Fire Resistance Ratings for Non-Load Bearing Walls and Partitions: (1973 Meeting)

Wall or Partition Assembly	Minimum Nominal Thickness for Fire Ratings Indicated (Inches)			
	4 Hr.	3 Hr.	2 Hr.	1 Hr.
Brick—Clay or Shale			6	O.S.U. #2
Brick Hollow—Plastered or Unplastered		8 (1)		N.B.S. #4 Vol. 35

(1) Units at least 71% solid.

TABLE 3 — FIRE RESISTANCE RATINGS FOR LOAD-BEARING WALLS AND PARTITIONS — Page B - 8

Add the following fire ratings to Appendix B, Table 3—Fire Resistance Ratings for Load-Bearing Walls and Partitions: (1973 Meeting)

Wall or Partition Assembly	Members Framed Into Wall or Partitions							
	Combustible				None or Non-Combustible			
	4 Hr.	3 Hr.	2 Hr.	1 Hr.	4 Hr.	3 Hr.	2 Hr.	1 Hr.
Solid Brick Walls								
Solid - (clay or shale)							6	4 O.S.U. T-1971 T-1972
Hollow Brick Walls								
Hollow units (clay or shale)								
Unplastered			8 (19)			8 (19)		N.B.S. #4 Vol.35
Plastered		8 (19)			8 (19)			

(19) Units at least 71% solid.

TABLE 7 — FIRE RESISTANCE RATINGS FOR MATERIALS AND CONSTRUCTION — Page B - 45

Add a new one-hour assembly to the Wood Joist Construction section. (1973 Meeting)

Flooring—Wood floor consisting of 23/32" Interior Group 1 Underlayment grade plywood (Exterior glue) with tongue and groove side joints. Butt joints of plywood centered over 2 x 10 joists spaced 24 inches on center (no bridging required). Underlayment bonded to joists with adhesive meeting the requirements of AFG-01. Plywood joints are protected with 5/8" Type X wallboard 6" wide centered over underside of T&G joint and stapled in position.

Ceiling—5/8" Type X gypsum wallboard installed on resilient furring channels formed of 25 MSG galvanized steel, spaced 16" o.c. perpendicular to joists using 1" wallboard screws spaced 12" o.c. in the field and 8" o.c. at ends. UL Design L 513, R 5229-2.

APPENDIX "B" — TABLE 7

Table 7 — Revise footnote (i) to read as follows: (1974 Meeting)

(i) Appropriate fire resistance ratings as listed in "Performance Data Acoustical Materials, Bulletin 1974", as published by the Acoustical and Insulating Materials Association may be accepted as if herein listed.

TABLE 600 NOTES — APPENDIX H

(q) Change phrase "over one story" to read "two (2) stories or more". (1974 Meeting)

APPENDIX "K"

Add a new Appendix K to read as follows: (1973 Meeting)

APPENDIX "K"

ACCESSIBILITY TO BUILDINGS AND FACILITIES WITHIN BUILDINGS FOR USE BY THE PHYSICALLY HANDICAPPED

(a) Toilet Facilities. Toilet facilities and water closet compartments shall have a clear and unobstructed access of not less than 32 inches, and a clear space, unobstructed by door swing, grab bars and similar items, of not less than 32 inches shall be provided in front of the water closet bowl. Grab bars shall be provided on two sides or one side and the back of each toilet compartment.

EXCEPTIONS: 1. In One and Two Family Dwellings the water closet compartments may be 30 inches in width and may have a clear space in front of the water closet bowl of not less than 24 inches. Grab bars need not be provided in such occupancies nor in apartment houses.

2. Where more than one toilet compartment is provided, only one such compartment need comply within each toilet room for each sex. All others need not have grab bars and clearances may be as specified in Exception No. 1.

(b) Toilet Room Facilities. In other than One and Two Family Dwellings and apartment house occupancies, toilet room facilities shall be as follows:

1. Except for the projection of bowls and waste piping, a clear unobstructed space 26 inches in width, 24 inches in height and 12 inches in depth shall be provided under at least one lavatory.

2. Where mirrors are provided, at least one shall be installed so that the bottom of the mirror is within 40 inches of the floor.

3. Where towel and disposal fixtures are provided, at least one shall be within 40 inches of the floor.

(c) Water Fountains. Where water fountains are provided, at least one shall have a spout within 33 inches of the floor and shall have up-front, hand operated controls. When fountains are located in an alcove, the alcove shall be not less than 32 inches in width.

Telephones. Where public telephones are provided, at least one shall be installed so that the headset, dial and coin receiver are within 40 inches of the floor. Unobstructed access within 12 inches of the telephone shall be provided. Such access shall be not less than 32 inches in width and depth.

(d) Ramps and/or elevators shall be required for the following uses:

Assembly Areas
Auction Rooms
Auditoriums
Bowling Alleys
(Assembly Areas)
Churches and Chapels
Dance Floors
Lodge Rooms
Stadiums
Conference Rooms
Dining Rooms
Bars
Exhibit Rooms
Gymnasiums
Library
Locker Rooms
Offices
Schools
Lounges
Skating Rinks
Stages
Children's Home
Homes for the Aged
Classrooms
Dormitories
Parking Garage
Hospitals
Sanitoriums
Nursing Homes
Hotels
Apartment Houses
(over 3 stories)
Nurseries for
Children (Day Care)
Stores (Retail)
Warehouses

(e) Change in Floor Level at Doors. Regardless of the occupant load, there shall be a floor or landing on each side of a door. The floor or landing shall be not more than 1 inch lower than the threshold of the doorway. Where doors open over landings, the landing shall have a length of not less than 5 feet.

EXCEPTIONS: 1. Where the door opens into a stair tower, the landing need not have a length of 5 feet.

2. In Group A Occupancies, a door may open on the top step of a flight of stairs or on an exterior landing provided the door does not swing over the top step or exterior landing and the landing is not more than 7½ inches below the floor level.

(f) Access to Exits. When more than one exit is required, they shall be so arranged that it is possible to go in either direction from any point in a corridor to a separate exit, except for dead ends permitted by this Code. When a corridor or exterior exit balcony is accessible to an elevator, changes in elevation of the floor shall be made by means of a ramp.

(g) Handrails shall be placed not less than 30 inches nor more than 34 inches above the nosing of treads, and ends of handrails shall extend 6 inches beyond stair or step treads and shall be returned or shall terminate in newel posts or safety terminals.

(h) Ramps. 1. General. Ramps used as exits shall conform to the provisions of this Section.

2. Width. The width of ramps shall be as required for stairways.

3. Slope. Ramps required by paragraph (d) shall not exceed a slope of one vertical to 10 horizontal. The slope of other ramps shall not exceed one vertical to 8 horizontal.

4. Landings. Ramps having slopes greater than one vertical to 15 horizontal shall have landings at the top and bottom and at least one intermediate landing shall be provided for each 5 feet of rise. Top landings and intermediate landings shall have a dimension measured in the direction of ramp run of not less than 5 feet. Landings at the bottom of ramps shall have a dimension in the direction of ramp run of not less than 6 feet.

Doors in any position shall not reduce the minimum dimension of the landing to less than 42 inches and shall not reduce the required width by more than 3½ inches when fully open.

5. Handrails. Ramps having slopes exceeding one vertical to 15 horizontal shall have handrails as required for stairways, except that intermediate handrails shall not be required.

6. Construction. Ramps shall be constructed as required for stairways.

7. Surface. The surface of ramps shall be roughened or shall be of nonslip materials.

APPENDIX "L" — THERMAL PERFORMANCE

(1973 Meeting)

SECTION 1. GENERAL

1.1 — SCOPE

All buildings and structures in Group A (Residential); Group B (Business); Group C (Schools); Group D (Institutional); and Group E (Assembly) which are heated and cooled mechanically shall be constructed to comply with the maximum "U" values contained in this chapter. A building which is both heated and cooled shall be insulated against the most severe climatic condition for the location and type of structure involved.

1.2 — CEILINGS AND WALL SECTIONS IN HEATED STRUCTURES

For ceilings and wall sections in heated structures only, the thermal transmittance shall be no more than the values shown in Table 1-A.

TABLE 1-A — HEATED BUILDINGS

Maximum "U" Values of Ceiling and Wall Sections of Various Constructions

Winter Degree Days	Flat Roof Deck	Masonry Wall Construction		Frame Wall Construction		Glazing
		Ceilings	Walls	Ceilings	Walls	
4500 or Less	.12	.05	.17	.08	.07	Single
4501 or More	.10	.05	.12	.05	.07	Double See Note(1)

(1) Double glazing requires that all glass surfaces in the structure shall be insulating glass or storm sash and that all door openings shall have insulated doors or storm doors.

1.3 — CEILING AND WALL SECTIONS IN HEATED MECHANICALLY COOLED STRUCTURES

Ceilings and wall sections which are heated and mechanically cooled, the thermal transmittance shall be no more than the values shown in Table 1-B.

TABLE 1-B — MECHANICALLY COOLED BUILDINGS

Maximum "U" Values for Ceilings and Wall Sections of Various Constructions

Summer Cooling-Hours Over 80° F	Flat Roof Deck	Masonry Wall Construction Ceilings	Masonry Wall Construction Walls	Frame Wall Construction Ceilings	Frame Wall Construction Walls
400 or Less	.12	.08	.17	.08	.07
401 or More	.10	.05	.12	.05	.07

Summer cooling hours may be obtained from the "NAHB Insulation Manual for Homes and Apartments." Manuals are available from NAHB Research Foundation, Inc., Rockville, Md. 20850 or National Mineral Wool Insulation Association, 211 E. 51st St., New York, New York 10022. Other recognized sources of degree day and summer cooling data may be used.

1.4 — FLOORS OVER UNHEATED SPACES

a. For floors over unheated basements, crawl spaces, breezeways and garages, the maximum coefficient of heat transfer (U-Value for heat flow down) shall not exceed:

Winter Degree Days	Structural Slab	Wood and Steel Framing
2500 or Less	No Requirement	No Requirement
2501 to 4500	.15	.10
4501 or More	.12	.07

Note 1. A basement shall be considered unheated unless it is provided with a positive heat supply equivalent to at least 15% of the total calculated heat loss of the living unit, or contains the heating unit and uninsulated ducts or piping, or is provided with a positive heat supply to maintain a minimum temperature of 50F.

Note 2. A crawl space with ventilation louvers is considered unheated. A crawl space without ventilation louvers is considered unheated unless it is (a) provided with positive heat supply equivalent to at least 10% of the total calculated heat loss of the living unit, or (b) contains uninsulated ducts or piping, or (c) is used as a supply or return plenum.

Note 3. A garage is considered unheated unless provided with a positive heat supply to maintain a minimum temperature of 50F.

1.5 — HEATED BASEMENTS AND CRAWL SPACES

When a basement or crawl space is heated the following insulation R Values shall be used for the perimeter walls and insulation may be omitted from floors over the heated area.

Winter Degree Days	Insulation R-Value
2500 or Less	R-3.75
2501 or More	R-7

1.6 — CONCRETE SLABS

Edge heat loss of concrete slabs around the perimeter of heated spaces shall not exceed a maximum value per lineal foot of exposed edge of 42 Btuh for unheated slabs and 50 Btuh for heated slabs. Calculations of heat loss through concrete slabs shall be made using the following formula:

$$H = F \times P$$

Where H = Heat Loss of the floor slab (Btuh)

Where F = Heat loss coefficient from the following Table (Btuh per lineal foot of exposed slab edge)

P = Perimeter or exposed floor edge (Lineal Feet)

SLAB EDGE HEAT LOSS FACTORS

Heat Loss Per Foot of Exposed Slab Edge (Btuh)

Winter Design Temperature (°F)	Total Width of Insulation Inches	F for unheated slab[1]			F for heated slab[1]		
		R=5.00	R=3.75	R=2.50	R=5.00	R=3.75	R=2.50
—30 and colder	24	34			46		
—25 to —29	24	32			44		
—20 to —24	24	30			41		
—15 to —19	24	28			39		
—10 to —14	24	27	40		37		
— 5 to — 9	24	25	38		35		
Zero to — 4	24	24	36		32	48	
+ 5 to + 1	24	22	33		30	45	
+ 10 to + 6	18	21	31	42	25	38	50
+ 15 to + 11[2]	12	21	31	42	25	38	50

Note 1. Insulation R-Values shown are for thermal insulation resistance (R = 1/C).

Note 2. Where Winter Design Temperatures are warmer than +15F, perimeter insulation is not required. If installed in these areas (edge only use values shown for +15 to +11 above. If not installed, use value of F = 45 for unheated and F = 60 for heated slabs.

SECTION 2 — DESIGN PARAMETERS

(a) Inside design temperature shall not be less than 70F. Heat loss and heat gain calculations shall be made using the winter design dry-bulb at 99% and summer design dry-bulb at 1% shown in current ASHRAE Handbook of Fundamentals.

(b) For the purpose of heat loss calculations air infiltration shall be assumed to be one (1) air change per hour for the space to be heated to 70° F.

SECTION 3 — LOOSE FILL

Blowing and poured type loose fill may be used in attic spaces where the pitch in roof design is not less than 2½ on 12 feet and there is at least 30 inches of clear headroom at the roof ridge. ("Clear Headroom" is defined as the distance from the top of the bottom chord of the truss or ceiling joists to the underside of the roof sheathing.) When eave vents are installed, adequate baffling of the vent opening must be provided so as to deflect the incoming air above the surface of the installed blown or poured insulation. Baffles shall be made of wood or other durable material and shall be installed at the soffit on a 60 degree angle. Baffles shall be in place at the time of framing inspection.

When loose fill insulation is proposed, the "R" value of the material shall be shown on the building plans together with the total number of bags and net coverage per bag.

SECTION 4. GLAZING REQUIREMENTS

(a) Where glass areas do not exceed 20% of the gross exposed wall area in high rise construction or 25% in medium and low rise the requirements of Paragraph 1.2, Table 1-A apply.

(b) Where in high rise construction the glass area exceeds 20% of the gross exposed wall area or 25% in low and medium rise construction the following tables apply.

HEATED BUILDINGS

Winter Degree Days	High Rise % Glass	Low-Medium Rise % Glass	Glazing
4500 or Less	21-35	26-40	Heat Absorbing
4501 or More	21 or more	25 or more	Reflective Insulating

MECHANICALLY COOLED BUILDINGS

Summer Cooling-Hours Over 80° F	High Rise % Glass	Low-Medium Rise % Glass	Glazing
400 or Less	21-35	26-40	Single Heat Absorbing
	36 or more	40 or more	Single Reflective
401 or More	21 or more	25 or more	Reflective Insulating

APPENDIX "M"

RECOMMENDED SCHEDULE OF PERMIT FEES

(1973 Meeting)

(a) Permit Fees

Schedule

Total Valuation	Fee
$100.00 and less	No fee, unless inspection required, in which case case a $5.00 fee for each inspection shall be charged.
$101.00 to $2000.00	$5.00 per thousand or fraction thereof.
$2001.00 to $15,000.00	$10.00 for the first $2000.00 plus $3.00 for each additional thousand or fraction thereof, to and including $15,000.00.
$15,001.00 to $50,000.00	$49.00 for the first $15,000.00 plus $2.50 for each additional thousand or fraction thereof, to and including $50,000.00.
$50,001.00 to $100,000.00	$136.50 for the first $50,000.00 plus $2.00 for each additional thousand or fraction thereof, to and including $100,000.00.

$100,001.00 to $500,000.00	$236.50 for the first $100,000.00 plus $1.25 for each additional thousand or fraction thereof, to and including $500,000.00.
$500,001.00 and up	$736.50 for the first $500,000.00 plus $.75 for each additional thousand or fraction thereof.

(b) Moving of Building or Structures

For the moving of any building or structure, the fee shall be $50.00.

(c) Demolition of Building or Structures

For the demolition of any building or structure, the fee shall be $25.00.

(d)

Where work for which a permit is required by this Code is started or proceeded with prior to obtaining said permit, the fees herein specified shall be doubled, but the payment of such double fee shall not relieve any persons from fully complying with the requirements of this Code in the execution of the work nor from any other penalties prescribed herein.

(e) Plan-checking Fees

When the valuation of the proposed construction exceeds $1,000.00 and a plan is required to be submitted by Section 105.3, a plan-checking fee shall be paid to the Building Official at the time of submitting plans and specifications for checking. Said plan-checking fee shall be equal to one-half of the building permit fee as set forth in Section 107.4.

Spotted something that isn't quite right?

Let us know and we'll fix it for future reprints.
And as a thank you for your diligence, we'll
send you a free Sothis book of your choice.

publisher@sothis.press

This page intentionally left blank

Made in the USA
Coppell, TX
05 November 2026

11289015R10360